# Lecture Notes in Business Information Processing 459

Series Editors

Wil van der Aalst
*RWTH Aachen University, Aachen, Germany*

John Mylopoulos
*University of Trento, Trento, Italy*

Sudha Ram
*University of Arizona, Tucson, AZ, USA*

Michael Rosemann
*Queensland University of Technology, Brisbane, QLD, Australia*

Clemens Szyperski
*Microsoft Research, Redmond, WA, USA*

More information about this series at https://link.springer.com/bookseries/7911

Andrea Marrella · Raimundas Matulevičius ·
Renata Gabryelczyk · Bernhard Axmann ·
Vesna Bosilj Vukšić · Walid Gaaloul ·
Mojca Indihar Štemberger · Andrea Kő ·
Qinghua Lu (Eds.)

# Business Process Management

## Blockchain, Robotic Process Automation, and Central and Eastern Europe Forum

BPM 2022 Blockchain, RPA, and CEE Forum
Münster, Germany, September 11–16, 2022
Proceedings

Springer

*Editors*
Andrea Marrella ⓘD
Sapienza University of Rome
Rome, Italy

Raimundas Matulevičius ⓘD
University of Tartu
Tartu, Estonia

Renata Gabryelczyk ⓘD
University of Warsaw
Warsaw, Poland

Bernhard Axmann ⓘD
Technische Hochschule Ingolstadt
Ingolstadt, Germany

Vesna Bosilj Vukšić ⓘD
University of Zagreb
Zagreb, Croatia

Walid Gaaloul ⓘD
Télécom SudParis
Evry, France

Mojca Indihar Štemberger ⓘD
University of Ljubljana
Ljubljana, Slovenia

Andrea Kő ⓘD
Corvinus University of Budapest
Budapest, Hungary

Qinghua Lu ⓘD
Data61
Eveleigh, NSW, Australia

ISSN 1865-1348 ISSN 1865-1356 (electronic)
Lecture Notes in Business Information Processing
ISBN 978-3-031-16167-4 ISBN 978-3-031-16168-1 (eBook)
https://doi.org/10.1007/978-3-031-16168-1

This Springer imprint is published by the registered company Springer Nature Switzerland AG
The registered company address is: Gewerbestrasse 11, 6330 Cham, Switzerland

# Preface

The International Conference on Business Process Management (BPM) was established about 20 years ago as the conference where people from academia and industry meet and discuss the latest developments in the area of business process management. In 2022, the conference was organized in Münster, Germany. This year's BPM also featured three specialized forums. This volume contains the proceedings of the Blockchain Forum, the Robotic Process Automation (RPA) Forum, and the Central and Eastern Europe (CEE) Forum, which took place during September 11–16, 2022.

A blockchain is a distributed data structure providing guarantees of immutability and integrity protection, delivering a practical solution to hard problems in coordination. Blockchain-based systems open up diverse opportunities in the context of the BPM lifecycle to redesign business activities in a wide range of fields, including healthcare, supply chain, logistics, and finance. However, these opportunities come with challenges to security and privacy, and to scalability and performance. The fourth edition of the Blockchain Forum provided a platform for the discussion of ongoing research and success stories on the use of blockchain, including techniques for, and applications of, blockchain and distributed ledger technology.

The concept of robotic process automation (RPA) has gained relevant attention in both industry and academia. RPA raises a way of automating mundane and repetitive human tasks requiring a lower level of intrusiveness with the IT infrastructure. The RPA Forum aimed to bring together researchers from various communities and disciplines to discuss challenges, opportunities, and new ideas related to RPA and its application to business processes in private and public sectors. The forum solicited contributions related to three main RPA areas: management, technology, and applications. The keynote given by Tathagata Chakraborti from IBM Research (USA) was focused on process automation from natural language inputs. The RPA Forum took place for the third time, after its previous appearances at BPM 2020, in Seville (Spain), and at BPM 2021, in Rome (Italy).

The main objective of the CEE Forum was to provide a discussion platform for BPM academics from Central and Eastern Europe to disseminate their research, compare results, and share experiences. This second CEE Forum was an opportunity for novice and advanced BPM researchers who have not yet had the chance to attend the International Conference on Business Process Management to get to know each other, initiate research projects, and join the international BPM community.

The Blockchain Forum received 15 submissions, of which seven papers were selected for presentation at the venue and for inclusion to this proceedings. The RPA Forum received 16 submissions, which led to the acceptance of the top nine as full papers. The CEE Forum received 9 submissions, and the top four high-quality papers were selected for presentation and publication. The overall acceptance rate was 50%. For the three forums, each submission was reviewed by at least three members of the respective Program Committees.

We hope that the reader of these proceedings will enjoy the papers presented at the forums. We would like to congratulate both the authors of the accepted papers and those who submitted their work that, unfortunately, was not accepted despite its quality. We also thank our colleagues who acted as reviewers in the selection process and provided the authors with meaningful and constructive comments. Finally, special thanks go to Katrin Bergener and Armin Stein (organizing chairs of BPM 2022) for organizing such an outstanding conference, despite the challenges that came with the COVID-19 pandemic.

September 2022

Andrea Marrella
Raimundas Matulevičius
Renata Gabryelczyk
Bernhard Axmann
Vesna Bosilj Vukšić
Walid Gaaloul
Mojca Indihar Štemberger
Andrea Kő
Qinghua Lu

# Organization

## Blockchain Program Chairs

Raimundas Matulevičius     University of Tartu, Estonia
Walid Gaaloul     Télécom SudParis, France
Qinghua Lu     CSIRO, Australia

## RPA Program Chairs

Andrea Marrella     Sapienza University of Rome, Italy
Bernhard Axmann     Technische Hochschule Ingolstadt, Germany

## CEE Program Chairs

Vesna Bosilj Vukšić     University of Zagreb, Croatia
Renata Gabryelczyk     University of Warsaw, Poland
Mojca Indihar Štemberger     University of Ljubljana, Slovenia
Andrea Kő     Corvinus University of Budapest, Hungary

## Combined Program Committee

Simone Agostinelli     Sapienza University of Rome, Italy
Aleksandre Asatiani     University of Gothenburg, Sweden
Agnieszka Bitkowska     Warsaw University of Technology, Poland
Edyta Brzychczy     AGH University of Science and Technology, Poland
Tathagata Chakraborti     IBM Research AI, USA
Marco Comuzzi     Ulsan National Institute of Science and Technology, South Korea
Adela Del Río Ortega     University of Seville, Spain
Carmelo Del Valle     University of Seville, Spain
Claudio Di Ciccio     Sapienza University of Rome, Italy
Peter Fettke     German Research Center for Artificial Intelligence (DFKI) and Saarland University, Germany
Christian Flechsig     Dresden University of Technology, Germany
José María García     University of Seville, Spain
Luciano García-Bañuelos     Monterrey Institute of Technology and Higher Education, Mexico

| Mark Staples | Data61, CSIRO, Australia |
| Burkhard Stiller | University of Zurich, Switzerland |
| Dalia Suša Vugec | University of Zagreb, Croatia |
| Rehan Syed | Queensland University of Technology, Australia |
| Martina Tomicic Furjan | University of Zagreb, Croatia |
| Horst Treiblmaier | Modul University Vienna, Austria |
| Inge van de Weerd | Utrecht University, The Netherlands |
| Maximilian Völker | Hasso Plattner Institute, Germany |
| Jonas Wanner | University of Würzburg, Germany |
| Barbara Weber | University of St. Gallen, Switzerland |
| Ingo Weber | Technical University of Berlin, Germany |
| Moe Thandar Wynn | Queensland University of Technology, Australia |
| Kaiwen Zhang | École de technologie supérieure de Montréal, Canada |

# Contents

**Blockchain Forum**

Blockchain for Business Process Enactment: A Taxonomy and Systematic
Literature Review .................................................... 5
*Fabian Stiehle and Ingo Weber*

Pupa: Smart Contracts for BPMN with Time-Dependent Events
and Inclusive Gateways ............................................... 21
*Rodrigue Tonga Naha and Kaiwen Zhang*

A Systematic Local Fork Management Framework for Blockchain
Sandbox Environments ................................................ 36
*Antreas Pogiatzis and Georgios Samakovitis*

Fine-Grained Data Access Control for Collaborative Process Execution
on Blockchain ....................................................... 51
*Edoardo Marangone, Claudio Di Ciccio, and Ingo Weber*

Challenges and Opportunities of Blockchain for Auditable Processes
in the Healthcare Sector ............................................. 68
*Walid Fdhila, Nicholas Stifter, and Aljosha Judmayer*

Measuring the Effects of Confidants on Privacy in Smart Contracts ........... 84
*Julius Köpke and Michael Nečemer*

Threshold Signature for Privacy-Preserving Blockchain .................... 100
*Sara Ricci, Petr Dzurenda, Raúl Casanova-Marqués, and Petr Cika*

**Robotic Process Automation (RPA) Forum**

From Natural Language to Workflows: Towards Emergent Intelligence
in Robotic Process Automation ....................................... 123
*Tathagata Chakraborti, Yara Rizk, Vatche Isahagian, Burak Aksar,
and Francesco Fuggitti*

Towards an Integrated Platform for Business Process Management
Systems and Robotic Process Automation ............................. 138
*Christian Flechsig, Maximilian Völker, Christian Egger,
and Mathias Weske*

Rolling Back to Manual Work: An Exploratory Research on Robotic
Process Re-Manualization .............................................. 154
  Artur Modliński, Damian Kedziora, Andrés Jiménez Ramírez,
  and Adela del-Río-Ortega

Steering the Robots: An Investigation of IT Governance Models
for Lightweight IT and Robotic Process Automation ....................... 170
  Vincent Borghoff and Ralf Plattfaut

Identifying the Socio-Human Inputs and Implications in Robotic Process
Automation (RPA): A Systematic Mapping Study ........................ 185
  Harmoko Harmoko, Andrés Jiménez Ramírez, José González Enríquez,
  and Bernhard Axmann

A Human-in-the-Loop Approach to Support the Segments Compliance
Analysis ............................................................ 200
  Simone Agostinelli, Giacomo Acitelli, Michela Capece,
  and Massimo Mecella

Recommending Next Best Skill in Conversational Robotic Process
Automation ......................................................... 215
  Avi Yaeli, Segev Shlomov, Alon Oved, Sergey Zeltyn, and Nir Mashkif

Process Discovery Analysis for Generating RPA Flowcharts ................ 231
  Fabian Rybinski and Selina Schüler

Can You Teach Robotic Process Automation Bots New Tricks? .............. 246
  Yara Rizk, Praveen Venkateswaran, Vatche Isahagian,
  Vinod Muthusamy, and Kartik Talamadupula

API as Method for Improving Robotic Process Automation .................. 260
  Petr Průcha and Jan Skrbek

**Central and Eastern Europe (CEE) Forum**

Business Process Management in CEE Countries: A Literature-Based
Research Landscape .................................................. 279
  Renata Gabryelczyk, Edyta Brzychczy, Katarzyna Gdowska,
  and Krzysztof Kluza

Process and Project Oriented Organization: The Essence and Maturity
Measurement ........................................................ 295
  Piotr Sliż

The Competencies for Knowledge Management and the Process
Orientation in the Post-Covid Economy. Generation Z Students Perspective .... 310
  *Waldemar Glabiszewski, Szymon Cyfert, Roman Batko, Piotr Senkus,*
  *and Aneta Wysokińska-Senkus*

The Future Development of ERP: Towards Process ERP Systems? ............ 326
  *Marek Szelągowski, Justyna Berniak-Woźny, and Audrone Lupeikiene*

**Author Index** ........................................................ 343

# Blockchain Forum

# Preface

## Blockchain Forum

The BPM 2022 Blockchain Forum provided a venue for discussion and introduction of new ideas related to research directions within techniques for, and applications of, blockchain and distributed ledger technology. While being associated to the BPM conference, this year's Blockchain Forum in Münster followed on from previous fora held in Rome (2021), Seville (2020), and Vienna (2019). Blockchain technology has already seen both academic interest and practical adoption, and several approaches exist which combine BPM and blockchain. This year, the forum attracted 15 submissions from which the seven top papers were accepted for presentation at the Blockchain Forum and inclusion in the proceedings.

Enactment of the blockchain-based business processes requires consideration of various guarantees and capabilities. In their paper on "Blockchain for Business Process Enactment: A Taxonomy and Systematic Literature Review", Stiehle and Weber discuss the challenges and opportunities of the inter-organizational processes. The authors highlight that blockchain technology could ensure traceability and the correctness of the process execution.

In the paper "Pupa: Smart Contracts for BPMN with Time-Dependent Events and Inclusive Gateways", Tonga Naha and Zhang propose an engine to support time-dependent events and inclusive gateways in Ethereum-based workflow execution. The findings show that the engine is similar to baseline solutions in terms of performance and cost, but it improves the decentralization and model semantics.

Pogiatzis and Samakovitis describe tools to support the soft forking of Ethereum blockchains. In the paper on "A Systematic Local Fork Management Framework for Blockchain Sandbox Environments", the authors discuss the proposed framework and its application in some DevOps and security-oriented use cases.

The topic of security and privacy continues in the paper on "Fine-grained Data Access Control for Collaborative Process Execution on Blockchain". Here, Marangone et al. propose to use attribute-based encryption to control read and write permissions in the public storage systems coordinated using public ledgers. This approach allows the users to maintain data integrity and guarantee data confidentiality by managing attributes in the transactions.

The contribution by Fdhila et al. illustrates how to apply principles of the self-sovereign identity in the healthcare sector. In the paper "Challenges and Opportunities of Blockchain for Auditable Processes in the Healthcare Sector", the authors illustrate how blockchain is applied in cross-organizational business processes to fulfil privacy requirements and constraints.

In their paper "Measuring the effects of Confidants on Privacy in Smart Contracts", Köpke and Nečemer define modeling constructs for privity requirements and confidants' inclusion. The proposal results in an approach to measure the impact of additional decision actors on privity. It potentially helps to resolve goal conflicts and compare alternative solutions.

In their paper on "Threshold Signature for Privacy-preserving Blockchain", Ricci et al. present a scheme for splitting a blockchain wallet into multiple devices so that a threshold of them is needed for signing. The approach increases security because more user devices must be compromised when signing blockchain transactions.

We wish to thank all those who contributed to making the BPM 2022 Blockchain forum a success: the authors who submitted papers, the members of the Program Committee who carefully reviewed the submissions, and the speakers who presented their work at the forum. We also express our gratitude to the BPM 2022 chairs and organizers for their support in preparing the Blockchain Forum.

September 2022
Raimundas Matulevičius
Walid Gaaloul
Qinghua Lu

# Organization

## Program Chairs

Raimundas Matulevičius      University of Tartu, Estonia
Walid Gaaloul      Télécom SudParis, France
Qinghua Lu      CSIRO, Australia

## Program Committee

Marco Comuzzi      Ulsan National Institute of Science
     and Technology, South Korea
Claudio Di Ciccio      Sapienza University of Rome, Italy
José María García      University of Seville, Spain
Luciano García-Bañuelos      Tecnológico de Monterrey, Mexico
Julien Hatin      Orange Labs, France
Inma Hernandez      University of Seville, Spain
Marko Hölbl      University of Maribor, Slovenia
Sabrina Kirrane      Vienna University of Economics and Business,
     Austria
Kais Klai      Université Paris 13 Nord, France
Julius Köpke      Alpen-Adria-Universität Klagenfurt, Austria
Agnes Koschmider      Kiel University, Germany
Nassim Laga      Orange Labs, France
Giovanni Meroni      Politecnico di Milano, Italy
Alex Norta      Tallinn University of Technology, Estonia
Pierluigi Plebani      Politecnico di Milano, Italy
Matti Rossi      Aalto University, Finland
Stefan Schulte      TU Hamburg, Germany
Madhusudan Singh      University of Tartu, Estonia
Volker Skwarek      Hamburg University of Applied Sciences,
     Germany
Tijs Slaats      University of Copenhagen, Denmark
Mark Staples      CSIRO, Australia
Burkhard Stiller      University of Zurich, Switzerland
Horst Treiblmaier      Modul University Vienna, Austria
Ingo Weber      TU Berlin, Germany
Kaiwen Zhang      École de technologie supérieure de Montréal,
     Canada

# Blockchain for Business Process Enactment: A Taxonomy and Systematic Literature Review

Fabian Stiehle[(✉)] and Ingo Weber

Software and Business Engineering, Technische Universitaet Berlin, Berlin, Germany
stiehle@campus.tu-berlin.de, ingo.weber@tu-berlin.de

**Abstract.** Blockchain has been proposed to facilitate the enactment of interorganisational business processes. For such processes, blockchain can guarantee the enforcement of rules and the integrity of execution traces—without the need for a centralised trusted party. However, the enactment of interorganisational processes pose manifold challenges. In this work, we ask what answers the research field offers in response to those challenges. To do so, we conduct a systematic literature review (SLR). As our guiding question, we investigate the guarantees and capabilities of blockchain-based enactment approaches. Based on this SLR, we develop a taxonomy for blockchain-based enactment. We find that a wide range of approaches support traceability and correctness; however, research focusing on flexibility and scalability remains nascent. For all challenges, we point towards future research opportunities.

**Keywords:** Blockchain · Business process enactment · Business process execution · Interorganisational processes · Taxonomy · SLR

## 1 Introduction

The enactment of a process is a central part of the business process management (BPM) lifecycle. Enactment comprises instantiation, execution, and monitoring of a process [1, Chapter 1.2]. Business process management systems (BPMS), also known as workflow management systems, have long been used in *intra*organisational processes to automate the enactment of business processes [1, Chapter 2.4][2, Chapter 9.1.2]. However, in an *inter*organisational setting, without central control, this is far more complex. To capture the complexity surrounding multiple autonomous distributed actors, Breu et al. denote such processes as *living*. Such processes, they argue, make traceability, scalability, flexibility, and correctness aspects far more challenging to address [3]. Similarly, Pourmirza et al. find that only 30% of BPMS consider interorganisational aspects. For these systems, the "autonomy of organisations" becomes an issue. This requires trust mechanisms, dynamism, and flexibility. Furthermore, they identify standardization and interoperability issues [4]. In this setting, blockchain has been proposed

© Springer Nature Switzerland AG 2022
A. Marrella et al. (Eds.): BPM 2022, LNBIP 459, pp. 5–20, 2022.
https://doi.org/10.1007/978-3-031-16168-1_1

to serve as a neutral ground between participants, by facilitating trust and enforcing conformance and integrity—without the introduction of a centralised trusted party [5]. In this work, we ask what answers the research field of blockchain-based enactment offers in response to the challenges posed by interorganisational challenges. To do so, we develop a taxonomy capable of describing and classifying blockchain-based enactment approaches. We derive this taxonomy from a comprehensive systematic literature review (SLR), based on 36 selected primary studies. We find that, while blockchain is a natural fit to ensure traceability and correctness of process execution, research focusing on flexibility and scalability remains nascent. For all challenges, we point out possible future research directions. Following open science principles, and to enable replicability, we make the data from our SLR available—see Footnote 2.

## 1.1  Blockchain-Based Business Process Enactment

In an interorganisational setting, process control crosses organisational boundaries. Without central control, properties such as traceability or correctness are hard to address, e.g., how to ensure integrity and availability of event data across organisations, or how to enforce control-flow when control is distributed [3]. With central control, the question arises which party is to host a hub or mediator component, i.e., a centralised trusted party must be introduced [5]. Blockchain technology can distribute this trust by offering "a single *logically-centralised* ledger of cryptocurrency transactions operated in an *organisationally-decentralised* and *physically-distributed* way" [6, p. 7]. The blockchain's ledger is in practice immutable, non-repudiable, fully transparent, and highly available [7, Chapter 5]. Smart contracts can be used to perform arbitrary computations on the blockchain. As conceptualized in the first work in the field [5], blockchain can assume control of the process, enforcing or monitoring process rules and providing an immutable process trace.

## 1.2  Related Work

Pourmirza et al. [4] presented a SLR of BPMS architectures; they have found that only 30% consider interorganisational aspects. Mendling et al. [8] formulated the possibilities and challenges of blockchain for BPM. Their seminal work can be seen as charting the research direction in BPM and blockchain. For enactment, they discussed the approach as outlined in Weber et al. [5]. Di Ciccio et al. [9] discussed the possibilities of business process monitoring using blockchain, which is part of the enactment lifecycle. For blockchain and BPM as a whole, Garcia-Garcia et al. [10] conducted a SLR investigating blockchain support for the different BPM lifecycles. In contrast, we present a taxonomy and classification of enactment approaches. This allows us to provide in-depth analysis specific to enactment. To the best of the authors' knowledge, this is the first work to present a SLR and taxonomy on blockchain-based business process enactment.

## 2    Methodology

A taxonomy is a classification system that produces groupings of objects based on common characteristics [11]. Such a classification is integral to scientific method. In a complex field, a taxonomy can facilitate understanding and analysis. It can help navigate the research field and identify research gaps. In the field of design science, Williams et al. [12] note that the classification of differences provides insights into the design—and design process—of artefacts. For taxonomy development, we follow the definitions and guidelines as outlined in Nickerson et al. [11]. A taxonomy has different dimensions that can be derived inductively (i.e., empirically) or deductively (i.e., conceptually). Induction requires empirical evidence (i.e., cases to investigate), while deduction requires sound knowledge to deduce dimensions through logical reasoning. Nickerson et al. recommend the application of both methods in an iterative manner. We did so, but relied mostly on induction. To collect empirical evidence, we conducted a SLR of the field as per Kitchenham et al. [13], interleaved with the methods outlined by Nickerson et al. for taxonomy development. That is, the identified primary studies were used to inductively derive our taxonomy. Afterwards, we classified our primary studies according to the taxonomy.

### 2.1    Taxonomy Development

Following Nickerson et al. [11], at first, we have defined the users, purpose, and the meta-charateristic for our taxonomy. As **users**, we identified design science researchers. For these researchers, the **purpose** of this taxonomy is to enable the assessment of the current state of the art and future research opportunities. More specifically, which challenges of interorganisational processes have been solved by integrating blockchain, and which are still unaddressed. A meta-characteristic is the most general and complete characteristic from which all dimensions are derived [11]. This characteristic can be thought of as the starting point for taxonomy development. Our **meta-characteristic** is comprised of the *guarantees and capabilities of blockchain-based process enactment*. Distributing trust is the central reason for introducing blockchain technology to process enactment. Blockchain establishes trust by providing certain guarantees, such as the immutability of the ledger. Therefore, the offered guarantees were of central interest to our research. In addition, we investigated the capabilities of approaches, such as resource allocation and process flexibility. The meta-characteristic also served as the guiding research question for our SLR.

### 2.2    Systematic Literature Review

Through early exploratory searches, we could not deduce a concise common terminology for blockchain-based business process enactment. Thus, we decided to conduct a search with a set of broad search keywords, connecting terms of business process management with blockchain, and then apply more restrictive exclusion criteria. To limit the search results (given that blockchain constitutes a buzzword mentioned in many works), we restricted the search to the title of studies. The search string is presented in Listing 1.

**Listing 1.** Search string. Note, in order to save space, here we implicitly mean both singular and plural versions of each search keyword.

```
("blockchain" OR "smart contract" OR "DLT" OR
"Distributed Ledger Technology") AND
("bpm" OR "business process" OR "choreography" OR
"workflow"))
```

To account for the fast research pace in which blockchain is evolving, we also considered pre-prints and conference papers. There is evidence that *Google Scholar*[1] performs especially well in such scenarios [14], which made it our tool of choice. The initial search was conducted on the 2022-03-10 and yielded 186 entries. A full list of applied inclusion and exclusion criteria is given in Table 1 below. In the first pass, we excluded works based on publication type and title; in the second we examined the abstract. Finally, we conducted a full reading. After applying our exclusion criteria, we obtained 30 studies. We then performed backward snowballing. To limit the scope of the study, we did not conduct a full forward snowballing. Due to our broad search keywords, we expected forward snowballing to only yield a large set of irrelevant or already reviewed studies. We confirmed this expectation for our two most cited primary studies, and indeed found no relevant additional studies. Trough backward snowballing, we obtained an additional six studies, leading to a final primary study set of 36 studies. The full process, each pass, and the application of the exclusion criteria is made transparent in our published data set.[2]

**Table 1.** Inclusion and exclusion criteria.

| Inclusion | The study presents an approach in the field of blockchain-based business processes enactment |
|-----------|----------------------------------------------------------------------------------------------|
| Exclusion | 1. The study presents a domain specific application (not meant for general business processes) |
|           | 2. The study is a theoretical work, or a non-technical work, it does not present and evaluate a research artefact such as a execution or monitoring engine |
|           | 3. The study is a tertiary study, i.e., it is a review or overview of other contributions |
|           | 4. The study is illegible, i.e, not written in English or containing heavy spelling mistakes |

---

[1] https://scholar.google.com, accessed 2022-05-30.

[2] Replication package available at: https://github.com/fstiehle/SLR-blockchain-BP-execution; for convenience, we also include a hosted interactive spreadsheet of our SLR at https://tubcloud.tu-berlin.de/s/M8JQtaRX5JkjXXZ.

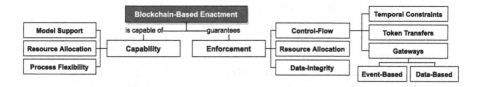

**Fig. 1.** Overview of our taxonomy of blockchain-based process enactment.

# 3   A Taxonomy of Blockchain-Based Enactment

## 3.1   Overview

By investigating our primary studies and following the methodology as per Sect. 2, we arrived at the taxonomy depicted in Fig. 1. When assessing blockchain-based applications, it is important to differentiate between application-specific properties and properties inherited from the employed blockchain. In our taxonomy, dimensions are kept independent from the chosen blockchain; this allows an independent assessment. The decision which blockchain platform to use is a different, but interrelated design decision [7, Chapter 6.3]. We have structured our taxonomy into supported capabilities and enforced guarantees. To improve readability, we defer the detailed introduction of our dimensions to the presentation of our classification results in Sect. 3.2. For capabilities, we capture with model support, which notation was chosen to represent the business process; with resource allocation capability, we differentiate between resource allocation strategies; with process flexibility capability, we capture how studies approach flexibility. For enforcement, we find that control-flow, resource allocation, and data-integrity aspects are enforced on-chain. Given the prevalence of control-flow, we subdivide this aspect further (see Sect. 3.2 below).

## 3.2   Dimensions and Classification Results

We detail the dimensions of our taxonomy and present the results of the classification our our primary studies.

**Capabilities.** First, we explore the supported capabilities, which are summarised in Table 2.

*Model Support.* The vast majority (70%) of studies are BPMN based. Notably, there is no close second. In terms of BPMN, 13 studies support the process, eight the choreography, and four the collaboration diagram.

**Table 2.** Classification of capabilities.

| Capability | Number (% of total) | Reference list |
|---|---|---|
| **Model Support** | **36 (100%)** | |
| BPMN process diagram | 13 (36%) | [15–27] |
| BPMN choreography diagram | 8 (22%) | [5, 28–34] |
| BPMN collaboration diagram | 4 (11%) | [35–38] |
| Undescribed model | 4 (11%) | [39–42] |
| YAWL | 2 (5%) | [43, 44] |
| Petri net | 2 (5%) | [45, 46] |
| Other (DCR, DEMO, and GSM) | 3 (8%) | [47–49] |
| **Resource Allocation** | **21 (58%)** | |
| Role-based | 13 (36%) | [5, 22, 28, 29, 31, 33–39, 48] |
| Direct | 5 (14%) | [25, 30, 42, 47, 49] |
| Dynamic | 3 (8%) | [17, 20, 21] |
| **Process Flexibility** | **5 (13%)** | |
| Looseness | 3 (8%) | [21, 47, 49] |
| Adaptation | 1 (2%) | [21] |
| Evolution | 1 (2%) | [35] |

*Resource Allocation.* Resource allocation assigns a process resource to a task [2, Chapter 10.5]. In blockchain-based enactment, a resource is typically identified by a blockchain account address. As blockchain transactions must be signed, a resource's involvement in a task cannot be repudiated (assuming the secrecy of their private key). 21 studies support resource allocation in general. Of these, we can differentiate between direct (five studies) and role-based (13 studies). A direct allocation binds a blockchain address directly to a task. Role-based allocation allows some indirection by assigning addresses to roles and roles to tasks. Only three allow a more dynamic strategy, which was first presented in [17]. These dynamic variants allow to specify the conditions for resource allocation in a so-called binding policy. For a given role, a participant can be nominated. Constraints can require that a participant must (or must not) already be bound to certain other roles. Endorsement constraints specify when and which other participants can vote on a nomination.

*Process Flexibility.* Process flexibility is essential for supporting less predictable processes. Reichert et al. characterised four flexibility needs: variability, looseness, adaptation, and evolution [50, Chapter 3]. We find that only five studies support flexibility needs. Looseness is supported by three studies. Two [47, 49] support looseness by using declarative models; these are loosely specified, providing more flexibility by only modelling constraints [50, Chapter 12]. López-Pintado et al. [21] support looseness and adaptation. Their approach allows late modelling: subprocesses can be modelled during run time. Furthermore, certain process elements can be adapted during run time. This is accompanied by an agreement policy, which allows to specify the participants that are allowed to adapt process elements and the conditions that must be met. For the adapted process, they can guarantee deadlock freeness. Klinger et al. [35] support process evolution. They implement blockchain design patterns (registry and proxy patterns) that enable the versioning of processes. These patterns decouple logic from data and allow logic to be updated. The approach includes a voting mechanism, which allows participants to vote on new process versions.

**Enforcement Guarantees.** Different perspectives of a process can be enforced on the blockchain. We find that control-flow, data-integrity, and resource allocation are enforced on-chain. Table 3 gives an overview of our classification result. We can observe a clear focus on control-flow enforcement (31 studies) over monitoring (5 studies).[3] Control-flow enforcement can be *data-based* (15 studies) or *event-based* (16 studies). Data-based enforcement enables, based on instance data, the evaluation of gateway conditions to automatically allow or disallow certain branches in the model. Event-based enforcement, on the other hand, can only enforce the semantics of the gateway (e.g., branching semantics of AND or XOR gateways), but not evaluate dynamic conditions.

Additionally, public blockchains enable the enforcement of token transfers (six studies). That is, certain behaviour may prompt automatic transfer of crypto tokens. Beyond fungible tokens (five studies), only Lu et al. [18] support the modelling and transfer of non-fungible tokens, which are integral for asset management.

Finally, only Ladleif et al. [28] and Abid et al. [27] allow to enforce temporal constraints. These constraints are based on the block timestamp. A blockchain network has no strong notion of a synchronised clock, the close world assumption and the transaction-driven nature of blockchain do not allow to access external time information or to continuously monitor an internal clock [51]. The block timestamp is the only readily available traditional notion of time on the chain; it is, however, of limited accuracy and can—to a certain extend—be manipulated by the block creator [51].

We listed 21 studies that support the allocation of resources to tasks. Most studies (19 studies) enforce this allocation by implementing authorisation mechanisms: only the allocated blockchain address can perform the task. In contrast, Prybila et al. [30] and Meroni et al. [49] present monitoring approaches that only guarantee the authenticity of the resource that has performed the task, they do not enforce authorisation. Lastly, all approaches make use of the integrity guarantee of blockchain to store the process trace. 27 studies allow the storage of instance data and five store a serialised version of the (original) process model on the blockchain.

**Methods.** Beyond our taxonomy, we investigate the evaluation methods employed. A summary is presented in Table 4. Most works evaluate their approach using Ethereum (24 studies). Nine studies use Hyperledger Fabric. Four

---

[3] We define monitoring as approaches where the control-flow is not enforced, but the process trace is still committed to the blockchain to ensure the integrity of the trace.

**Table 3.** Classification of enforcement guarantees.

| Enforcement Guarantee | Number (% of total) | Reference list |
|---|---|---|
| **Control-Flow** | **31 (86%)** | |
| Event-based gateways | 16 (44%) | $[5, 19, 25, 33, 35, 37\text{–}42, 44\text{–}48]$ |
| Data-based gateways | 15 (42%) | $[15\text{–}18, 20\text{–}22, 24, 27\text{–}29, 31, 32, 34, 36]$ |
| Token transfers | 6 (17%) | $[5, 18, 26, 29, 31, 34]$ |
| Temporal constraints | 2 (6%) | $[27, 28]$ |
| **Resource Allocation** | **19 (53%)** | $[5, 17, 20\text{–}22, 25, 28, 29, 31, 33\text{–}39, 42, 47, 48]$ |
| **Data-Integrity** | **36 (100%)** | |
| Execution trace | 36 (100%) | $[5, 15\text{–}49]$ |
| Instance data | 27 (75%) | $[5, 15\text{–}18, 20\text{–}24, 27\text{–}32, 34, 39\text{–}46, 48, 49]$ |
| Process model | 5 (14%) | $[39, 44\text{–}46, 49]$ |

**Table 4.** Employed methods.

| Method | Number (% of total) | Reference list |
|---|---|---|
| **Blockchain Selection** | **36 (100%)** | |
| Ethereum | 24 (67%) | $[5, 15\text{–}18, 20\text{–}22, 24\text{–}29, 31\text{–}38, 47, 49]$ |
| Hyperledger Fabric | 9 (25%) | $[19, 23, 26, 34, 39\text{–}41, 43, 48]$ |
| Custom implementation | 4 (11%) | $[42, 44\text{–}46]$ |
| Bitcoin | 2 (6%) | $[26, 30]$ |
| **Evaluation Criteria** | **33 (92%)** | |
| Cost | 23 (64%) | $[5, 15\text{–}22, 25, 28\text{–}35, 37, 38, 45, 47, 49]$ |
| Qualitative discussion | 18 (50%) | $[5, 23, 25, 26, 28, 30, 31, 33, 34, 38, 40, 41, 43\text{–}46, 48, 49]$ |
| Correctness | 9 (25%) | $[5, 15, 16, 18, 30, 33, 36, 42, 49]$ |
| Throughput | 3 (8%) | $[15, 42, 46]$ |
| Finality | 3 (8%) | $[5, 30, 34]$ |

present a custom blockchain implementation and only two studies consider Bitcoin. Finally, Corradini et al. [34] and Falazi et al. [26] present artefacts for multiple blockchains. In terms of evaluation, cost is the most frequently regarded metric (23 studies). For ten studies, cost is the sole focus of the evaluation.

Indeed, this concern is understandable when considering the cost of public blockchain compared to more traditional computing [52]. Most (19 studies out of 23) report cost based on Ethereum gas[4]. Transaction fees on Ethereum have increased significantly along with the popularity of the network. While Weber et al. [5] were still able to conduct evaluation experiments on the main network of Ethereum, more recent works resort to test networks or private deployments. We show this development in Fig. 2, where we compare publication year and execution cost for a repeated instance execution, excluding the cost that only incur once (e.g., deployment or configuration). As a point, we depict the original cost in US\$ at the time of publication. This price is based on the average gas

---

[4] See *Gas and fees*, https://ethereum.org/en/developers/docs/gas/, accessed 2022-05-30.

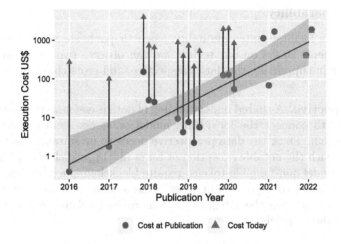

**Fig. 2.** Execution cost on Ethereum at publication and today.

cost and exchange rate of the publication year.[5] The point is connected vertically with a triangle, which represents the cost the same execution would incur taken the average gas cost and exchange rate of 2021. We can observe that the cost has risen significantly over time. The result is a projected mean cost of $1010 for a singular instance run in 2021. No approach would lie significantly below $100 for one run. As a result, most recent works argue for the use of a private network in almost all scenarios.

Besides cost, 18 studies discuss qualitative aspects (e.g., privacy or trust) and nine studies evaluate the correctness of their approach. For correctness, they investigate whether non-conforming traces are prevented (enforcing approaches) or detected (monitoring approaches) correctly.[6] Notably, finality[7] and throughput are only investigated by three studies each. Lastly, 18 studies have published their code and eleven have made a replication package available.

## 4    Discussion: Challenges and Future Research Directions

We discuss the guarantees and capabilities of blockchain-based enactment in the light of the challenges of interorganisational processes (see Sect. 1): interoperability [4], traceability, scalability, flexibility, and correctness [3].

---

[5] Our script and data sets used for calculation can be found at: https://github.com/fstiehle/SLR-blockchain-BP-execution. Historical data was taken from Etherscan (https://etherscan.io/chart/gasprice) and Yahoo Finance (https://finance.yahoo.com/quote/ETH-USD/, both accessed 2022-06-01.

[6] We here need to remark that [25] is not handling the claimed subset of BPMN correctly; as noted in [16, Sect. 2.2], the OR join is handled incorrectly.

[7] We consider finality as the time it takes a transaction to be durably committed with a certain probability $e$.

## 4.1   Interoperability

Blockchain can facilitate interoperability, as participants share the same execution environment. Based on our review, we can observe two opportunities for future research: supporting the data perspective, and cross-chain compatibility.

**Data Perspective.** A shared understanding of data is essential for participants; for example, to connect the local data model or assess security and privacy implications [53]. Thus, the data perspective needs to be suitably modelled. All approaches, with the notable exception of Lu et al. [18], which use a UML class diagram to model non-fungible-tokens, require blockchain-specific code snippets (e.g., solidity code) to express data types or data-based gateway conditions. In the future, we envision the integration of more comprehensive and platform-independent data models.

**Cross-chain Compatibility.** Only [26,34] present artefacts for multiple blockchains. However, the choice of the blockchain platform is application-specific [7, Chapter 6.3]. In terms of cross-chain compatibility, we envision three lines of research. First, when and how are multi-chain deployments a suitable implementation choice, and which basic requirements arise (e.g., cross-chain guarantees on integrity)? Second, the creation of artefacts for different blockchain platforms from one process model. Third, as discussed in [54], the execution of parts of the same process instance on different blockchain platforms.

## 4.2   Traceability and Correctness

Blockchain is a natural fit to ensure traceability and correctness of execution. Event data stored on the blockchain ledger is immutable and globally traceable. The integrity of this data can be enforced without introducing a centralised trusted party. To ensure correctness, the model-driven engineering paradigm is typically applied. This allows to generate well-tested artefacts following best practices. Consequently, all approaches either enforce the control-flow on-chain or allow the monitoring of the control-flow by committing the process trace to the ledger. In the following, we outline two opportunities for future research: dispute resolution and the extension of enforcement guarantees.

**Dispute Resolution.** Based on the immutable process trace, blockchain is envisioned to enable the resolution of contractual disputes between process participants (see e.g., [5]). However, no approach details a dispute resolution process, nor is its facilitation supported. It is unclear in which state the process remains once a dispute is raised. We expect that research conducted in this direction could provide real benefit to organisations. However, integrating the resolution of disputes may prove to be challenging and require different escalation levels [55]. Furthermore, it remains unclear whether blockchain traces would be accepted in a litigation process. A dispute resolution process could also include incentive mechanisms, facilitating the honest behaviour of participants and penalising malicious behaviour. Such research, would have to be conducted in an interdisciplinary context, including law and economic disciplines.

**Enforcement Guarantees.** While control-flow, resource allocation, and integrity aspects are supported, we expect that organisations would benefit from the blockchain-based enforcement of other process related rules. Our taxonomy remains extensible to further enforcement dimensions. For example, resource allocation is a complex decision problem into which many characteristics can be factored in [2, Chapter 10.5]. Enforcing these rules on-chain would make the allocation process transparent and globally enforceable. Currently, most works focus on role-based allocation, but it remains intransparent why a certain participant was allocated to a role.

Lastly, while most primary studies support on-chain enforcement, only nine evaluate the correctness of this enforcement capability. A more stringent evaluation, or even formal correctness proofs of the enforcement capability should be a central concern for the field, as this is the basis for all guarantees offered.

### 4.3  Flexibility and Scalability

While traceability and correctness aspects are already well supported, enabling flexibility and scalability remains a challenge. We see three major research opportunities: controlled flexibility, comprehensive performance studies, and enactment on public blockchains.

**Controlled Flexibility.** Only five studies address flexibility challenges. Introducing flexibility in blockchain applications is a challenge due to the ledger's immutability. Furthermore, introducing flexibility capabilities may lead to trust concerns and correctness issues [21]. Participants must be convinced that flexibility will not introduce uncertainties beyond their control, otherwise it will undermine traceability and correctness guarantees. Recharting the development of traditional enactment approaches, we can observe that the main focus, so far, has been on predictable processes. Addressing the challenges surrounding unpredictable processes and integrating different techniques to support variability, looseness, adaptation, and evolution [50]—but in a controlled manner—remains a line for future research.

**Comprehensive Performance Studies.** Across all studies, the most prominent evaluation goal is to demonstrate (low) gas cost. Gas cost can give an indication on throughput scalability on a public or private blockchain. The notion of gas has been introduced in Ethereum to calculate transaction fees. The goal was to control network propagation and storage requirements.[8] However, other factors

---

[8] Blockchain and Mining, Ethereum Whitepaper, https://ethereum.org/en/whitepaper/blockchain-and-mining, accessed 2022-06-01.

also play a crucial role. In our set of primary studies, scalability factors beyond gas cost are rarely explored. While a lot of performance properties depend upon the underlying blockchain platform, many use cases, especially private deployments, would benefit from more comprehensive performance studies. For example, in a private blockchain, with a few participants, gas cost may be not of paramount importance. The choice and configuration of a blockchain network is a complex trade-off between different parameters [7, Chapter 3] and can be optimised for a specific use-case [7, Chapter 6.3]. We envision future work to go beyond reporting gas cost and contribute to a discussion on the assumptions, advantages and drawbacks of different deployment options. Here, the question remains what properties must be investigated for a specific use case and which can be simulated or deduced from existing benchmarks, e.g. of the underlying blockchain platform.

**Enactment on Public Blockchains.** From our cost analysis (Sect. 3.2), we see that current transaction fees render the public Ethereum mainnet prohibitively expensive for the presented approaches in our primary studies. When considering blockchain, less quantifiable requirements play an important role also. Many of the guarantees a blockchain offers are a result of decentralisation [7, Chapter 3.2]. Certain high-risk use cases (e.g., the transportation of dangerous goods, as in Meroni et al. [49]) may benefit from decentralised and resilient public blockchains. Beyond Ethereum, we see a lot of promise in exploring alternative public blockchains. Next generation proof-of-stake blockchains like Algorand[9] or Avalanche[10] promise a more sustainable operation and low cost. Studying enactment approaches on different public platforms would produce valuable insights. However, identifying advantages and drawbacks, and comparing different blockchain setups could prove to be challenging. We envision a first step in connecting our presented taxonomy to a taxonomy of blockchain platforms.

Beyond exploring alternative public blockchains, a different line of research has opened around *Layer-2 technologies*.[11] These technologies reduce the involvement of the blockchain and perform most tasks off-chain; these off-chain tasks remain verifiable on the blockchain. We expect monitoring approaches to benefit in the short term from this line of research, as storing process traces becomes significantly cheaper. Long-term, we believe that enforcement approaches can make use of verifiable off-chain computations to significantly reduce cost.

## 5   Conclusion and Outlook

We performed a SLR on blockchain-based business process enactment. We identified a final set of 36 primary studies. Based on these primary studies, we developed a taxonomy capable of describing the guarantees and capabilities of

---

[9] https://www.algorand.com, accessed 2022-05-30.
[10] https://www.avax.network, accessed 2022-05-30.
[11] Layer 2 scaling, Ethereum development documentation, https://ethereum.org/en/developers/docs/scaling/layer-2-scaling, accessed 2022-05-30.

enactment approaches, and classified each study accordingly. We discussed our results in relation to the challenges of interorganisational processes.

We find that blockchain is a natural fit to ensure traceability and enforce correctness of process execution. However, in terms of research focusing on flexibility and scalability, the field of blockchain-based enactment remains nascent. For all challenges, we have pointed out a range of research opportunities. We have not addressed privacy or security concerns as these are often a result of, or strongly dependent on, the employed blockchain technology. However, our taxonomy remains open for extensions—e.g., towards security or privacy properties. In the future, we envision the development of a decision model, based on our taxonomy, to support stakeholders considering blockchain-based enactment.

# References

1. Weske, M.: Business Process Management: Concepts, Languages. Architectures, 3rd edn. Springer, Heidelberg (2019). https://doi.org/10.1007/978-3-642-28616-2
2. Dumas, M., La Rosa, M., Mendling, J., Reijers, J.A.: Fundamentals of Business Process Management. Springer, Heidelberg (2018). https://doi.org/10.1007/978-3-662-56509-4
3. Breu, R., Dustdar, S., Eder, J., et al.: Towards living inter-organizational processes. In: IEEE 15th Conference on Business Informatics, pp. 363–366 (2013)
4. Pourmirza, S., Peters, S., Dijkman, R., Grefen, P.: A systematic literature review on the architecture of business process management systems. Inf. Syst. **66**, 43–58 (2017)
5. Weber, I., Xu, X., Riveret, R., Governatori, G., Ponomarev, A., Mendling, J.: Untrusted business process monitoring and execution using blockchain. In: La Rosa, M., Loos, P., Pastor, O. (eds.) BPM 2016. LNCS, vol. 9850, pp. 329–347. Springer, Cham (2016). https://doi.org/10.1007/978-3-319-45348-4_19
6. Weber, I., Staples, M.: Programmable money: next-generation conditional payments using Blockchain - keynote paper. In: CLOSER (2021)
7. Xu, X., Weber, I., Staples, M.: Architecture for Blockchain Applications, 1st edn. Springer, Cham (2019). https://doi.org/10.1007/978-3-030-03035-3
8. Mendling, J., Weber, I., Aalst, W.V.D., et al.: Blockchains for business process management - challenges and opportunities. ACM Trans. Mange. Inf. Syst. **9**(1) (2018)
9. Di Ciccio, C., Meroni, G., Plebani, P.: Business process monitoring on blockchains: potentials and challenges. Enterpr. Bus. Process Inf. Syst. Model. **387**, 36–51 (2020)
10. Garcia-Garcia, J.A., Sanchez-Gomez, N., Lizcano, D., Escalona, M.J., Wojdynski, T.: Using Blockchain to improve collaborative business process management: systematic literature review. IEEE Access **8**, 142312–142336 (2020)
11. Nickerson, R.C., Varshney, U., Muntermann, J.: A method for taxonomy development and its application in information systems. Euro J. Inf. Syst. **22**(3), 336–359 (2013)
12. Williams, K., Chatterjee, S., Rossi, M.: Design of emerging digital services: a taxonomy. Euro J. Inf. Syst. **17**(5), 505–517 (2008)
13. Kitchenham, B., Charters, S.: Guidelines for performing systematic literature reviews in software engineering. Technical Report EBSE 2007–001 Version 2.3, Keele University and Durham University Joint Report (2007)

14. Martín-Martín, A., Orduna-Malea, E., Thelwall, M., Delgado López-Cózar, E.: Google Scholar, Web of Science, and Scopus: a systematic comparison of citations in 252 subject categories. J. Informet. **12**(4), 1160–1177 (2018)
15. García-Bañuelos, L., Ponomarev, A., Dumas, M., Weber, I.: Optimized execution of business processes on blockchain. In: Carmona, J., Engels, G., Kumar, A. (eds.) BPM 2017. LNCS, vol. 10445, pp. 130–146. Springer, Cham (2017). https://doi. org/10.1007/978-3-319-65000-5_8
16. López-Pintado, O., García-Bañuelos, L., Dumas, M., Weber, I., Ponomarev, A.: Caterpillar: A business process execution engine on the Ethereum blockchain. Softw. Pract. Exp. (2019) spe.2702
17. López-Pintado, O., Dumas, M., García-Bañuelos, L., Weber, I.: Dynamic role binding in blockchain-based collaborative business processes. In: Giorgini, P., Weber, B. (eds.) CAiSE 2019. LNCS, vol. 11483, pp. 399–414. Springer, Cham (2019). https://doi.org/10.1007/978-3-030-21290-2_25
18. Lu, Q., et al.: Integrated model-driven engineering of blockchain applications for business processes and asset management. Softw. Pract. Exp. **51**(5), 1059–1079 (2021)
19. Nakamura, H., Miyamoto, K., Kudo, M.: Inter-organizational business processes managed by blockchain. In: Hacid, H., Cellary, W., Wang, H., Paik, H.-Y., Zhou, R. (eds.) WISE 2018. LNCS, vol. 11233, pp. 3–17. Springer, Cham (2018). https:// doi.org/10.1007/978-3-030-02922-7_1
20. López-Pintado, O., Dumas, M., García-Bañuelos, L., Weber, I.: Interpreted execution of business process models on blockchain. In: EDOC, pp. 206–215. IEEE (2019)
21. López-Pintado, O., Dumas, M., García-Bañuelos, L., Weber, I.: Controlled flexibility in blockchain-based collaborative business processes. Inf. Syst. **104** (2022)
22. Mercenne, L., Brousmiche, K.L., Hamida, E.B.: Blockchain studio: a role-based business workflows management system. In: IEMCON, pp. 1215–1220. IEEE (2018)
23. Alves, P.H.C., et al.: Exploring Blockchain technology to improve multi-party relationship in business process management systems. In: ICEIS, vol. 2, pp. 817–825 (2020)
24. Brahem, A., Messai, N., Sam, Y., Bhiri, S., Devogele, T., Gaaloul, W.: Running transactional business processes with blockchain's smart contracts. In: 2020 IEEE International Conference on Web Services (ICWS), pp. 89–93. IEEE (2020)
25. Sturm, C., Szalanczi, J., Schönig, S., Jablonski, S.: A lean architecture for blockchain based decentralized process execution. In: BPM Workshops (2018)
26. Falazi, G., Hahn, M., Breitenbücher, U., Leymann, F., Yussupov, V.: Process-based composition of permissioned and permissionless blockchain smart contracts. In: EDOC, pp. 77–87. IEEE (2019)
27. Abid, A., Cheikhrouhou, S., Jmaiel, M.: Modelling and executing time-aware processes in trustless blockchain environment. In: Kallel, S., Cuppens, F., Cuppens-Boulahia, N., Hadj Kacem, A. (eds.) CRiSIS 2019. LNCS, vol. 12026, pp. 325–341. Springer, Cham (2020). https://doi.org/10.1007/978-3-030-41568-6_21
28. Ladleif, J., Weske, M., Weber, I.: Modeling and enforcing blockchain-based choreographies. In: Hildebrandt, T., van Dongen, B.F., Röglinger, M., Mendling, J. (eds.) BPM 2019. LNCS, vol. 11675, pp. 69–85. Springer, Cham (2019). https:// doi.org/10.1007/978-3-030-26619-6_7
29. Corradini, F., Marcelletti, A., Morichetta, A., Polini, A., Re, B., Tiezzi, F.: Engineering trustable choreography-based systems using blockchain. In: Symposium on Applied Computing, pp. 1470–1479. ACM (2020)

30. Prybila, C., Schulte, S., Hochreiner, C., Weber, I.: Runtime verification for business processes utilizing the Bitcoin blockchain. Fut. Gene. Comput. Syst. **107**, 816–831 (2020)
31. Corradini, F., Marcelletti, A., Morichetta, A., Polini, A., Re, B., Tiezzi, F.: Engineering trustable and auditable choreography-based systems using blockchain. In: SAC 2020: The 35th ACM/SIGAPP Symposium on Applied Computing, vol. 13, pp. 1–53 (2022)
32. Lichtenstein, T., Siegert, S., Nikaj, A., Weske, M.: Data-driven process choreography execution on the blockchain: a focus on blockchain data reusability. In: Abramowicz, W., Klein, G. (eds.) BIS 2020. LNBIP, vol. 389, pp. 224–235. Springer, Cham (2020). https://doi.org/10.1007/978-3-030-53337-3_17
33. Loukil, F., Boukadi, K., Abed, M., Ghedira-Guegan, C.: Decentralized collaborative business process execution using blockchain. World Wide Web **24**(5), 1645–1663 (2021). https://doi.org/10.1007/s11280-021-00901-7
34. Corradini, F., et al.: Model-driven engineering for multi-party business processes on multiple blockchains. Blockchain: Res. Appl. **2**(3) (2021)
35. Klinger, P., Nguyen, L., Bodendorf, F.: Upgradeability concept for collaborative blockchain-based business process execution framework. In: ICBC (2020)
36. Morales-Sandoval, M., Molina, J.A., Marin-Castro, H.M., Gonzalez-Compean, J.L.: Blockchain support for execution, monitoring and discovery of inter-organizational business processes. Peer J. Comput. Sci. **7**, e731 (2021)
37. Klinger, P., Bodendorf, F.: Blockchain-based cross-organizational execution framework for dynamic integration of process collaborations. In: Wirtschaftsinformatik (Zentrale Tracks), pp. 1802–1817 (2020)
38. Sturm, C., Scalanczi, J., Schönig, S., Jablonski, S.: A blockchain-based and resource-aware process execution engine. Fut. Gene. Comput. Syst. **100**, 19–34 (2019)
39. Bore, N., et al.: On using blockchain based workflows. In: ICBC, pp. 112–116. IEEE (2019)
40. Nagano, H., Shimosawa, T., Shimamura, A., Komoda, N.: Reliable architecture of cross organizational workflow management system on blockchain. Fut. Gene. Comput. Syst. **15**(2), 29–43 (2020)
41. Nagano, H., Shimosawa, T., Shimamura, A., Komoda, N.: Blockchain Based Cross Organizational Workflow Management System. AC 97–104 (2020)
42. Osterland, T., Rose, T., Putschli, C.: On the Implementation of Business Process Logic in DLT nodes. In: Asia Service Sciences and Software Engineering Conference, pp. 91–99. ACM (2020)
43. Adams, M., Suriadi, S., Kumar, A., ter Hofstede, A.H.M.: Flexible integration of blockchain with business process automation: a federated architecture. In: Advanced Information Systems Engineering, pp. 1–13 (2020)
44. Evermann, J.: Adapting workflow management systems to BFT blockchains-The YAWL example. In: EDOCW, pp. 27–36. ACM (2020)
45. Evermann, J., Kim, H.: Workflow management on proof-of-work blockchains: implications and recommendations. SN Comput. Sci. **2**(1), 1–22 (2021). https://doi.org/10.1007/s42979-020-00387-6
46. Evermann, J., Kim, H.: Workflow Management on BFT Blockchains. Enterprise Modelling and Information Systems Architectures (EMISAJ) **15**, 14–18 (2020)
47. Madsen, M.F., Gaub, M., Høgnason, T., Kirkbro, M.E., Slaats, T., Debois, S.: Collaboration among adversaries: Distributed workflow execution on a blockchain. In: Symposium on Foundations and Applications of Blockchain (2018)

48. Silva, D., Guerreiro, S., Sousa, P.: Decentralized enforcement of business process control using blockchain. In: Advances in Enterprise Engineering, pp. 69–87 (2019)
49. Meroni, G., Plebani, P., Vona, F.: Trusted artifact-driven process monitoring of multi-party business processes with blockchain. In: Di Ciccio, C., Gabryelczyk, R., García-Bañuelos, L., Hernaus, T., Hull, R., Indihar Štemberger, M., Kő, A., Staples, M. (eds.) BPM 2019. LNBIP, vol. 361, pp. 55–70. Springer, Cham (2019). https://doi.org/10.1007/978-3-030-30429-4_5
50. Reichert, M., Weber, B.: Enabling Flexibility in Process-Aware Information Systems. Springer, Berlin (2012). https://doi.org/10.1007/978-3-642-30409-5
51. Ladleif, J., Weske, M.: Time in blockchain-based process execution. In: EDOC, pp. 217–226. IEEE (2020)
52. Rimba, P., Tran, A.B., Weber, I., Staples, M., Ponomarev, A., Xu, X.: Quantifying the cost of distrust: comparing blockchain and cloud services for business process execution. Inf. Syst. Front. **22**(5), 1–19 (2018). https://doi.org/10.1007/s10796-018-9876-1
53. Meyer, A., Pufahl, L., Batoulis, K., Fahland, D., Weske, M.: Automating data exchange in process choreographies. Inf. Syst. **53**, 296–329 (2015)
54. Ladleif, J., Friedow, C., Weske, M.: An architecture for multi-chain business process choreographies. In: Abramowicz, W., Klein, G. (eds.) BIS 2020. LNBIP, vol. 389, pp. 184–196. Springer, Cham (2020). https://doi.org/10.1007/978-3-030-53337-3_14
55. Migliorini, S., Gambini, M., Combi, C., La Rosa, M.: The rise of enforceable business processes from the hashes of blockchain-based smart contracts. In: Reinhartz-Berger, I., Zdravkovic, J., Gulden, J., Schmidt, R. (eds.) BPMDS/EMMSAD - 2019. LNBIP, vol. 352, pp. 130–138. Springer, Cham (2019). https://doi.org/10.1007/978-3-030-20618-5_9

# Pupa: Smart Contracts for BPMN with Time-Dependent Events and Inclusive Gateways

Rodrigue Tonga Naha[(✉)] and Kaiwen Zhang

École de technologie supérieure, Université du Québec, 1100 Notre-Dame St W,
Montreal, QC H3C 1K3, Canada
rodrigue.tonga-naha.1@ens.etsmtl.ca, kaiwen.zhang@etsmtl.ca

**Abstract.** The digital transformation of business processes faces a major hindrance due to the lack of trust and transparency. As blockchain and other distributed ledger (DLT) are considered key enabling technologies, there is a need for supporting tools which can deploy business models over smart contracts in order to leverage these decentralized platforms. Existing blockchain-based Business Process Management (BPM) solutions support various blockchain platforms and different types of modelling language, i.e., Ethereum and Business Process Model Notation (BPMN). However, the majority of these methods do not support processes with time events and inclusive gateways due to severe limitations imposed by smart contract programming languages. In other words, mainstream blockchain platforms do not offer a straightforward way to execute a transaction at a later time. To overcome these aforementioned issues, we propose an engine called Pupa, a blockchain-based decentralized protocol to translate business processes with time events and inclusive gateways to smart contracts. Pupa accomplishes this by adding task feature to time events, check function on top of activities succeeding time events and, listening variables to sequence flow forking or joining inclusive gateways. We implemented Pupa by extending Caterpillar, an existing BPMN solution using Solidity and Ethereum, and evaluated the performance of our proposed engine and its generated smart contracts with a baseline solution. Our results show that Pupa is competitive with baseline solutions in terms of cost and performance, while offering additional advantages in terms of decentralization and supporting additional BPMN semantics.

**Keywords:** Blockchain · Business process management · Smart contract · BPMN · Timer · Inclusive gateway

## 1 Introduction

In the context of collaborative business processes, blockchain and distributed ledger technologies (DLTs) allow mutually untrusted parties to cooperate without need of a central authority. In order to leverage DLTs to provide transparency

© Springer Nature Switzerland AG 2022
A. Marrella et al. (Eds.): BPM 2022, LNBIP 459, pp. 21–35, 2022.
https://doi.org/10.1007/978-3-031-16168-1_2

and traceability to business processes, smart contracts must be deployed which can manage and validate information generated by the business processes. While these contracts can be implemented manually using specialized programming languages, it is desirable to rely on a translation tool to automatically translate models (e.g., written in BPMN) to smart contracts (e.g., written in Solidity), effectively bridging the gap between the two domains.

To this end, several approaches have been proposed, such as Caterpillar [13] and Lorikeet [27], and others [4,20,25]. These tools are able to translate many of the core functionalities of the BPMN language directly into smart contracts that can be readily deployed into a blockchain. However, they do not support two key BPMN features: timer events and inclusive gateways. The former is used to represent event triggered by a defined time while, the latter is used to create a combination of alternative and parallel paths of process flow where all paths are evaluated.

These two BPMN expressions are particularly challenging to implement in smart contracts because of temporal constraint restriction coming from smart contracts. On the other hand, the combination of parallel and exclusive gateway behaviour covered by inclusive gateway complexifies his usage and request a higher flexibility while designing.

In this paper, we want to address these limitations with our proposed solution, Pupa, which is a new BPMN engine capable of generating Solidity code which can be deployed on Ethereum. Our contributions in this paper are:

1. We provide support for time events generation in smart contracts. Our solution allows the user to specify a time duration input, which is verified while the smart contract is executed.
2. We provide support for inclusive gateways by employing a marking variable inside the generated smart contract which can properly execute the forking and joining behaviours according to BPMN semantics.
3. We implemented our solution by extending Caterpillar. We evaluated Pupa and its generated sample code and compared it against baseline solutions in terms of cost and performance.

The remainder of this paper is organized as follows: Sect. 2 introduces background concepts and related works for blockchains and BPMN. Section 3 provides details of our solution Pupa. Section 4 illustrates our approach with an order management use case. Section 5 contains the results of our experimental evaluation. And finally, Sect. 6 concludes the paper.

## 2      Background and Related Works

This section is subdivided in two parts. First, we describe background knowledge regarding smart contracts and BPMN. Then, we review works related to supporting timer events and inclusive gateways in BPMN smart contracts.

## 2.1  Background on Smart Contracts

The idea of smart contracts was presented initially in 1997 [26] and popularized with Ethereum blockchain. Smart contracts are computer programs deployed directly on the blockchain, which execute autonomously in response to certain triggers or transactions [21]. Smart contracts help blockchain technology in establishing trust between untrusted parties. Among the blockchain platforms implementing Smart contract, Ethereum is the most globally used [29]. Ethereum Smart contracts are code written in the Solidity language, and executed on the Ethereum Virtual Machine (EVM) once deployed. EVM is runtime component embedded within each full Ethereum node. To execute specific tasks, EVM uses a small set of low-level machine instructions (also called opcodes) but, sufficient enough to allow EVM to be Turing-complete. In order to store efficiency opcodes, they are encoded to bytecode. Every opcode is allocated a byte; therefore, the maximum number of opcodes is 256 ($16^2$). The EVM works as a stack-based virtual machine with a depth of 1024 items. Each item is a 256-bit word and only the top 16 items are accessible at given moment in the execution. Due to these limitations some opcodes use contract memory which is a not persistent memory, to retrieve or pass data. Since everyone running an Ethereum node can perform contract execution, an attacker could try to spam the network by creating contracts including lots of computationally expensive operations. In order to avoid accidental or hostile computation wastage code, a limit on computational steps of code execution to use for each transaction is required. The base unit of computation is called gas. Since a transaction include a gas limit, any gas not used in a transaction is returned to the user; however, any transaction with missing gas will be reverted, and the limited gas provided is consumed. The deployment cost is based on the Smart contract size, and measured in gas. The truthful execution of the code on Ethereum can serve to establish trust among untrusted parties.

## 2.2  Background on BPMN

To execute and perform processes inside a business process management system (BPMS), a standard language called BPMN, have been defined. BPMN is a communication tool used to represent business process flows. BPMN elements are classified in five basic categories [19]: flow Objects, data, connecting Objects, swimlanes, and artifacts. BPMS and BPMN have been used amply by companies to simplify and automate intra-organizational processes. But for inter-organizational processes, one major challenge remains: the lack of mutual trust [17]. Some existing BPMS have been able to combine blockchain and BPMN in order to, support the execution of collaborative business processes, between mutually untrusted participants. However, the vast majority of these solution does not support business process with timer event or inclusive gateway.

## 2.3    Background on Timer Events

A timer event is a type of event which can be used to influence the commence-
ment of an activity execution based on temporal constraints. Timer events can
be used as start event, intermediate event or boundary event. Figure 1 gives an
overview of timer events. A timer start event serves to create process instance
at a given time this means that a process should start only once and should
start in specific time intervals (example: every Tuesday) [5]. A timer intermediate catching event works as a stopwatch. This means that, when an execution is
triggered by a timer intermediate catching event, a timer begins. When the time
duration is over, the sequence flow outgoing of the timer intermediate catching
event is followed. A timer boundary event acts as a combination of stopwatch
and an alarm clock. When an execution reaches the activity to which the boundary event is attached, a timer is started. After a specified duration, the activity
is interrupted and the sequence flow leaving the timer event is followed. When
the execution of an activity is interrupted after a deadline, the timer event is
considered as an interrupting event in the other hand, when the original task or
sub process is not interrupted, the timer event is a non-interrupting event.

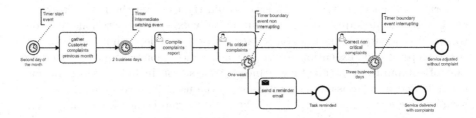

**Fig. 1.** Different types of timer events

## 2.4    Background on Inclusive Gateways

Inclusive gateway, equally called inclusive decision, is used to create alternative
and also, parallel paths inside a process flow. It is the combination of exclusive
and parallel gateway. With inclusive gateway, all conditions are evaluated, Unlike
the exclusive gateway, the true test of one condition expression does not exclude
the test of other condition. All paths produced by sequence flows with a true
evaluation, are taken. It should be designed such that all paths may be taken
(similarly to parallel gateway) or at least one path is taken [18] (OMG 2011).
Inclusive gateway can support fork behaviour and join behaviour [2]. In the fork
behaviour, all sequence flows going out of the gateway are evaluated; the sequence
flows with a true evaluation will be followed in parallel and the activities attached
to these sequence flows will be executed concurrently. In the join behaviour, all
the concurrent sequence flows incoming to the inclusive gateway will be checked
and the inclusive gateway will only wait for the incoming sequence flows that are
executed. After that, the process execution can move to the next step. Figure 2
shows the fork and join concept.

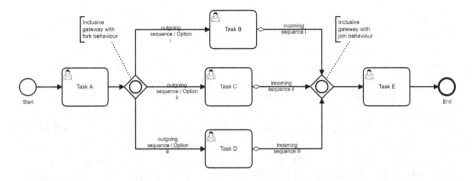

**Fig. 2.** Inclusive gateway - fork & join view

## 2.5   Related Works on Process-Oriented Smart Contract Solution

Thanks to its properties, Blockchain technology becomes more and more attractive in the field of process-centric solution. Therefore, there exist many approaches of Blockchain-based business process management systems. These approaches can be classified according to the type of blockchain used. Concerning the public blockchain, the majority of blockchain-based BPMS are made on top of Ethereum. [25] provides a light framework supporting the process execution on Ethereum. [27] uses the features of model-driven engineering to simplify the development of blockchain-based process-oriented applications. [28] has presented a collaborative approach of blockchain-based process execution with, a major drawback about expensive gas cost for data storage on chain. In order to improve the previous work, [7] has presented an optimized solution consisting of converting BPMN process into petri net and, compiling the output petri net into a smart contract written in Solidity. However, this solution covers a small set of BPMN items and does not care about access control of process participants on chain. Next, [13] designed a mature solution called Caterpillar which implements the translation of BPMN 2.0 constructs into smart contract. Caterpillar supports a large set of BPMN item, process state and execution handling. In order to improve this first version of Caterpillar, [14] and [11], provide a compiled version of Caterpillar with integration of role-based access policy. Despite the fact that Caterpillar is a remarkable solution, it does not accept timer events and inclusive gateways.

## 2.6   Related Works on Timer Events

According to [6], and [3], design and management of temporal specifications in the business process area is definitely a critical topic of research. In blockchain-based BPMS field, the integration of temporal constraints is supported by a very limited research works. However, some remarkable framework such as Lorikeet [27], Caterpillar [14], Chorchain [4,25], and the interpreted version of Caterpillar [12], do not support at all time events. [8] discuss about improving execution time from off chain component without implementing timer events.

[10] admits some challenges on timer implementation and, put it as a potential future improvement. [1] provides an approach which extends Caterpillar and consists of adding time-guards inside the functions of activities implementing temporal constraints such as task duration, absolute start/end times. However, this approach doesn't provide a clear link between temporal constraints implemented and the different variants of timer events as described by OMG 11 [18]. Mavridou and Laszka [16] use finite-state machine modelling language to generate Solidity code while taking into consideration delayed processes, and block timestamps on Ethereum. Ladleif et al. [9] present a good discussion paper about challenges and alternatives solutions available regarding the integration of time constraints in blockchain-based BPMS. After comparing properties of these solutions, they present some hints which can be helpful to choose the right answer for specific scenarios.

### 2.7   Related Works on Inclusive Gateways

Large number of works regarding blockchain-based BPMN engine, do not support inclusive gateway. Among them we can find some notable frameworks like Caterpillar [12,14], Lorikeet [20,27]. On the other hand, [23] and [24] have integrated inclusive gateways in a blockchain-based business process. Their approach consists of formalizing execution of semantics of BPMN inclusive gateways on blockchain. However, these implementations face some limitations such as the deviation of the execution semantics from the BPMN standard, and the lack of support of non-block-structured process. Schinle et al. [22] present an approach for the integration, execution and monitoring of modelled business processes based on Hyperledger Fabric's chaincode. This approach supports inclusive gateways in theory, but no implementation is provided. Loukil et al. [15] provide a decentralized collaborative business process solution, builds on top of Ethereum, which supports gateway elements. However, no clear implementation of inclusive gateway is presented.

## 3   Details of the Proposed Solution

In this section, we present details of our proposed solution Pupa. First, we provide details of our solution for support timer events. Then, we explain how to support inclusive gateways. Finally, we demonstrate how Pupa functions using a case study modelled after customer order management process.

### 3.1   Handling Time-Dependent Events

Our proposed timer solution has the following properties:

1. **Timer event as user task.** Timer event acts similarly as User task or Service task. Unlike other implementations [14], the timer event supports input data, function, and role access. Therefore, a user can input delay time in second

on timer event while executing the deployed business process smart contract. Only users having the same role than the role linked to timer, can perform operation on timer.

2. **Time check done on next node.** Since it is not possible to trigger execution of solidity code at a given time with EVM, authorized user will perform this check. Execution of activity succeeding timer event in business process model is done in two steps. The first step is to check if the duration time previously specified on timer event is over. If the duration is completed then execution flow will call the second step which is the real execution function of the activity.

3. **Timer variables and timer functions.** Timer variables and timer functions are created dynamically meaning that for a business process with no timer, a smart contract with no timer variables and no timer functions is generated.

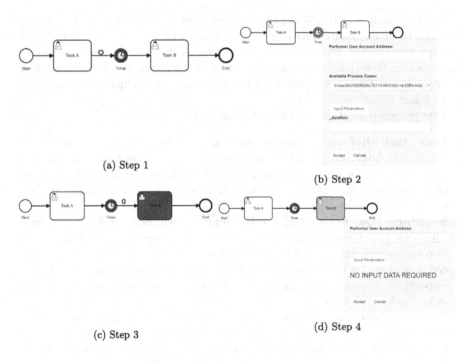

(a) Step 1

(b) Step 2

(c) Step 3

(d) Step 4

**Fig. 3.** User flow for timer events

The implementation of timer event follows a token mechanism previously used by Caterpillar. Firstly, the incoming sequence flow of timer receives process token, after that, timer event is executed. At this step, an authorized user will provide the timer duration in second ($d$) and validate. This transaction is saved on blockchain and the timestamp of associated block ($T_b$) is used as the starting point of the timer stopwatch. Choosing block timestamp as time reference, helps to prevent time variation coming from various Ethereum node. The third step will move process token to the outgoing sequence flow of timer. In the last step, the token arrives on node following timer event. At this phase, a test condition consisting to verify authorized role attached to node. During this transaction, the timestamp of the current block ($T_{cb}$) is captured and If the duration previously set is over then the step is completed. In order words, if the remaining time is equal or less than 0, then the execution process will move if not, the check will take place again. In order to ease the interaction with user and reduce gas, in case of the result of test condition is false, the remaining waiting time ($T_r$) is output. Expression 1 describes the value of remaining time ($T_r$). Figure 3 illustrates the entire process for a user.

$$T_r = (T_b + d) - T_{cb} \tag{1}$$

### 3.2  Supporting Inclusive Gateways

Inclusive gateway is a BPMN element classified under the category Flow objects. It is used to create an alternative and parallel path inside a process flow. Inclusive gateway can support multiple outgoing sequences flow (fork), multiple incoming sequences flow (join) or, both fork and join behaviours. Our implementation covers inclusive gateways supporting fork behaviour, and inclusive gateways supporting join behaviour.

Our proposed solution has the following characteristics:

1. **User task before inclusive gateway.** Inclusive gateway with fork behaviour should be preceded by a user task. The user task will offer the opportunity to choose the path to follow. All available choices can be selected or at least one. If no choice is selected, an error is raised and the process token will not move.

2. **Sequence flow associated to Boolean expression.** Inclusive gateway with fork behaviour should have each outgoing sequence flow linked to a Boolean expression. Also, inclusive gateway with join view and which is preceded near or far by an inclusive gateway with fork behaviour, should have respectively incoming sequence flow and outgoing sequence flow linked to the same Boolean variable.

3. **Inclusive gateway variables and functions.** The presence of inclusive gateway in a business process will create customized variables in Smart contracts. There are not specific functions for Inclusive gateway. Operations about inclusive decisions are handled inside the workflow component. These instructions have two major purposes. Firstly, to ensure that in fork behaviour, process token moves only to the selected sequence flow. If no choice is made, nothing will happen. Secondly, these instructions should ensure that in join behaviour, process token moves to the next node following inclusive gateway only if, all the incoming chosen paths of this gateway have been completed.

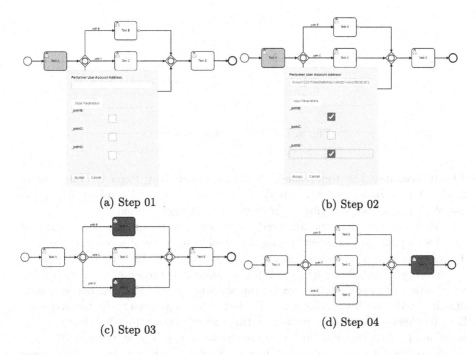

(a) Step 01          (b) Step 02

(c) Step 03          (d) Step 04

**Fig. 4.** User flow for inclusive gateways

Our solution consists of four steps, as shown in Fig. 4. In Step 1 and Step 2, an activity is inserted before inclusive gateway forking. This activity is used to choose path to follow. As soon as paths are chosen, the process token will move to the corresponding path and, variables associated to outgoing and incoming sequence flows are updated. In Step 3, activities found in the paths holding process token, are performed. When process token arrives to the inclusive gateway joining, the gateway will compare the number of tokens received with the number of paths previously chosen. If there are equals, process execution flow will move to activity next to inclusive gateway. This move represents Step 4.

## 4   Use Case Study

We demonstrate our approach with a walk-through example shown in Fig. 5, which models a business process managing a customer order for an Internet Service Provider company. When a customer asks for service from ISP, a quotation including the installation equipment and based on customer location, is sent to customer. Customer has three days to accept or refuse the quotation. After accepting the quotation, the next step is the preparation of customer order. Order can support at least one of the following services: incentive, shipment and, sending order only. After all the selected service have been completed, data from customer installation are recorded.

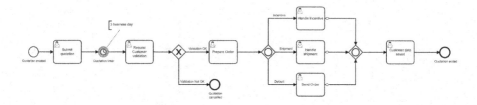

**Fig. 5.** Use case - customer order

**Timer Events:** The timer contract code generated by Pupa is presented in Listing 1.1. In order to better understand this code, we will explain briefly the use of *step* function, *marking* variable and bit-wise operations.

*step* is an internal function used to update the business process state. It handles step of sequence flow during the whole execution of a business process. To achieve this, *step* function uses *marking* variable and bit-wise operations.

*marking* is a global variable (256-bits unsigned integers) in charge of the distribution of process token across the sequence flows. Each sequence flow is mingled with one bit in this variable: 1 if a token is present in the sequence flow, 0 otherwise Values supported by this variable is equal to $2^i$ (where $i$ is the position of a sequence flow inside the business process flow starting at 0).

Regarding bit-wise operations, they are used to handle queries on the process state. *AND* (&) is used to verify if an element is started or enabled and allows testing set inclusion. *OR* (|) operator provides a method to encode the set union as an integer. Finally, the combination of *NOT* (~) and *AND* (&) serves to replace the old token from the variable *marking* by a new one.

In line 1 and 2, the step function is declared and while loop is invoked. After crossing tasks before timer event (line 3), the process token arrives at timer event node. After a successful check done on *marking* variable, the *start* function of timer node is called, the index of corresponding timer will be used to initialize the mapping between timer node and its incoming and outgoing sequence flow id and, the token will be replaced (Lines 4–10). The *start* function is associated to each activity and is used to register activity index and the authorized role address for this activity. In lines 11–20, the token moves to the node just after timer. If the timer event has not yet been initialized, only the *start* function of this node is called otherwise, the *start* function is called and the process token is removed from the node. Note that the start function of this node will check always if node is already registered to avoid duplicated records.

**Inclusive Gateways:** Listing 1.2 shows the generated code for forking inclusive gateways. The *marking* variable will move token only on sequence flow with a Boolean variable set true (Lines 1–9). In lines 10–27 code of activity located on outgoing paths of the fork inclusive gateway is represented. Only activities located on path holding process token will be executed.

When the token arrives at the joining inclusive gateway (Listing 1.3), an initializing check is done. A test is done on each incoming path to know if it holds or not process token. After this test, the number of expected incoming path holding token is linked with the index of inclusive gateway (Lines 1–11). If the gateway have been already initialized as shown in lines 12–21, the number of expected incoming path holding process token is compared to the number of incoming path already completed. If they are equals token will move to activity next to the gateway and the global variable called *inclusiveGatewayCounter* representing counter of incoming path performed, is set to 0. Otherwise, the token will only be removed on this inclusive gateway. Lines 22–39 represent the generated code of all incoming sequence flow joining to the inclusive gateway. For each sequence, the Boolean variable attached to is tested. If it is true, *inclusiveGatewayCounter* is incremented and the token move to inclusive gateway position.

```
1    function step(uint tmpMarking, uint tmpStartedActivities) internal {
2        while (true) {
3            ...
4            if (tmpMarking & uint(4) != 0) {
5                Use_case_AbstractWorlist(worklist).Quotation_timer_start(2);
6                timerFlows[uint(2)] = timerFlow(uint(4), uint(8));
7                tmpMarking &= uint(~4);
8                tmpStartedActivities |= uint(4);
9                continue;
10           }
11           if (tmpMarking & uint(8) != 0) {
12               if (TimerNodes[uint(3 -1)].timerRoleAddr == address(0)) {
                     Use_case_AbstractWorlist(worklist).Customer_validation_start(3);
13               } else {          Use_case_AbstractWorlist(worklist).Customer_validation_start(3);
14                   tmpMarking &= uint(~8);
15                   tmpStartedActivities |= uint(8);
16               }
17               continue;
18           }
```

**Listing 1.1.** Step Function - Timer Instructions

```
1        if (tmpMarking & uint(128) == uint(128)) {
2            if (handleInc)
3                tmpMarking = tmpMarking & uint(~128)| uint(256);
4            if (handleShip)
5                tmpMarking = tmpMarking & uint(~128)| uint(512);
6            if (sendOrder)
7                tmpMarking = tmpMarking & uint(~128)| uint(1024);
8            continue;
9        }
10       if (tmpMarking & uint(256) != 0) {
11           Use_case_AbstractWorlist(worklist).Handle_Incentive_start(8);
12           tmpMarking &= uint(~256);
13           tmpStartedActivities |= uint(256);
14           continue;
15       }
16       if (tmpMarking & uint(512) != 0) {
17           Use_case_AbstractWorlist(worklist).Handle_shipment_start(9);
18           tmpMarking &= uint(~512);
19           tmpStartedActivities |= uint(512);
20           continue;
21       }
22       if (tmpMarking & uint(1024) != 0) {
23           Use_case_AbstractWorlist(worklist).Send_Order_start(10);
24           tmpMarking &= uint(~1024);
25           tmpStartedActivities |= uint(1024);
26           continue;
27       }
```

**Listing 1.2.** Step Function - Inclusive Gateway Fork Behaviour

```
 1        if (tmpMarking & uint(14336) == uint(14336)) {
 2            if(SelecIncomInclNodes[uint(11)] == 0) {
 3                uint internalcter = 0;
 4                if (handleInc)
 5                    internalcter +=1;
 6                if (handleShip)
 7                    internalcter +=1;
 8                if (sendOrder)
 9                    internalcter +=1;
10                SelecIncomInclNodes[uint(11)] = internalcter;
11            }
12            else {
13                if(SelecIncomInclNodes[uint(11)] == inclusiveGatewayCounter) {
14                    tmpMarking = tmpMarking & uint(~14336) | uint(16384);
15                    inclusiveGatewayCounter = 0;
16                }
17                else
18                    tmpMarking = tmpMarking & uint(~14336);
19            }
20            continue;
21        }
22        if (tmpMarking & uint(2048) != 0 && (handleInc)) {
23            inclusiveGatewayCounter +=1;
24            tmpMarking &= uint(~2048);
25            tmpMarking |= uint(14336);
26            continue;
27        }
28        if (tmpMarking & uint(4096) != 0 && (handleShip)) {
29            inclusiveGatewayCounter +=1;
30            tmpMarking &= uint(~4096);
31            tmpMarking |= uint(14336);
32            continue;
33        }
34        if (tmpMarking & uint(8192) != 0 && (sendOrder)) {
35            inclusiveGatewayCounter +=1;
36            tmpMarking &= uint(~8192);
37            tmpMarking |= uint(14336);
38            continue;
39        }
```

**Listing 1.3.** Step Function - Inclusive Gateway Join Behaviour

The source code of Pupa can be downloaded under the BSD 3-clause License from https://github.com/rodrigueNTprojects/Pupa.

## 5    Result and Evaluation

This section presents an experimental evaluation of Pupa. The goal here is to compare execution cost between our solution and Caterpillar.

**Table 1.** Deployment cost comparison between Pupa and Caterpillar

| Contracts | Deployment cost | | |
|---|---|---|---|
| | Caterpillar | Pupa | Difference |
| Statics | | | |
| Process registry | 597895 | 599947 | (2052) |
| BindingPolicy (For 2 role) | 159859 | 158863 | 996 |
| Dynamic | | | |
| Process with 3 tasks | 1405839 | 1380197 | 25642 |
| Process with 10 tasks | 2379973 | 2368627 | 11346 |
| Process with 20 tasks | 3812180 | 3797209 | 14971 |
| Process with 35 tasks | 5991255 | 5967344 | 23911 |
| Process with 40 tasks | N/A | 6742984 | |
| Process with 45 tasks | N/A | 7435775 | |

## 5.1 Compiling and Deploying New Smart Contracts

Contracts generated from Pupa are written in Solidity 0.6.12. Security aspect have been taken in consideration. In particular, we run a smart contract security profiler tool called Slither against statics and dynamics contracts, in order to reduce any risk. Contracts are generated, compiled and deployed thanks to JavaScript components such as EJS (Embedded JavaScript template) and TS (TypeScript).

## 5.2 Gas and Performance Evaluation

To run our evaluation, we tested in series many processes. Our goal was to analyze gas consumption, and the application load. Because Ethereum blocks have limited sizes, we tried to understand how many basics tasks a process can support when it is generated once with Pupa. We conducted gas evaluation in order to ensure that Pupa is optimized and cheaper in use. We compared deployment cost of Pupa with Caterpillar. Tests are performed on local blockchain network called Ganache. Table 1 demonstrates that Pupa consumes less gas than Caterpillar. Also, processes with more than 35 tasks cannot be executed once on Caterpillar while Pupa is able to do so.

## 6 Conclusion

Time events and inclusive gateways are core features of BPMN which are difficult to support in smart contract generation tools due to the limitations of blockchain environments. In this paper, we present Pupa, a Solidity generator for BPMN which supports the missing features. Timer events are handled by adding check functions on top of the execution of activity succeeding to time event. Pupa supports inclusive gateways by attaching a marking variable to properly support joining and forking behaviours. We implemented our solution by extending Caterpillar and compared against a known baseline. For future work, we aim to implement the deferred choice timer event in order to provide a listenable feature which triggers when some events have not been selected. Furthermore, we plan to extend Pupa across multiple platforms, to facilitate the adoption of our tool on popular enterprise blockchain systems such as Hyperledger Fabric and Quorum.

## References

1. Abid, A., Cheikhrouhou, S., Jmaiel, M.: Modelling and executing time-aware processes in trustless blockchain environment. In: Kallel, S., Cuppens, F., Cuppens-Boulahia, N., Hadj Kacem, A. (eds.) CRiSIS 2019. LNCS, vol. 12026, pp. 325–341. Springer, Cham (2020). https://doi.org/10.1007/978-3-030-41568-6_21
2. Camunda.org: Camunda 7 docs, inclusive gateway (2022). https://docs.camunda.org/manual/7.16/reference/bpmn20/gateways/inclusive-gateway/

3. Cheikhrouhou, S., Kallel, S., Guermouche, N., Jmaiel, M.: The temporal perspective in business process modeling: a survey and research challenges. SOCA **9**(1), 75–85 (2014). https://doi.org/10.1007/s11761-014-0170-x
4. Corradini, F., Marcelletti, A., Morichetta, A., Polini, A., Re, B., Tiezzi, F.: Engineering trustable choreography-based systems using blockchain. In: Proceedings of the 35th Annual ACM Symposium on Applied Computing, pp. 1470–1479 (2020)
5. Dumas, M., La Rosa, M., Mendling, J., Reijers, H.A., et al.: Fundamentals of Business Process Management, vol. 1. Springer, Heidelberg (2013). https://doi.org/10.1007/978-3-642-33143-5
6. Eder, J., Panagos, E., Rabinovich, M.: Time constraints in workflow systems. In: Jarke, M., Oberweis, A. (eds.) CAiSE 1999. LNCS, vol. 1626, pp. 286–300. Springer, Heidelberg (1999). https://doi.org/10.1007/3-540-48738-7_22
7. García-Bañuelos, L., Ponomarev, A., Dumas, M., Weber, I.: Optimized execution of business processes on blockchain. In: Carmona, J., Engels, G., Kumar, A. (eds.) BPM 2017. LNCS, vol. 10445, pp. 130–146. Springer, Cham (2017). https://doi.org/10.1007/978-3-319-65000-5_8
8. Klinger, P., Bodendorf, F.: Blockchain-based cross-organizational execution framework for dynamic integration of process collaborations. In: Wirtschaftsinformatik (Zentrale Tracks), pp. 1802–1817 (2020)
9. Ladleif, J., Weske, M.: Time in blockchain-based process execution. In: 2020 IEEE 24th International Enterprise Distributed Object Computing Conference (EDOC), pp. 217–226. IEEE (2020)
10. Ladleif, J., Weske, M., Weber, I.: Modeling and enforcing blockchain-based choreographies. In: Hildebrandt, T., van Dongen, B.F., Röglinger, M., Mendling, J. (eds.) BPM 2019. LNCS, vol. 11675, pp. 69–85. Springer, Cham (2019). https://doi.org/10.1007/978-3-030-26619-6_7
11. López-Pintado, O., Dumas, M., García-Bañuelos, L., Weber, I.: Dynamic role binding in blockchain-based collaborative business processes. In: Giorgini, P., Weber, B. (eds.) CAiSE 2019. LNCS, vol. 11483, pp. 399–414. Springer, Cham (2019). https://doi.org/10.1007/978-3-030-21290-2_25
12. López-Pintado, O., Dumas, M., García-Bañuelos, L., Weber, I.: Interpreted execution of business process models on blockchain. In: 2019 IEEE 23rd International Enterprise Distributed Object Computing Conference (EDOC), pp. 206–215. IEEE (2019)
13. López-Pintado, O., García-Bañuelos, L., Dumas, M., Weber, I.: Caterpillar: a blockchain-based business process management system. In: BPM (Demos), vol. 172 (2017)
14. Lòpez-Pintado, O., García-Bañuelos, L., Dumas, M., Weber, I., Ponomarev, A.: Caterpillar: a business process execution engine on the Ethereum blockchain. Softw. Pract. Exp. **49**(7), 1162–1193 (2019)
15. Loukil, F., Boukadi, K., Abed, M., Ghedira-Guegan, C.: Decentralized collaborative business process execution using blockchain. World Wide Web **24**(5), 1645–1663 (2021). https://doi.org/10.1007/s11280-021-00901-7
16. Mavridou, A., Laszka, A.: Designing secure ethereum smart contracts: a finite state machine based approach. In: Meiklejohn, S., Sako, K. (eds.) FC 2018. LNCS, vol. 10957, pp. 523–540. Springer, Heidelberg (2018). https://doi.org/10.1007/978-3-662-58387-6_28
17. Mendling, J., et al.: Blockchains for business process management-challenges and opportunities. ACM Trans. Manag. Inf. Syst. (TMIS) **9**(1), 1–16 (2018)
18. Object Management Group BPMN Technical Committee: Business process model and notation, version 2.0. OMG Document Number Formal/2011-01-03 (2011)

19. Object Management Group Business Process Model: Notation (BPMN) version 2.0.2, Object Management Group, 2013 (2016). http://www.omg.org/spec/BPMN/2.0

20. Prybila, C., Schulte, S., Hochreiner, C., Weber, I.: Runtime verification for business processes utilizing the Bitcoin blockchain. Future Gener. Comput. Syst. **107**, 816–831 (2020)

21. Savelyev, A.: Contract law 2.0: 'smart' contracts as the beginning of the end of classic contract law. Inf. Commun. Technol. Law **26**(2), 116–134 (2017)

22. Schinle, M., Erler, C., Andris, P.N., Stork, W.: Integration, execution and monitoring of business processes with chaincode. In: 2020 2nd Conference on Blockchain Research & Applications for Innovative Networks and Services (BRAINS), pp. 63–70. IEEE (2020)

23. Sturm, C., Scalanczi, J., Schönig, S., Jablonski, S.: A blockchain-based and resource-aware process execution engine. Future Gener. Comput. Syst. **100**, 19–34 (2019)

24. Sturm, C., Szalanczi, J., Jablonski, S., Schönig, S.: Decentralized control: a novel form of interorganizational workflow interoperability. In: Grabis, J., Bork, D. (eds.) PoEM 2020. LNBIP, vol. 400, pp. 261–276. Springer, Cham (2020). https://doi.org/10.1007/978-3-030-63479-7_18

25. Sturm, C., Szalanczi, J., Schönig, S., Jablonski, S.: A lean architecture for blockchain based decentralized process execution. In: Daniel, F., Sheng, Q.Z., Motahari, H. (eds.) BPM 2018. LNBIP, vol. 342, pp. 361–373. Springer, Cham (2019). https://doi.org/10.1007/978-3-030-11641-5_29

26. Szabo, N.: Formalizing and securing relationships on public networks. First Monday (1997). https://journals.uic.edu/ojs/index.php/fm/article/view/548

27. Tran, A.B., Lu, Q., Weber, I.: Lorikeet: a model-driven engineering tool for blockchain-based business process execution and asset management. In: BPM (Dissertation/Demos/Industry), pp. 56–60 (2018)

28. Weber, I., Xu, X., Riveret, R., Governatori, G., Ponomarev, A., Mendling, J.: Untrusted business process monitoring and execution using blockchain. In: La Rosa, M., Loos, P., Pastor, O. (eds.) BPM 2016. LNCS, vol. 9850, pp. 329–347. Springer, Cham (2016). https://doi.org/10.1007/978-3-319-45348-4_19

29. Xu, X., et al.: A taxonomy of blockchain-based systems for architecture design. In: 2017 IEEE International Conference on Software Architecture (ICSA). IEEE, April 2017. https://doi.org/10.1109/icsa.2017.33

# A Systematic Local Fork Management Framework for Blockchain Sandbox Environments

Antreas Pogiatzis$^{(\boxtimes)}$ and Georgios Samakovitis

University of Greenwich, Old Royal Naval College, Park Row, London SE109LS, UK
a.pogiatzis@greenwich.ac.uk

**Abstract.** Blockchain technology presently permeates multiple industries, resulting in an increasing number of agents (primarily end-users and engineers) interacting with it in a variety of contexts. This, in turn, has introduced the practical need for a sufficient tooling ecosystem for blockchain solutions testing and evaluation under secure environment conditions. To that end, this study presents a robust framework for the creation and management of customisable, persistent, private, scalable blockchain environments that fulfill precisely the need for live sandbox platforms with on-chain interaction capabilities. We extend the concept of local chain forking, emphasising the limitations of existing tooling and methodologies and propose how our framework mitigates identified weaknesses and bridges some of the resulting gaps in desirable non-functional attributes. We offer a reference implementation for our framework and discuss how it can be applicable to a broad array of DevOps and security-oriented Use Cases. Finally, emerging challenges are discussed and potential directions for further research are drawn.

**Keywords:** Blockchain · Chain fork · Sandbox · Framework · DevOps · Security

## 1 Introduction

Since the original seminal work on Bitcoin and the introduction of blockchain as its underlying technology platform [26], the space has evolved at an unparalleled pace. Relevant literature [20,33] typically places strong emphasis on the technology's capability to facilitate distributed and verified execution of arbitrary logic through smart contracts [25], driving in turn the emergence of a new class of applications - also known as decentralised applications (dApps) [15]. Along with new opportunities, the advent of smart contracts comes with a multitude of previously unseen challenges [34]. Ironically, a part of these stem directly from the very virtues that earned blockchain its popularity, namely immutability and transparency.

For the past few years, decentralised applications were deemed applicable across numerous industry fields [23], including financial transactions, supply chain solutions, healthcare records management and IoT, to name a few. The massive

© Springer Nature Switzerland AG 2022
A. Marrella et al. (Eds.): BPM 2022, LNBIP 459, pp. 36–50, 2022.
https://doi.org/10.1007/978-3-031-16168-1_3

interest towards blockchain technology has given rise to an increasingly growing number of software engineers tuning their interest to that space. Yet with modern blockchains introducing a non-traditional platform for computation (in the sense that they operate on a distributed network, governed in its entirety by a decentralised community) traditional software development lifecycle practices do not always apply [24]. What is more, testing has increased significance in blockchain environments, since updating a smart contract is a non-trivial task due to the inherent immutability of the blockchain and, more importantly, faulty smart contracts can result to substantial loss of funds [17]. Developers are hence faced with significant challenges in testing their smart contracts as they often refrain from deploying on mainnets to avoid public exposure and costs.

Even though testnets offer sandbox environments for trialling smart contracts without incurring financial costs, they are still too transparent for production testing, potentially leaking information about the development of the contract. Furthermore, dApps with dependencies on third party smart contracts such as DeFi protocols, cannot be tested reliably on testnet due to the uncertainty of the state of their dependencies which may be outdated or nonexistent.

With these limitations in mind, a common approach is to resort to local mainnet forks which replicate the state of the mainnet in a local instance of the network that can be used for testing. This is an acceptable and convenient solution for short manual tests, unit or integration tests in isolation. Nonetheless, local forks are most often volatile and short-lived by design, and hence not be suitable in a multi-user setting that uses a forked version of a blockchain. In this work, we argue that maintaining synchronised long-lived forks of the mainnet chains supports tooling for more flexible and maintainable DevOps, multi-user smart contract testing, creation of sandbox environments for education and training, traceable auditing and in-depth security analysis of smart contracts.

Considering that, at the time of writing, more than 70% of the total value locked (TVL) is held by dApps that are built on blockchains powered by the Ethereum Virtual Machine (EVM) [9], we focus our work only on EVM-based blockchains. Consequently the presented framework is bounded by the functionality that EVM entails.

The contribution of this paper is manifold. For one, it introduces a systematic, scalable framework for managing multiple internal non-immutable blockchain environments through chain forking, hence offering a methodological approach for tooling, scarcely addressed in the literature so far. Second, it discusses an indicative set of real-life practical use cases for applying the framework, drawing on the versatile applicability of chain forks and hence stressing the extensive potential of the tool. Finally, it provides a reference implementation for this framework, precisely on the back of aforementioned use cases, and taking account of their common challenges and features.

The rest of the paper is structured as follows: Sect. 2 offers the conceptual and technical frame and rationale of chain forks and their operation and features. Section 3 provides the background rationale of our proposed design, followed by a complete presentation of the framework and its architecture, in Sect. 4.

Section 5 then offers two indicative practical use cases, also drawing on underlying particularities and challenges. The outline implementation follows (Sect. 6) and the discussion of limitations and future directions of our proposed approach in Sects. 7 and 8 close the paper.

## 2   Preliminaries

Blockchain's immutability stems from the cryptographic primitives of the consensus protocol being used. At the end of the mining or validation process, blocks are being appended to the main chain, thereby extending the block trail with new transactions that modify the global state of the network. The data in this trail is the single source of truth of the events that occurred since the genesis of the chain. It is often the case that two blocks are validated or mined at the same time, resulting to two versions of the same chain until that point, a structure commonly known as a chain fork (Fig. 1), also known as chain "forking". The chain inconsistencies that originally result to forking are resolved at later blocks by the principle of the longest chain.

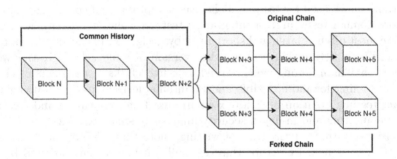

**Fig. 1.** Chain forking

Although forks can happen accidentally, by design of the consensus protocol, there are cases where forks are intentionally created. For instance, forks may be introduced as interim chain versions during protocol updates to facilitate gradual user migration to the new version of the chain. Alternatively, a fork may be used to settle a dispute or controversial decision amongst the community of the network, occasionally resulting to a different sustainable version of a blockchain sharing its history with the parent chain until the forked block number [21].

More importantly, chain forking is a crucial step in blockchain-based software development. Very often, dApps and protocols interact with other dApps already deployed on the mainnet. Given that determinism is an exigency of unit and integration tests, relying on mainnet, or even testnets, for executing the tests is unreliable and introduces confounding variables during the testing process. Instead, a widely used development practice is to create a short-lived local mainnet fork. That is, a forced chain fork of the mainnet chain which runs in a simulated environment to provide granular control over the state of the network.

Despite their recognised usefulness for unit testing and other short-term code verification, and however popular, intentional short-lived forks come with significant limitations as tools for blockchain software engineering [18]. Indicatively, they are unsuitable for maintaining a desired state when longer testing periods are required, as is the case with more complex scenarios in security testing for smart contracts; they are furthermore often unfit for multi-user interaction testing or when realistic network conditions are required.

# 3   Related Work

Systematic fork management and implementation frameworks have so far received a relatively sparse treatment in the related blockchain literature. Recent work emphasises on the challenges that emerge in Blockchain Oriented Software Engineering (BOSE) [18], with previous research focusing on tooling approaches that address BOSE challenges in isolation; of these, security testing appears to receive increased attention [29], especially given the recurrent financial impact associated with dApps exploits.

Conversely, more emphasis has been placed on advanced dedicated simulation tools (discussed below) which, while offering relatively effective evaluation instruments, do not directly support 'live' testing environments. That latter task is at the core of our proposed framework, while we discuss below a number of the more popular blockchain simulation tools.

BlockSim is an advanced blockchain simulation software tool by Alharby et al. [11], that builds on top of their relevant framework [12]. Although its flexible structure allows to address design and deployment concerns of blockchain solutions, it focuses on protocol rather than decentralised application evaluation. Its configuration allows the imitation of multiple blockchain architectures such as Bitcoin or Ethereum, yet, the absence of support for connecting to public mainnet networks prevents its use for the integration of dApps and replication of existing transactions.

On the other hand, EtherClue [14] and TxSpector [32] are security-oriented tools for post-factum detection and analysis of smart contract attacks. Their functionality is intended to be short-lived without any persistency or longevity capabilities with respect to the local blockchain state. Such qualities are central to our proposal: While our framework does not feature the detection and tracing capabilities of the aforementioned tools, it provides suitable tooling for post-factum attack analysis and allows for plugins to extend detection features.

A different contribution is that of Alsahan et al. [13], offering a Bitcoin network simulator to asses blockchain performance under various network conditions. Although their scope and objectives diverge from the research direction of this work, their data acquisition and performance analysis module is similar to our proxying architecture in the node layer.

Finally, Wang et al. [30] introduced a FinTech-focused blockchain-powered regulatory sandbox, which also consists of a *virtual sandbox*, conceptually similar to our framework without however offering any technical details.

To the best of our knowledge, while the above approaches offer some efficient tooling for mitigating smart contract design risks, they primarily operate under an ephemeral off-chain, simulated environment, and arguably lack the systematic structure necessary in cases where mutability and multi-user interaction is needed. In that sense, the present work is arguably one of the first attempts to address the limitations and challenges of existing approaches through offering an open framework for managing blockchain environments suited to applications in multiple areas, including security, training or development.

## 4    Proposed Framework

### 4.1    Scope

Although a multitude of approaches have been published by both the scientific community and industry to address the peculiarities of blockchain-based development, there still exists a gap in effectively managing internal mainnet-based blockchain environments. We acknowledge that decentralised governance, immutability and transparency may not be desired attributes in cases where granular control is essential. To that end, we propose a generic framework to build and maintain persistent, portable and controllable blockchain environments.

Private blockchain networks offer a tradeoff between immutability and control that is often useful in industrial applications due primarily to the challenges described above [22]. Nevertheless, such setups are subject to scrutiny with regards to centralisation and governance concerns. Additionally, private blockchain setups that serve as testbeds or sandboxes for applications aimed to be deployed on public networks suffer from inconsistencies in the global network state, which in turn prevents the integration with existing dApps hosted on the public mainnet. In contrast, this work introduces a methodology to setup a private version of the public blockchain which may be tuned to accommodate a breadth of features in accordance with the application context.

The proposed framework builds on top of widely used chain forking capabilities to offer a private version of a public blockchain that can be applied in various contexts such as blockchain-based software development, education (training), security and auditing. It is important to clarify, that this is not a replacement of a production permissioned blockchain setup, but rather a sandbox environment to be used in operations where the intrinsic security features of blockchain technology may impede the desired outcomes.

### 4.2    High Level Overview

Our framework is comprised of three distinct layers: (a) a Node layer (b) an Environment Layer and (c) a Control Layer. This tiered layer distinction dictates the grouping of components with respect to their responsibilities in the domain of the framework. In addition it defines the boundaries at which components can horizontally scale independently. Figure 2 depicts a high level overview of the framework. The three layers are explained below:

i **Node Layer:** Encapsulates the infrastructure required to access historical data on the blockchain in a highly available and scalable manner. The absence or failure of this layer would leave the environment layer functioning only with the local state.

ii **Environment Layer:** Comprised of the set of forked environments and their corresponding components. Each environment 'lives' in isolation and its composition can vary depending on the requirements that it attempts to satisfy. The Environment Layer is managed by the Control Layer.

iii **Control Layer:** Exposes an API for users to manage and control environments and entails all required components that are relevant for the API's operation and assist seamless user experience.

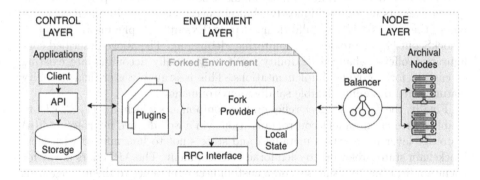

**Fig. 2.** Architecture of the proposed framework

## 4.3   Components

Each layer encompasses multiple components that work in tandem to accommodate the desired functionality. Starting at the right (Fig. 2), there are two main components that make up the *Node Layer*: (i) the load balancer and (ii) the archival nodes. In contrast with conventional nodes that store only the state of the most recent 128 blocks, archival nodes hold the state of the blockchain network since genesis [7]. Archival nodes are used instead of full nodes, as they offer a stable interface for accessing historical data on the blockchain, albeit coming with high storage costs; those may however be mitigated through third part service providers such as Moralis or QuickNode [6,8], providing on-demand access to archival nodes without the hustle of maintaining them.

The load balancer component provides a single node-agnostic access point between the Environment and Node Layer thereby in-house maintained archival nodes can be used interchangeably with nodes offered by third party service providers. Furthermore, the load balancer enables horizontal scalability of archival nodes by distributing traffic to multiple instances and, at the same time, provides observability by serving as a single point of traffic auditing and monitoring.

Moving into the *Environment Layer*, this comprises several independent forked environments, each of which 'lives' in isolation and can be customised according to the needs that it serves. In a minimal setup, it consists of a fork provider which is accompanied by a storage layer that keeps the local state of the forked chain and a Remote Procedure Call (RPC) interface. Generally the term 'fork provider' is an abstract term that refers to piece of software which supports chain forking from an archival node along with a local EVM engine and additional functionality to support state mutability. In fact the extend of features that the fork provider allows is unbounded and lies fully with the implementation of the fork provider. The fork provider component can be interchangeably served by already existing software such as Ganache, Hardhat, Anvil etc. [1–3] or custom implementations. Optionally, the forked environment's feature set can be broadened by the introduction of plugins. These are plug-and-play standalone software agents which would mostly—but not necessarily—utilise the RPC interface of the fork provider to fulfil their purpose. Examples of plugins are explorers (BlockScout [4]) or proxies for monitoring, debugging. The above-described configuration offers enhanced flexibility, as it may equally accommodate existing or custom fork provider implementations. This is arguably a critical usability feature when designing flexible sandbox environments

Finally, the *Control Layer* relies on two components which are: (1) an API to enable forked environment control operations via a well defined interface and (ii) a storage layer to maintain metadata and state - not to be confused with local blockchain state - relevant to each forked environment. The API offers a way for client applications to visualise, add a custom user experience to the management of the forked environments or integrate it with external systems. The amount or format of client application is left to the preference of the framework users. Examples of client applications are, but not limited to, CLIs or web applications.

### 4.4  Features

One of the core contributions of our model is its ability to mitigate shortcomings in the practice of using intentional short-lived local chain forks as blockchain testing tools. As such shortcomings may be best expressed, as deficiencies in a series of attributes (portability; longevity; observability; mutability; extensibility), we have introduced precisely these attributes as our primary features in our framework design and implementation. These are outlined below:

i **Portability:** With the local state of each forked environment stored in a persistency layer, one may easily extract a snapshot of the local state in a compressed file artefact with the help of client control applications. This can then be transferred to other hosts or shared with peers and reloaded in new environments, or stored externally for backup and future reference.

ii **Longevity:** Ephemeral blockchain forks are currently the norm in blockchain development especially for automated testing. However there are cases where actions performed in a local forked environment need to be persistent across multiple sessions (See use cases section for examples). Again, the persistency of local blockchain state enables the fork provider to import and reload that state thereby achieving longevity over disrupted fork sessions.

iii **Observability:** The load balancer and plugins components in the Node and Environment layers respectively enable observability capabilities at two levels: (a) at the environment level and (b) at the node level. As a result real-time monitoring and debugging of RPC requests becomes possible and can be used to provide insights into the RPC call sequence of specific operations.

iv **Mutability:** In cases where control is favoured over security, the intrinsic immutability of blockchains becomes irrelevant. Using our framework, the local state becomes mutable and enables mutations that can materialise as changes in storage addresses, modifications in already deployed bytecode or impersonation of existing wallets. Such features, conveniently serve sophisticated stateful testing cases for security or demonstration purposes.

v **Extensibility:** As discussed in Sect. 4, each layer separates the domain of this framework to self-sufficient entities. Although a single layer implementation in isolation may be very limited in terms of functionality, it can be replaced by, or integrated with, other layer implementations of this framework or even parts of larger systems. In addition, the framework provides abstract guidelines to achieve the desired outcomes; yet as highlighted in multiple occasions, each implementation of this framework is extensible.

Introducing those features into our framework comes with the obvious advantages of systematic management and feature control capabilities for designing local fork sandbox environments. In the next section we discuss example use cases for our framework, viewed in light of two distinct application contexts ((a) application development and (b) education/training) as we deem this connection necessary for underlining our framework's usability and informing future research directions.

## 5   Use Cases

To encourage adoption and provide clarity, we classify use cases by (i) operation and (ii) application context. This taxonomy allows a non-exhaustive but systematic treatment of example practical use cases, hence clearly illustrating the usefulness of our framework. These are centered around two application context areas: (a) *Decentralized application development* and (b) *Education/Training*.

**Setup of Persistent Environments for DevOps.** A common practice of the DevOps culture is to establish and maintain a streamlined process of releasing software to users, or in more technical terms, to *production* [27]. During this process, the software to be released passes through multiple stages and environments that are subject to packaging operations, automated or manual tests and deployment procedures. It is well- established that such practices result to more robust, bug-free and secure software to be delivered to users [27]. In the context of dApp development, similar DevOps procedures are followed, albeit with some twists due to the challenges that emerge by the distinct nature of blockchain development practices [31]. Automated testing does not suffice to capture bugs in complex hard-to-test interactions which involve multiple components and existing state. Generally, unit tests run on a short-lived local mainnet forks but manual quality assurance (QA) on staging environment is inevitably either done in testnet networks [31] or not done at all. Such testnet environments suffer from the problems discussed in previous sections in this paper.

Our framework addresses these issues by allowing a DevOps team to setup internal blockchain environments that can be used by QA teams to manually test dApps without leaking any information to public networks as to what features are currently in development and without seeking funds from testnet faucets. Furthermore, new features can be tested in an incremental manner with existing data since the local state is persistent across releases. This allows for manual regressions tests, a category of tests that is quite complex to automate. To emphasize how this persistent environment can outperform ephemeral forks, consider a DAO smart contract with numerous roles, contingent to a multivariate state or external protocols. Testing all possible internal state combinations to verify the soundness of the DAO can become intractable. Alternatively, a closed version of the DAO can be launched and run for an extended duration with multi user interactions, thereby uncovering potential bugs that could have been missed from the test suite. The recent governance attack on Yam finance illustrates such complex scenarios where the dApp does not malfunction in anyway, but rather the game theory dynamics of the DAO undermine the security of the system [10]. That is not to say that this methodology renders normal testing using ephemeral forks obsolete. On the contrary, it extends the arsenal of developers to conduct more realistic testing. Finally these environments can be replicated to support internal demos and Proof of Concept (PoC) implementations.

**DApp Assessment for Exploits and Transaction Trail Tracing.** Security is a concern of utmost importance in dApp development with devastating financial impact in the event of malicious exploitation [17]. Recent literature suggests that attacks on smart contracts tend to have a complex trail of transactions [28]. The in-depth comprehension and identification of the scope, origin and on-chain interactions of dApp exploits is a cumbersome task, forcing security analysts to resort to tools that aid the disambiguation of such transactions. There are several publicly available tools for performing transaction tracing and debugging [29], however, despite their advanced debugging capabilities, none of the address multi-tenant use, or simulating what-if scenarios by mutating the local blockchain's state.

Our proposed framework enables adopters to create a blockchain environment, forked at the block number just prior to an attack. Henceforth, the malicious transactions can be extracted from the public blockchain history and replayed in the simulated environment either externally by in-house tools or by replay plugins. During this process, analysts can debug and trace transactions, annotate changes on the local state, invite more experts to inspect the local state or share it with colleagues. Alternatively, all transactions of the next block can be replayed incrementally in order to identify which transactions comprised the attack's kill-chain in the first place. The framework therefore provides a flexible, versatile training and risk assessment tool suitable for exploring smart contract attacks and the conditions under which these occurred. In light of the rising popularity of flash-loan based attacks, the use of this work as an analysis tool can offer transparency through each stage of an attack via interactive state transitions and synergise with other tools such as Flashot [16].

Our classification of potential Use Cases by operation/context may be used to identify numerous additional scenarios for our framework; this however falls beyond the scope of this work and is left for future research. At this stage, the above use cases provide ample evidence on how and why our suggested local fork management framework may be practically used.

## 6   Implementation

So far in this work, we have introduced the architecture, structure and operating principles of a versatile systematic blockchain local fork management framework and have presented the case for its use under different conditions to support diverse requirements, primarily in testing and cyber-risk mitigation environments. In this section, we provide a reference implementation of this framework, a high-level view of which is provided in Fig. 3.

Starting from the Control Layer, a Python CLI has been developed to facilitate the interaction with the layer's API which is built on a FastAPI web framework. Moreover, a Postgres DB is acting as the storage layer for the Environment Layer's metadata. The CLI can be used to create/modify and delete environments which fork the local mainnet chain using Ganache v7 (Environment Layer). Ganache by design stores the local state in a LevelDB database on-disk. On top of that an HTTP debugging proxy is added as a plugin to intercept and monitor the JSON-RPC requests to Ganache's JSON-RPC server. Any of the requests that query data on-chain from previous states are forwarded to the Node Layer where HAProxy [5] distributes them equally to a Moralis and Alchemy archival nodes via an HTTP JSON-RPC interface. HAProxy was selected to serve as load balancer on the basis of its low latency and high throughput [19].

The provided tech stack was deployed on Amazon Web Services (AWS) public cloud provider for testing and evaluation. To allow for modularity and low maintenance, all components which require computational resources were deployed using Docker containers on Fargate and Elastic Container Service (ECS), for which more detailed specifications are provided in Table 1. Namely there are

**Table 1.** AWS ECS task specifications

| Image description | Image size (GB) | vCPUs | Memory |
|---|---|---|---|
| Control layer API | 0.578 | 2 | 4 |
| HAProxy load balancer | 0.097 | 1 | 2 |
| Forked environment | 1.56 | 4 | 8 |

HAProxy, Python FastAPI, CLI, and each forked environment as a single image. Note that the Docker image for the environment comprises of the fork provider, plugins and LevelDB for simplicity and single point deployment. To maintain and persist local state of the environments, an Elastic Filesystem (EFS) Volume was mounted to the environments' container. Further, Amazon Relational Database Service (RDS) was used for hosting the PostgresDB in the Control Layer, on a db.t4g.micro Elastic Compute Cloud (EC2) instance with a storage of 20 GB. Finally the Python CLI was run locally and connected to the API to manage the environments.

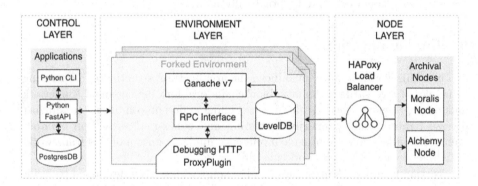

**Fig. 3.** Reference implementation of the proposed framework

# 7   Discussion, Limitations and Future Work

In this paper, we have introduced a systematic framework for the creation and management of customisable, persistent, scalable blockchain environments with the use of local forking. We have demonstrated an approach and architecture for developing long-lived adaptable forks which can be well-suited as tools for DevOps and activities like security testing, applied in training or even direct development of smart contracts in production. We have furthermore listed a feature set for our framework that ensures the development of sandbox environments that can be portable, long-lived, observable, mutable and extensible. Our reference implementation was delivered with these principles in mind, while we acknowledge the need for further research to accommodate non-EVM features.

Evaluating the effectiveness and scalability of the framework is paramount for informing the decision of practical application in specific scenarios, however, the evaluation metrics are highly contingent to different implementation strategies. To accommodate this particularity, we have chosen to concentrate our work on the high-level framework that offers a blueprint for more concrete implementation instances. An indicative implementation was presented in Sect. 6; however, a single implementation instance would not suffice for a sound experimental evaluation. Alternatively, we propose a non-exhaustive list of metrics that can be taken into account. As the tiered architecture presented in this paper distinguishes the framework into three standalone layers, in the same token, a complete evaluation must assess each layer independently. Namely, one can examine the control plane by looking into metrics such as requests per second, time required to deploy infrastructure for a new environment, or whether it supports multi-tenancy. In order to correctly evaluate throughput, transactions must be categorised based on the number of state writes produced. Therefore, the environment layer requires evaluation of other metrics such as transactions per minute (for each class of transactions, and consensus approach i.e. transaction batching) and max number of live environments without performance degradation. Finally the node layer is primarily concerned with the latency from the proxy to the archival nodes.

Even where applied under the guidelines presented, the framework comes with certain limitations that are most often contingent to the application context, which must be taken into consideration. In fact, when considering the framework in its abstract form, limitations only materialise naturally at the implementation stage. Nevertheless, such impediments lay down the foundations and draw directions for future work.

The evolving nature of the field of blockchain research evidently dictates that any novel sandbox platform is bound to come with restrictions, and our proposed framework is no exception. That notwithstanding, we identify two main broad limitations to our work, which are outlined and discussed in detail below.

**The Fork Provider Throughput Bottleneck.** A critical component which can be the origin of many system constraints is the fork provider. The fork provider is the a core component and the entry point of all of the RPC requests to the framework at the Environment Layer of our architecture (Fig. 2). Large volumes of traffic can place the fork provider under stress, as requests must be served and update or read the local state accordingly. Therefore, the throughput of such component relies heavily on the implementation of the fork provider. A prominent example is the currently available fork provider implementations (HardHat, Ganache etc.), which operate with the launch of a single-threaded JSON-RPC server, thereby limiting the throughput of the framework. In cases where local chain forks are only used for unit tests, a single threaded JSON-RPC server provides sufficient throughput, but it can become a bottleneck for some of the proposed use cases discussed in this work.

There are in fact multiple potential solutions to address the throughput bottleneck challenge, however, each one comes with its challenges. In particular, the same architectural pattern that is presented in the framework for load-balancing traffic to the archival nodes can be utilised here to load-balance RPC requests to multiple instances of the fork provider. Yet, this approach may not be effective with the common provider implementations as most, if not all of them employ LevelDB [25] to store the local state which does not allow for multiple writer processes. Alternatively, application-level customisations, such as a custom multi-threaded fork provider or a multi-threaded re-engineering on existing implementations, would reasonably improve the throughput but these are all non-trivial solutions. To further explore and quantify the strengths and weaknesses of each fork provider, we aim to perform a benchmarking study across multiple dimensions on existing fork providers as an extension of this research.

**Lack of Guidelines for Infrastructure Topology.** The scope of this work extends only to the high-level components of the framework presented, pending any precise guidelines of infrastructure and network topology. We intentionally leave these details out of scope, as they are heavily dependent on the application context. For instance, considering the first use case presented in Sect. 5, this comes with the topological particularity that each forked environment created in the Environment Layer may 'live' in ephemeral machines, VMs or even containers in an internal network. On the other hand, the Control Layer and the Load Balancer component in the Node layer may be hosted in another server on a different network, but still be accessible by the remaining layers. Nevertheless, this is not the only possible topological configuration of this particular application. Similar application-specific contingencies apply in other use cases such as providing blockchain sandbox environments for participants in training workshops, which necessitates that relevant infrastructure or network topology details accompany each specific implementation. We therefore leave the investigation, evaluation and assessment of suitability of such configurations per Use Case, as directions for future work.

## 8   Conclusion

This work has introduced a novel systematic framework as a tool to develop and manage customisable sandbox blockchain environments. We outlined the architecture design and features, provided a reference implementation and presented a taxonomy of example use cases where the framework may be applied.

The contributions of our work is manifold: for one, to the best of our knowledge, this is one of the first attempts to offer such systematic treatment to testbed environments for blockchain-oriented development, allowing low-risk/low-cost extensive testing via long-lived forking. Secondly, we provide a robust platform for maintaining and managing multiple stateful local fork sandbox environments, which can be configured to support a multitude of practical use cases with any subset of requirements matching the feature-set in Sect. 4.4. Further we provided an extensive description of two such scenarios and conceptually placed in a taxonomy derived from this work. Thirdly, we offer a flexible design and tiered architecture, able to equally accommodate external fork provider implementations (Ganache, HardHat and others, as discussed in Sect. 4.3) and bespoke ones. As limitations of our framework are contingent to the application scope, the applicability to different cases remains to be investigated in detail. Doing so will allow us to define, at a more granular level, the framework parameters and propose architectural configurations which are potentially more attuned to specific categories of Use Cases.

# References

1. Anvil · foundry-rs/foundry—github.com (2022). https://github.com/foundry-rs/foundry/tree/master/anvil. Accessed 05 May 2022
2. Ethereum development environment for professionals by Nomic Foundation—hardhat.org (2022). https://hardhat.org/. Accessed 05 May 2022
3. Ganache - Truffle Suite—trufflesuite.com (2022). https://trufflesuite.com/ganache/. Accessed 03 May 2022
4. GitHub - blockscout/blockscout: blockchain explorer for Ethereum based network and a tool for inspecting and analyzing EVM based blockchains—github.com (2022). https://github.com/blockscout/blockscout. Accessed 30 May 2022
5. HAProxy - The Reliable, High Performance TCP/HTTP Load Balancer (2022). http://www.haproxy.org/. Accessed 10 May 2022
6. Moralis » The Ultimate Web3 Development Platform—moralis.io (2022). https://moralis.io/. Accessed 12 May 2022
7. Nodes and clients, May 2022. https://ethereum.org/en/developers/docs/nodes-and-clients/
8. QuickNode - Blockchain API and Node Infrastructure: Ethereum, Solana, Polygon, BSC + More (2022). https://www.quicknode.com/. Accessed 12 May 2022
9. Total value locked all chains, May 2022. https://defillama.com/chains/EVM
10. Yam finance governance attack attempt—gist.github.com (2022). https://gist.github.com/ethedev/248f931dbb29d054a9366fe43f37d42e. Accessed 12 July 2022
11. Alharby, M., van Moorsel, A.: BlockSim: an extensible simulation tool for blockchain systems. Front. Blockchain **3**, 28 (2020)
12. Alharby, M., Van Moorsel, A.: BlockSim: a simulation framework for blockchain systems. ACM SIGMETRICS Perform. Eval. Rev. **46**(3), 135–138 (2019)
13. Alsahan, L., Lasla, N., Abdallah, M.: Local bitcoin network simulator for performance evaluation using lightweight virtualization. In: 2020 IEEE International Conference on Informatics, IoT, and Enabling Technologies (ICIoT), pp. 355–360. IEEE (2020)
14. Aquilina, S.J., Casino, F., Vella, M., Ellul, J., Patsakis, C.: EtherClue: digital investigation of attacks on Ethereum smart contracts. Blockchain Res. Appl. **2**(4), 100028 (2021)
15. Buterin, V., et al.: Ethereum white paper. GitHub repository **1**, 22–23 (2013)
16. Cao, Y., Zou, C., Cheng, X.: Flashot: a snapshot of flash loan attack on DeFi ecosystem. arXiv preprint arXiv:2102.00626 (2021)
17. Chainanalysis: The 2022 crypto crime report. Chainalysis Inc., New York (2022). https://go.chainalysis.com/rs/503-FAP-074/images/Crypto-Crime-Report-2022.pdf

18. Destefanis, G., Marchesi, M., Ortu, M., Tonelli, R., Bracciali, A., Hierons, R.: Smart contracts vulnerabilities: a call for blockchain software engineering? In: 2018 International Workshop on Blockchain Oriented Software Engineering (IWBOSE), pp. 19–25. IEEE (2018)
19. Dillon, G.: Benchmarking 5 Popular Load Balancers: Nginx, HAProxy, Envoy, Traefik, and ALB (2018). https://www.loggly.com/blog/benchmarking-5-popular-load-balancers-nginx-haproxy-envoy-traefik-and-alb/. Accessed 10 May 2022
20. Golosova, J., Romanovs, A.: The advantages and disadvantages of the blockchain technology. In: 2018 IEEE 6th Workshop on Advances in Information, Electronic and Electrical Engineering (AIEEE), pp. 1–6. IEEE (2018)
21. Hall, G., Mansi, M., Makrant, I.: Novel method for handling Ethereum attack. arXiv preprint arXiv:1909.12934 (2019)
22. Hamida, E.B., Brousmiche, K.L., Levard, H., Thea, E.: Blockchain for enterprise: overview, opportunities and challenges. In: The Thirteenth International Conference on Wireless and Mobile Communications (ICWMC 2017) (2017)
23. Khan, S.N., Loukil, F., Ghedira-Guegan, C., Benkhelifa, E., Bani-Hani, A.: Blockchain smart contracts: applications, challenges, and future trends. Peer-to-Peer Netw. Appl. **14**(5), 2901–2925 (2021). https://doi.org/10.1007/s12083-021-01127-0
24. Marchesi, M., Marchesi, L., Tonelli, R.: An agile software engineering method to design blockchain applications. In: Proceedings of the 14th Central and Eastern European Software Engineering Conference Russia, pp. 1–8 (2018)
25. Mohanta, B.K., Panda, S.S., Jena, D.: An overview of smart contract and use cases in blockchain technology. In: 2018 9th International Conference on Computing, Communication and Networking Technologies (ICCCNT), pp. 1–4. IEEE (2018)
26. Nakamoto, S.: Bitcoin: a peer-to-peer electronic cash system. Decentralized Business Review, p. 21260 (2008)
27. Senapathi, M., Buchan, J., Osman, H.: DevOps capabilities, practices, and challenges: insights from a case study. In: Proceedings of the 22nd International Conference on Evaluation and Assessment in Software Engineering 2018, pp. 57–67 (2018)
28. Su, L., et al.: Evil under the sun: understanding and discovering attacks on Ethereum decentralized applications. In: 30th USENIX Security Symposium (USENIX Security 2021), pp. 1307–1324 (2021)
29. Vacca, A., Di Sorbo, A., Visaggio, C.A., Canfora, G.: A systematic literature review of blockchain and smart contract development: techniques, tools, and open challenges. J. Syst. Softw. **174**, 110891 (2021)
30. Wang, S., et al.: Blockchain-powered parallel fintech regulatory sandbox based on the ACP approach. IFAC-PapersOnLine **53**(5), 863–867 (2020)
31. Wöhrer, M., Zdun, U.: DevOps for Ethereum blockchain smart contracts. In: 2021 IEEE International Conference on Blockchain (Blockchain), pp. 244–251. IEEE (2021)
32. Zhang, M., Zhang, X., Zhang, Y., Lin, Z.: TxSpecTor: uncovering attacks in ethereum from transactions. In: 29th USENIX Security Symposium (USENIX Security 2020), pp. 2775–2792 (2020)
33. Zheng, Z., Xie, S., Dai, H.N., Chen, X., Wang, H.: Blockchain challenges and opportunities: a survey. Int. J. Web Grid Serv. **14**(4), 352–375 (2018)
34. Zou, W., et al.: Smart contract development: challenges and opportunities. IEEE Trans. Softw. Eng. **47**(10), 2084–2106 (2019)

# Fine-Grained Data Access Control for Collaborative Process Execution on Blockchain

Edoardo Marangone[1]([envelope]) [ORCID], Claudio Di Ciccio[1] [ORCID], and Ingo Weber[2] [ORCID]

[1] Sapienza University of Rome, Rome, Italy
{edoardo.marangone,claudio.diciccio}@uniroma1.it
[2] Software and Business Engineering, Technische Universitaet Berlin, Berlin, Germany
ingo.weber@tu-berlin.de

**Abstract.** Multi-party business processes are based on the cooperation of different actors in a distributed setting. Blockchains can provide support for the automation of such processes, even in conditions of partial trust among the participants. On-chain data are stored in all replicas of the ledger and therefore accessible to all nodes that are in the network. Although this fosters traceability, integrity, and persistence, it undermines the adoption of public blockchains for process automation since it conflicts with typical confidentiality requirements in enterprise settings. In this paper, we propose a novel approach and software architecture that allow for fine-grained access control over process data on the level of parts of messages. In our approach, encrypted data are stored in a distributed space linked to the blockchain system backing the process execution; data owners specify access policies to control which users can read which parts of the information. To achieve the desired properties, we utilise Attribute-Based Encryption for the storage of data, and smart contracts for access control, integrity, and linking to process data. We implemented the approach in a proof-of-concept and conduct a case study in supply-chain management. From the experiments, we find our architecture to be robust while still keeping execution costs reasonably low.

**Keywords:** Attribute Based Encryption · Blockchain · Business process management · IPFS

## 1 Introduction

Blockchain technology is gaining momentum, among other reasons because it allows for the creation and enactment of business processes between multiple parties with low mutual trust [26,29]. The distributed nature of public permissionless blockchains allows every user in the network to have a copy of the ledger and therefore all the data is freely accessible. This transparency, together with the permanence of data and non-repudiability of transactions granted by the technology, motivate the use of blockchains as a reliable ground for verifiable and trustworthy interactions.

© Springer Nature Switzerland AG 2022
A. Marrella et al. (Eds.): BPM 2022, LNBIP 459, pp. 51–67, 2022.
https://doi.org/10.1007/978-3-031-16168-1_4

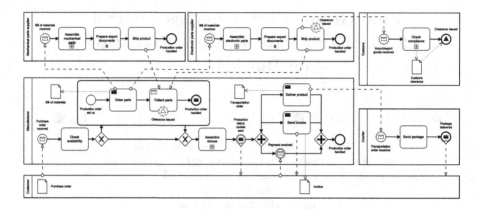

**Fig. 1.** BPMN collaboration diagram of a multi-party process

Especially in cases wherein the parties lack trust in one another, though, hiding some data from the majority of users can be useful. As a matter of fact, security and privacy are at the centre of the debate when considering blockchain technology [9,33]. For example, Corradini et al. [5] point to security and privacy aspects as relevant points. The authors underline that the encryption of the payload of messages (a solution already present in the literature) does not preserve the secrecy of information. Sharing a decryption key among process participants does not allow data owners to selectively control the access to different parts of a single message. Using the public key of a recipient forces the sender to create multiple copies of every message (one per intended reader) and severely hampers the traceability of the process. Another proposed solution is the usage of permissioned blockchains. However, this scheme entails strong complexity and management issues.

Our work aims to close the gap by proposing a technique that guarantees data privacy among parties. With this architecture, the parties can exchange information in a secure way and can also hide data (or parts thereof) from other players with whom they do not want to share it. As such, this paper introduces a novel approach to address security and privacy problems by presenting an architecture that allows for the ciphering of selected data using Attribute-Based Encryption (ABE) [25] so as to control fine-grained read and write access to data.

In the following, Sect. 2 presents a running example, to which we will refer throughout the paper, and illustrates the problem we tackle. Section 3 outlines the fundamental notions that our solution is based upon. In Sect. 4, we describe our approach in detail. In Sect. 5, we present our proof-of-concept implementation and illustrate the results of the experiments we conducted therewith. Section 6 presents the related work in the literature. Finally, Sect. 7 concludes the paper and draws some avenues for future works.

# 2    Running Example and Problem Illustration

Figure 1 depicts a Business Process Model and Notation (BPMN) collaboration diagram representing the supply chain behind the production of drones. We will use this scenario as a running example throughout our paper. We assume process execution is backed by a blockchain-based infrastructure as illustrated in [8].

A new process instance begins when a *Customer* orders one or multiple drones from a *Manufacturer*. After checking the availability of the mechanical and electronic components in the warehouse, the *Manufacturer* orders the missing ones from a local *Mechanical parts supplier* and an international *Electronic parts supplier*, respectively. After the assemblage of the required parts, the suppliers prepare the shipment documents, the package, and send the products. *Customs* then check the documents of the international supplier and release the custom clearance after the verification of compliance concludes positively. Upon the receipt of the parts, the *Manufacturer* proceeds with their assemblage. After sending a notification about the stage reached by the production process, the *Manufacturer* sends an invoice to the paying *Customer*, and requests a *Courier* to deliver the package. The process concludes with the consignment of the ordered product.

We highlight the information artefacts we are going to primarily focus on for our examples as paper documents with twisted corners, namely *(i)* purchase order, *(ii)* bill of materials (BoM), *(iii)* customs clearance, *(iv)* invoice (for the customer), and *(v)* transportation order. First and foremost, we observe that the exchanged information in this process should not be fully accessible outside of the involved counterparts in the process execution. Notice that, instead, a non-encrypted communication through the blockchain allows every node (not necessarily involved in the process either) to disclose the full content of all data attached to transactions. If all parties knew a secret key, they could store the data on-chain once encrypted with that key to ensure nobody outside their circle can read through them, yet ensuring that the data are notarised by the blockchain. However, we remark that although the information exchanges involve multiple actors in collaborative processes, it is rare that *every* actor is supposed to read *all* the exchanged data in their entirety – particularly in this scenario, it never is the case. For example, the invoice details should be undisclosed to any other party that is not the *Customer* or the *Manufacturer*, just like the purchase order. Likewise, the transportation order should be fully accessible only to the *Manufacturer* and the *Courier*.

Whenever a message sender and recipient are single players who know one another in advance, the data producer could encrypt the message and give the access key to the sole expected consumer. However, this may not be a reasonable assumption in cases like the one we discuss here. The customs clearance, for instance, should be known to more than two parties, as the *Electronic parts supplier* and the *Customs* are directly involved but the *Manufacturer* should also be made aware of the result at the end of the border controls. Besides, not only the operators in the *Customs* office involved in the first inspection should be granted access – this restriction would impede future checks.

**Table 1.** Requirements and corresponding actions in the approach

| | Requirement | Approach |
|---|---|---|
| R1 | Access to parts of messages should be controllable in a fine-grained way (attribute level), while integrity is ensured | We use Attribute-Based Encryption (ABE) to encrypt messages, which are stored off-chain while their locator and hash are kept on chain. Access is mediated by a component that decrypts messages only if the requester has the necessary attributes |
| R1.1 | Access policies should be linked to individual (parts of) information artefacts | Access policies associate granted classes of users to the sole messages or sections (slices) thereof that pertain to them |
| R1.2 | Access policies should control access levels for authenticated users | The policies are fine-grained, and the component that decrypts messages does so as per on-chain information |
| R1.3 | Non-authorised access is prevented | Data is kept in an encrypted form, and only authorised requests allow for decryption; salting prevents leakage of information through hashes |
| R2 | Information artefacts should be written in a permanent, tamper-proof and non-repudiable way | We use hashed, permanent off-chain storage in combination with hashes on-chain |
| R3 | The system should be independently auditable with low overhead | On-chain information is publicly available to users of the system, and through hashes integrity of off-chain data becomes auditable |

Another example of non-binary communication channel pertains to the bill of materials. The section of the BoM for the *Mechanical parts supplier* should not be read by any other party but the recipient of the production order and the *Manufacturer*. Notice that, albeit the *Electronic parts supplier* is also a producer of basic components for the *Manufacturer*, it should access the sole part of the BoM referred to its area of competence. Therefore, different parts of a shared data artefact should be accessible to different players. In contrast, the section of the BoM with the identifying data of the *Manufacturer* should be visible to both suppliers.

In the last few years, research work flourished for blockchain-based control-flow automation and decision support for processes like the one in this section [16,17,27, 29]. Our investigation complements this body of research by focussing on the secure information exchange among multiple parties in a collaborative though partially untrusted scenario. We list the key requirements for our approach in Table 1. Next, we focus on the background knowledge that to which our approach resorts.

## 3   Background

In this section, we give an overview of the fundamental notions upon which our approach is built. The fundamental building blocks of our work are Distributed Ledger Technology (DLT), particularly programmable blockchain platforms, and Attribute-Based Encryption (ABE). Next, we outline the basic notions they build upon and relate them to our running example.

Distributed Ledger Technologies (DLTs) realise protocols for the storage, processing and validation of transactions among a network of peers. Their distributed nature entails that no central authority or intermediary are involved in the management of the data. To all these transactions a timestamp and a unique cryptographic signature are attached. To produce signatures, a public/private key scheme is adopted. Every user holds an account with a unique address to which the public and private keys are associated. The shared database is public so all participants in the network can have access to the data. **Blockchain** is a type of DLT, wherein segments of the ledger are collated into blocks and those blocks are backward-linked together forming a chain. DLTs in general and the blockchain in particular cannot be tampered with thanks to a blend of cryptographic techniques, including the hashing of blocks themselves, the inclusion in every block of the hash of the previous one, and the distributed validation of transactions. Public blockchain platforms such as Bitcoin [20], Ethereum [31] and Algorand [4] require fees to be paid in order to let transactions be submitted and processed by the platform. More recent blockchain protocols such as Ethereum and Algorand include the opportunity to run **Smart Contracts**, namely programs deployed, stored and executed in the blockchain [7]. Smart contracts are invoked via transactions. The execution is spread among the nodes without the involvement of a trusted third party so that the overall behaviour can be verified and trusted. Moreover, smart contracts can also trigger the next steps of a workflow when some conditions are met [29]. As with transactions, the execution of smart contract code is subject to costs that in the Ethereum nomenclature fall under the name of *gas*. These costs depend on the complexity of the invoked code and on the amount of data exchanged and stored. To reduce the invocation costs of smart contracts, external Peer-to-peer (P2P) systems are typically employed to save larger bulks of data [32]. One such system is the **InterPlanetary File System (IPFS)**, a distributed system for the storage and access to files. Having it a Distributed Hash Table (DHT) at its core, the stored files are scattered among several nodes. Akin to DLTs, no central authority or trusted organisation retaining the whole bulk of data is thus involved. IPFS makes use of content-addressing to uniquely identify each file in the network. The data saved on IPFS are hash-linked by resource locators that are then sent to contracts that store them on chain [15]. Thereby, the hash of external data together with their remote handle are permanently stored on chain to link them to the ledger.

In a multi-party collaboration scenario like the one we described in Sect. 2, the blockchain creates a layer of trust: the ledger operates as an auditable notarisation infrastructure to certify the occurrence of transactions among the involved actors (e.g., the purchase orders or custom clearances), the smart contracts guarantee that the workflow is followed as per the agreed behaviour, as illustrated in [8,18]. Documents such as purchase orders, bill of materials and custom clearances can be stored on IPFS and linked to the transactions that report on their submission. However, those data are accessible to all peers on chain. Techniques to cipher the data and control their accessibility to predefined users become necessary so as to take advantage of the security and traceability guarantees of blockchain while managing read and write grants on the stored information.

**Attribute-Based Encryption (ABE)** is a type of public-key encryption in which the *ciphertext* (i.e., an encrypted text derived from a *plaintext*) and the corresponding private key to decipher it are linked together by attributes [25].

In particular, the Ciphertext-Policy ABE (CP [3]) is such that every potential user is associated with a number of *attributes* over which *policies* are expressed. Attributes, in particular, are propositional literals that are affirmed in case a user enjoys a given property. In the following, we shall use the teletype font to format attributes and policies. For example, user `0xE756[...]b927` is associated with the attributes `Supplier`, to denote their role, and `14548487`, to specify their involvement in process instance number 14548487. For the sake of brevity, we omit from the attribute name that the former is a role and the latter a process instance identifier (e.g., `Supplier` in place of `RoleIsSupplier` or `14548487` instead of `InvolvedIn14548487`) as we assume it is understandable from the context. Policies are associated to messages and expressed as propositional formulae on the attributes (the literals) to determine whether a user is granted access (e.g., `Courier or Manufacturer`).

All users can attain a unique *secret key* (*sk*). The *sk* is a fixed-length numeric sequence (typically of 512 bits) generated on the basis of the user attributes and a pair of keys, namely a *master public key* (*mpk*) and a *master key* (*mk*). In turn, *mpk* and *mk* are generated through a cryptographic parametric algebraic structure (e.g., a pairing group). A message is encrypted via the *mpk* and the policy. The users can decrypt the ciphertext by using the *mpk* and their own *sk*. It follows that an unauthorised user would not have the suitable *sk* as per the policy. Furthermore, without knowledge of the *mpk*, the user cannot read the encrypted data either. Notice that *mpk* alone would not allow for the generation of new *sks* as the master key (*mk*) is also necessary. To conclude, we remark that the generation of keys, the encryption of plaintexts, and the deciphering thereof, are operations that are algorithmically handled and thus no trusted party is needed – any peer with access to the required credentials could run the necessary code.

In our setting, intuitively, users are process participants, messages are the data artefacts exchanged during the process execution, ciphertexts are the encrypted data artefacts, policies determine which artefacts can be access by whom, and keys are the instruments that are granted to the process parties to try and access the artefacts. Next, we explain how we combine the use of blockchain and CP-ABE to build an access-control architecture for data exchanges in blockchains that meet the requirements listed in Table 1.

## 4    The CAKE Approach

In this section, we describe our approach, named Control Access via Key Encryption (CAKE). Figure 2 illustrates the main components of our architecture alongside their interaction by means of a UML collaboration diagram. The involved parties are:

1. the Data Owner, who wants to cipher the information artefacts (henceforth also collectively referred to as *plaintext*) with a specific access policy (e.g., the *Manufacturer* who wants to restrict access to the bill of materials to the sole intended parties, i.e., the suppliers); we assume Data Owner is equipped with a public/private key pair;

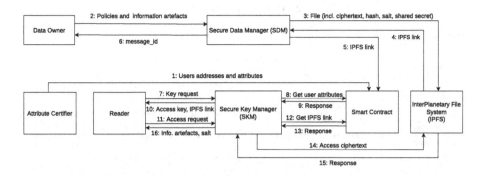

**Fig. 2.** The key component interactions in the CAKE approach

2. one or more Readers, interested in some of the information artefacts (e.g., the *Manufacturer*, the *Electronic parts supplier*, the *Mechanical parts supplier*); we assume every Reader to keep their own public/private key pair;
3. the Attribute Certifier, specifying the attributes characterising the potential readers of the information artefacts; we assume the Attribute Certifier to hold a blockchain account;
4. the Secure Data Manager (SDM), a stateless software component ciphering the plaintext with the policy received from Data Owner; we assume the Data Owner to hold a blockchain account;
5. the Secure Key Manager (SKM), a stateless software component generating access keys for Readers and that the Readers invoke to decrypt messages; we assume the SKM comes endowed with a pair of public and private encryption keys and to hold a blockchain account;
6. IPFS, used to store the ciphertext (i.e., the ciphered plaintext); and finally
7. the Smart Contract, used to safely store the resource locator to the ciphertext saved on IPFS and the information about potential readers of the information artefacts; at deployment time, the Smart Contract is associated with the blockchain account addresses of the SDM, of the SKM, and of the Attribute Certifier, so as to accept invocations only by those components.

Using the enumeration schema of Fig. 2, action (1) is a preliminary operation in which the Attribute Certifier transmits the attributes and the identifying blockchain account addresses of the Readers to the Smart Contract so as to make them publicly verifiable on chain. To this end, the Attribute Certifier operates as a push-inbound oracle [19]. The Attribute Certifier stores on chain the attributes that determine the role of the Reader and, optionally, the list of process instances in which they are involved. For example, the Attribute Certifier stores on chain that 0x906D[...]Dba8 is the address of a user that holds the **Manufacturer** attribute, determining the role, alongside the numeric identifier 14548487 for the running process (the so-called *case id*), specifying the participation of the manufacturer in that particular process instance. Also, it registers that 0xE756[...]b927 and 0xE2C8[...]A2810 are Readers endowed with the

**Table 2.** Message policy examples

| Message | Slice | Policy |
|---|---|---|
| Purchase order | 1 | `14548487 and (Customer or Manufacturer)` |
| Bill of materials | 1 | `14548487 and (Manufacturer or (Supplier))` |
|  | 2 | `14548487 and (Manufacturer or (Supplier and Electronics))` |
|  | 3 | `14548487 and (Manufacturer or (Supplier and Mechanics))` |
| Customs clearance | 1 | `Customs or (14548487 and (Manufacturer or (Supplier and Electronics)))` |
| Invoice | 1 | `14548487 and (Manufacturer or Client)` |
| Transportation order | 1 | `14548487 and (Manufacturer or Courier)` |

Supplier and 14548487 attributes, and that the Electronics and Mechanics attributes belong to the first and the second Reader, respectively.

Thereafter (2), the Data Owner sends the plaintext (i.e., an information artefact such as the bill of material) and the access policies to the SDM, so that the latter can make use of the ABE algorithm to cipher the plaintext with the policy. The access policy declares the conditions according to which a user can be granted access to the ciphered information.

Notice that a message can be separated into multiple *slices*, and each of those can be associated to a different policy. For example, the bill of materials of process instance 14548487 is partitioned as follows (see Table 2): a slice is accessible to all suppliers and manufacturers involved in the process instance, as the policy reads 14548487 and (Manufacturer or Supplier); another one pertains to the sole production order of mechanical parts – i.e., 14548487 and (Manufacturer or (Supplier and Mechanics)); a third slice is specific for the electronic parts supplier – i.e., 14548487 and (Manufacturer or (Supplier and Electronics)). Notice that actors who are granted access to the data do not necessarily need to be directly involved in a process instance. It is the case of *Customs*, e.g., in our running example: the policy reads, indeed, Customs or (14548487 and ([...])). Therefore, *Customs* are authorised to access data across all instances with their key, unlike *Manufacturers*. In other words, the inclusion of a case_id as an attribute in the policy determines a design choice on whether a Reader can use the *sk* across different process instances or not. If the case_id is specified, different access keys are generated for separate instances.

We assume every slice to be associated with a unique identifier (henceforth, slice_id). Table 2 lists the policies used in our running example. The semantics of access policies meets **R1.1**, as they are at the fine-grain level of slices within messages.

Then (3), the SDM runs the algorithm for the generation of the ABE master public key (*mpk*) and master key (*mk*). It uses the master key (*mk*) and the policies to encrypt the plaintext and attain the ciphertext. Thereafter, it generates a

**Table 3.** Examples of messages encoded by the CAKE system.

| Message | Original data | | File header | File body (slices) |
|---|---|---|---|---|
| Purchase order | Company_name:<br>Address:<br>E-mail:<br>Quantity:<br>Price: | Alpha<br>34, Alpha street<br>cpny.alpha@mail.com<br>5<br>$5000 | sender:    0x989a [...] FeaD,<br>message_id: 2206302810394556865,<br>pk:    {"g": "\\u00af [...] 00f4},<br>mk:    {"beta": "\\u009b [...] 009a} | slice_id: 1032292967141064041,<br>hash:    0x2958 [...] fb611,<br>salt:    "\u008d [...] 01bd",<br>metadata: {"c1": [...] 4E1s},<br>cipherText: "qp21 [...] 7Ue9Q" |
| Bill of materials | Manufacturer_company:<br>Address:<br>E-mail: | Beta<br>82, Beta street<br>mnfctr.beta@mail.com | sender:    0x906D [...] Dba8,<br>message_id: 17071949511205323542,<br>pk:    {"g": "\\u0087 [...] 00ca},<br>mk:    {"beta": "\\u00b2 [...] 00fb} | slice_id: 7816105805828306901,<br>hash:    0x953a [...] f8d8,<br>salt:    "Zu00 [...] u004",<br>metadata: {"c1": [...] 00a0},<br>cipherText: "oT2W [...] MQ--" |
| | Frames_quantity:<br>Propeller_quantity:<br>PropellerGuard_quantity:<br>Camera_quantity:<br>Controller_quantity:<br>Amount_paid: | 8<br>80<br>63<br>30<br>4<br>$12000 | | slice_id: 6847895862959863592,<br>hash:    0x12es [...] 1g23,<br>salt:    "bw32 [...] b464",<br>metadata: {"c1": [...] asq2},<br>cipherText: "AS2w [...] btvd" |
| | IMU_quantity:<br>ESC_quantity:<br>Engines_quantity:<br>Batteries_quantity:<br>Amount_paid: | 6<br>40<br>9<br>25<br>$9850 | | slice_id: 3147899764966459866,<br>hash:    0xj4rs [...] ne3d,<br>salt:    "ns1w [...] mey4",<br>metadata: {"c1": [...] 23rs},<br>cipherText: "ht3r [...] asf3" |
| Customs clearance | Date:<br>Sender:<br>Receiver: | 2022-05-10<br>Beta<br>Alpha | sender:    0x0182 [...] 8DC0,<br>message_id: 5757578887823057098,<br>pk:    {"g": "\\uKu00 [...] 00b9},<br>mk:    {"beta": "\\u004d [...] 0d2r} | slice_id: 4807011054135544290,<br>hash:    0x4ee6 [...] 2386,<br>salt:    \u0010 [...] 0013,<br>metadata: {"c1": [...] 00c2},<br>cipherText: "udBA [...] 1A==" |
| Invoice | Gross_total:<br>Company_VAT:<br>Issue_date: | $5000<br>U12345678<br>2022-05-12 | sender:    0x906D [...] Dba8,<br>message_id: 6796003701952936428,<br>pk:    {"g": "\\u00dc [...] 00a2},<br>mk:    {"beta": "\\u00be [...] 00c0} | slice_id: 1264178261449339594 9,<br>hash:    0xad46 [...] 0f79,<br>salt:    "o9\u [...] 01e5,<br>metadata: {"c1": [...] u09a},<br>cipherText: "7QaM [...] KVRS" |
| Transportation order | E-mail:<br>Arrival_date: | cpny.alpha@mail.com<br>2022-06-06 | sender:    0x906D [...] Dba8,<br>message_id: 9846697684368436866,<br>pk:    {"g": "\\ur25d [...] 3a7s},<br>mk:    {"beta": "\\u001q [...] 08q2} | slice_id: 8655357017007860466,<br>hash:    0ze1de [...] 3f9f,<br>salt:    "\bvRA [...] 01in",<br>metadata: {"c1": [...] 00b4},<br>cipherText: "opBK [...] J709" |

unique identifier for the message (`message_id`), such as 17071949511205323542. For every slice, it builds a unique identifier (`slice_id`) and a random number (named *salt*) to be additively used for hashing. Finally, it stores on IPFS the `message_id` and, for each slice, the `slice_id`, ciphertexts, hash of the slice's plaintext combined with the corresponding salt, and the following data encrypted with the public key of the SKM, which we collectively refer to as *shared secret*: *(i)* the *mpk*, *(ii)* the *mk*, *(iii)* additional parametric metadata for the cryptographic algebraic structure (for every slice). In our approach, the SDM forgets both *mpk* and *mk* after storing them as it is stateless. Also, notice that a new pair of keys is created for every message (i.e., IPFS file) to address **R1.2**. As a result (4), the IPFS returns the resource locator (i.e., the link to the IPFS file) to the SDM, which the SDM stores in the Smart Contract (5). Next (6), the SDM returns the `message_id` to the Data Owner. The Data Owner can send the `message_id` to the interested parties to let them know the content is ready for retrieval. For example, the *Manufacturer* sends the suppliers the information that 17071949511205323542 is the identifier to use to fetch the bill of material.

As said, the SDM stores the association between the message and the resource locator on chain via the Smart Contract (5). Thus, we have data stored off-chain that is linked with the blockchain ledger, as per **R2**. Table 3 illustrates the messages we described in our running example in Sect. 2 as saved on IPFS by the SDM. Every IPFS file in our approach consists of a header with the address of the sender (i.e., the Data Owner), the `message_id`, and the encrypted pair of keys (*mpk* and *mk*). The body consists of slices, each with its identifier (`slice_id`), hash, ciphertext, salt and metadata. We recall that salt and metadata are encrypted with the public key of the SKM. Furthermore, notice that the plaintext is encrypted, and albeit being stored semi-publicly on IPFS, it is

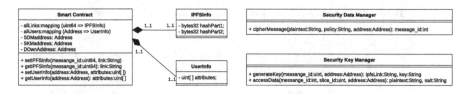

**Fig. 3.** The implemented components of the CAKE system

unreadable even to the Data Owner (unless a party obtains a suitable key, which can be granted only by the SKM). Thereby, we meet **R1.3**.

When the Reader (e.g., the *Electronic parts supplier*) wants to read the data of a message (e.g., the section of interest in the bill of materials), it requests a key from the SKM (7). Then, the SKM retrieves the Reader data (the blockchain address and attributes) from the Smart Contract (8, 9). Notice that these pieces of information were previously stored by the Attribute Certifier at step (1). Equipped with these pieces of information and with the shared secret (including the pk and mk), it produces an ABE secret key ($sk$) for the Reader and sends it back (10) together with the IPFS link corresponding to the requested message (e.g., the one identified by **17071949511205323542**). Notice that the shared secret (including the $mpk$ and $mk$) is saved on IPFS encrypted with the public key of the SKM, so that only the SKM can use its private key to retrieve the necessary information and produce the $sk$. Also, we remark that the SKM is stateless, so it retains no information after it responds to the Reader.

Equipped with their own access key ($sk$), the Reader can begin the message decryption procedure. As per the ABE paradigm, the $sk$ alone is not sufficient to decipher messages though. The $mpk$ is also necessary, though it is encrypted in the IPFS file with the public key of the SKM. Therefore, the Reader makes an access request to the SKM (11). In turn, the SKM asks for the IPFS link from the Smart Contract (12, 13). Then, the SKM retrieves the ciphertext from IPFS (14) and decrypts it with the $sk$ of the user and the shared secret, extracted and deciphered from the requested message. If the decryption is successful, the SKM component sends the information artefacts back (15). Otherwise the Reader request is denied.

Recall that a message can be composed of multiple slices. In the case of the bill of material, e.g., message **17071949511205323542** consists of three slices (see Tables 2 and 3). The first slice contains information available to all suppliers, the second one only for *Electronic parts supplier*, and the third one only for *Mechanic parts supplier*. Therefore, with the $sk$ of the *Electronic parts supplier*, its attributes and the shared secret kept by the SKM, the latter can decipher only the first and second slice, but not the third one – as per the specified policies. The SKM component thus returns those slices only (16). The controlled, fine-grained data access in CAKE is designed to meet the requirements regarding auditability (**R3**), integrity and control (**R1**) and specifically granularity (**R1.1**).

We conclude this section with a few more remarks about security and integrity. When a Reader has received the information artefacts, they may want to verify that the data is not counterfeit. This is the reason why the SKM component returns the (decrypted) `salt` along with the information artefacts to the Reader (16). With the received deciphered data and the salt, the Reader can compute the hash and check if it is equal to the one stored on IPFS by the SDM at step (3) or not. We remark that the Reader had received the IPFS link along with the key at step (8), so that they could directly access the data on IPFS to check the integrity of the information artefacts received from the SKM later on. This design contributes to meeting **R1.3**. The data on IPFS is ciphered and only the SKM can decipher it. The usage of the salt prevents leakage of information, like dictionary attacks.

Also, we remark that the communication backbone outside of blockchain and IPFS for the information exchanges between components is based on the Secure Sockets Layer (SSL) protocol, so as to avoid packet sniffing from malicious third parties that could intercept the data. Furthermore, we assume that the communication from Data Owner to SDM, and from Reader to SKM, are preceded by an initial authentication phase to address **R1.2**. During a preliminary handshake, the SDM and the SKM send a random value to the callers. The callers responds with that value signed with their own private key, so as to let the invoked components verify their identity. Notice that, without this measure, any malicious peer could request the *sk* in place of the real Reader by knowing their address and guessing a file they could be granted access to.

## 5    Implementation and Evaluation

This section describes the proof-of-concept implementation of our approach and the test runs we conducted to assess its affordability for data access control and audits.

Figure 3 depicts the core CAKE components in the form of a UML class diagram. The code of our prototype can be found at https://github. com/apwbs/CAKE together with the detailed results of our experiments. We implemented the SDM, SKM and the communication channels in Python. We encoded the Smart Contract in Solidity as we employ the Ethereum testnet Ropsten for the deployment of our blockchain components: all transactions directed to the CAKE Smart Contract instance we used for our tests can be freely inspected at https://ropsten.etherscan.io/ address/0x2D9EAe20E1E7515d47fBB9A5d454Ce7Be59cA03f. To manage the public/private key pair system for the Data Owner, Readers and SKM, we resort to the Rivest-Shamir-Adleman (RSA) algorithm [24]. In our software prototype, the length of the pair of keys amounts to 2048 bits.

To test our system, we called the methods of the deployed Smart Contract to measure gas consumption. More in the detail, we focussed on the invocations that require the payment of gas fees, namely *(i)* the storage of the address and attributes of Readers (`setUserInfo(...)` in Fig. 3), and *(ii)* the storage of the IPFS link associated to a message (`setIPFSInfo(...)`). The data we used to run our experiments are taken from our running example (see Table 3). We executed

**Table 4.** Gas consumption and total cost of test transactions

| | setIPFSInfo | | | setUserInfo | | | ETH/EUR exchange |
|---|---|---|---|---|---|---|---|
| | gasUsed [unit] | gasPrice [wei] | Total cost [EUR] | gasUsed [unit] | gasPrice [wei] | Total cost [EUR] | |
| Avg. | 67 486.52 | 1399 400 015 | 0.164 38 | 40 755 | 1370 810 611 | 0.097 34 | 1746.35 |
| Min | 67 484.6 | 1000 000 007 | 0.123 78 | 40 755 | 1000 000 007 | 0.074 75 | 1650.68 |
| Max | 67 487 | 1649 000 034 | 0.185 87 | 40 755 | 1644 053 019 | 0.111 91 | 1834.14 |

fifteen calls per day in five consecutive days. Out of the fifteen calls, ten were directed to setUserInfo and five to setIPFSInfo. The higher numerosity of the former is due to the fact that the latter has a rather fixed input format as the length in bits of IPFS locators is constant. As setUserInfo takes as input arrays, the variability of inputs is potentially higher.

Table 4 summarises the results. For every call we provide the average, minimum and maximum of (i) units of gas used to run the code, (ii) price in wei paid for the gas consumption (using the Ropsten default setting), (iii) total cost in Euros based on the daily exchange rate with an Ether. The costs are relatively limited and range between ten and twenty Euro cents, which can be considered a reasonably low amount in light of the permanency and security guarantees provided by the system. Most importantly, the size of the information artefacts do not have a significant impact on the price paid to store them.

To save on gas expenditures, we have adopted a few mechanisms that reduced the size of input and output data. Among them, we recall the following two. First, we have turned IPFS links from their native base-58 encoded format in strings of 46 characters to pairs of bytes32 elements (the IPFSInfo struct in Fig. 3). This allowed for a saving of approximately 30 000 units of gas per call of setIPFSInfo. Secondly, we have encoded attributes into numeric identifiers to avoid the usage of strings for denumerable entities, thereby saving more gas units as the length of the attribute array increases. Further gas-consumption optimisation techniques may be achievable especially for the attribute checking. This challenge paves the path for future work.

# 6    Related Work

Over the last few years, several research endeavours have been dedicated to the automation of collaborative processes based on blockchain. Weber et al. [29] present a technique that resorts to blockchain technology to execute business process between parties who do not trust each other. In their seminal work, they show how the actors can find a mutual agreement on the enacted behaviour without the need to trust a central authority for its enforcement. López Pintado et al. [16] present Caterpillar, a blockchain-based BPMN execution engine. Caterpillar allows users to create instances of a process and to monitor their status. Tran et al. [27] introduce Lorikeet, a model-driven engineering (MDE) tool to implement business processes on chain for the management of assets (e.g., cars, houses), thereby proposing a solution for a scenario that traditionally relies on central authorities. Di Ciccio et al. [8] describe how to design and run business processes where several parties are involved, present the building blocks of model-driven approaches for blockchain-based collaborative business processes with a comparison between Caterpillar and Lorikeet. López Pintado et al. [15] present a model to dynamically bind the actors in a multi-party business process to roles and a specification language for binding policies. CAKE can handle dynamic role binding as the attributes are set by the Attribute Certifier possibly at run time or deploy time. Access keys are generated upon request and not before the process starts. Madsen et al. [17] investigate distributed declarative workflow execution where the collaboration is among adversaries. In such settings, the involved parties do not trust each other and they can also suspect that a party might not act like established. In this work, the authors demonstrate that the execution of the distributed declarative workflow could be implemented as a Smart Contract while ensuring the enforcement of workflow semantics and notarisation of the execution history. Corradini et al. [5] present ChorChain. It takes a BPMN choreography model as input and outputs its translation into a Solidity Smart Contract. ChorChain also allows auditors to obtain ex-post and runtime information on the process instances. These works undoubtedly contribute to the integration of blockchain and process management thus unlocking security and traceability opportunities. However, they do not include mechanisms to ensure fine-grained access control to data saved on a public platform. In contrast, our work precisely focuses on this aspect in a collaborative business process scenario.

Another branch of research work that pertains to our investigation area is the privacy and integrity of data stored on chain. Several papers in the literature document the adoption of encryption to this extent. Hawk [12] is a decentralised system that automatically implements cryptographic devices based on user-defined private Smart Contracts. We take inspiration from this work in that we resort to policies backed by Smart Contracts to cipher messages. Bin Li et al. [13] present RZKPB, a privacy protection system for transactions in shared economy built upon blockchain. This method does not require third trusted parties and preserves transaction privacy as it does not store the financial transactions publicly on chain. Their methodology relates with ours in that we resort to external data stores to save data too, yet we link it with transactions on the ledger. In [14], the authors describe FPPB, a fast privacy protection method based on licenses. It uses zero-knowledge proof, secret address and encryption primitives in the blockchain. Thanks to these features, it grants consistency without disclosing data. This architecture can be used in several shared economic applications. Rahulamathavan et al. [23] propose a new privacy-preserving blockchain architecture for IoT applications based on Attribute-Based Encryption techniques. We employ ABE too, yet with the objective of enhancing existing architectures with our approach. In contrast, this model aims at changing the blockchain protocol at its core. Benhamouda et al. [2] present a solution that allows a public blockchain to act as a repository of secret data. In their system, at first, a secret is stored on chain, then the conditions under which to release it are specified and, finally, the secret is disclosed if and only if the conditions are met. In our approach, we employ shared secrets among components but we do not use the blockchain as a storage for secret data nor expect to disclose the secret. Differently from the techniques above, we tackle the problem of controlled data access in a multi-party process scenario, wherein several information artefacts are exchanged and different actors can access (parts of) messages based on fine-grained policies.

Wang et al. [28] present a secure electronic health record system wherein they combine ABE, Identity-Based Encryption (IBE) and Identity-Based Signature (IBS) with the blockchain technology. This architecture differs from CAKE because in this case the hospital owns the data about patients, and patients specify the policies. In our case, no authority is intended to manage the data except the data owners themselves – in healthcare processes, e.g., they would be the patients. Pournaghi et al. [22] provide a scheme based on blockchain technology and attribute-based encryption, named MedSBA. Their architecture differs from ours for two main reasons. Firstly, MedSBA makes use of two private blockchains, whereas we consider a public-blockchain scenario. Secondly, they cipher the data with AES symmetric cryptography with a random key and then they cipher that random key via ABE. By ciphering with the AES encryption scheme, MedSBA does not allow different users to read the same message, or slices thereof.

# 7  Conclusion and Future Remarks

In this work, we have proposed CAKE, an approach that combines blockchain technology and Attribute-Based Encryption (ABE) to control data access in the context of a multi-party business process. Our approach also makes use of IPFS to store information artefacts, access policies and meta-data. We employ Smart Contracts to store the user attributes, determining the access granted to the process actors, and the link to IPFS files. CAKE provides a fine-granular specification of access grants, data integrity, permanence and non-repudiability, allowing for auditability with minor overheads.

An important aspect to analyse in future studies is the integration with alternative encryption methods. For example, Odelu et al. [21] propose an RSA-based CP-ABE scheme with constant-size secret keys and ciphertexts (CSKC). Their approach targets high efficiency for limited-battery devices. The adoption of CSKC could be of help to integrate IoT devices in the management of blockchain-based processes. Key-Policy Attribute-Based Encryption (KP-ABE) [10] seems a promising asset for a more agile management of the process instance identifiers (case ids). With KP-ABE, attributes are associated with the ciphertext while the policy is associated with users, so the latter can decrypt the ciphertext only if the attributes of the encrypted text satisfy the user policy.

We also plan to overcome existing limitations of our approach. If a Data Owner wants to revoke access to data for a particular Reader, e.g., they can change the policy and cipher the messages again. However, the old data on IPFS would still be accessible. To overcome this limitation, we are considering the usage of InterPlanetary Name System (IPNS), as it allows for the replacement of existing files, hence the substitution of a message with a new encryption thereof. Furthermore, we plan to turn the SDM and SKM into distributed components in order to make our architecture more robust. We are investigating the adoption of secure multi-party computation schemes [6] to this end.

The integration of CAKE with existing blockchain-based process automation toolkits such as Caterpillar [16], Lorikeet [27] and ChorChain [5] is an interesting research avenue as well. CAKE can complement the control-flow-centric perspectives of the above tools with the data access control facilities it provides. To this end, the automated translation of task-based authorisation constraints to policies would be part of the endeavour [30]. Lorikeet specifically includes methods for on-chain data management, which CAKE can complement for confidential off-chain data. As we resort to IPFS to store data, though, the integration should include oracles that permit Smart Contracts to interact with off-chain data [1,19]. The system designer would then be able to determine the trade-off between full transparency on the decision process and access control, by balancing the on-chain and off-chain storage of data as discussed in [11]. Finally, we aim to implement this system with other public blockchains in the future (e.g., Algorand [4]) and test this system with real-world multi-party business processes in production.

**Acknowledgements.** The work of E. Marangone and C. Di Ciccio was partially funded by the Cyber 4.0 project BRIE and by the Sapienza research projects SPECTRA and "Drones as a service for first emergency response".

# References

1. Basile, D., Goretti, V., Di Ciccio, C., Kirrane, S.: Enhancing blockchain-based processes with decentralized oracles. In: González Enríquez, J., Debois, S., Fettke, P., Plebani, P., van de Weerd, I., Weber, I. (eds.) BPM 2021. LNBIP, vol. 428, pp. 102–118. Springer, Cham (2021). https://doi.org/10.1007/978-3-030-85867-4_8
2. Benhamouda, F., et al.: Can a public blockchain keep a secret? In: Pass, R., Pietrzak, K. (eds.) TCC 2020. LNCS, vol. 12550, pp. 260–290. Springer, Cham (2020). https://doi.org/10.1007/978-3-030-64375-1_10
3. Bethencourt, J., Sahai, A., Waters, B.: Ciphertext-policy attribute-based encryption. In: SP, pp. 321–334 (2007)
4. Chen, J., Micali, S.: Algorand: a secure and efficient distributed ledger. Theor. Comput. Sci. **777**, 155–183 (2019)
5. Corradini, F., Marcelletti, A., Morichetta, A., Polini, A., Re, B., Tiezzi, F.: Engineering trustable and auditable choreography-based systems using blockchain. ACM Trans. Manag. Inf. Syst. **13**(3), 1–53 (2022)
6. Cramer, R., Damgård, I.B., et al.: Secure Multiparty Computation. Cambridge University Press, Cambridge (2015)

7. Dannen, C.: Introducing Ethereum and Solidity: Foundations of Cryptocurrency and Blockchain Programming for Beginners. Apress (2017)
8. Di Ciccio, C., et al.: Blockchain support for collaborative business processes. Informatik Spektrum **42**, 182–190 (2019). https://doi.org/10.1007/s00287-019-01178-x
9. Feng, Q., He, D., Zeadally, S., Khan, M.K., Kumar, N.: A survey on privacy protection in blockchain system. J. Netw. Comput. Appl. **126**, 45–58 (2019)
10. Goyal, V., Pandey, O., Sahai, A., Waters, B.: Attribute-based encryption for fine-grained access control of encrypted data. IACR Cryptology ePrint Archive, p. 309 (2006). http://eprint.iacr.org/2006/309
11. Haarmann, S., Batoulis, K., Nikaj, A., Weske, M.: Executing collaborative decisions confidentially on blockchains. In: Di Ciccio, C., et al. (eds.) BPM 2019. LNBIP, vol. 361, pp. 119–135. Springer, Cham (2019). https://doi.org/10.1007/978-3-030-30429-4_9
12. Kosba, A., Miller, A., Shi, E., Wen, Z., Papamanthou, C.: Hawk: the blockchain model of cryptography and privacy-preserving smart contracts. In: 2016 IEEE Symposium on Security and Privacy (SP), pp. 839–858 (2016)
13. Li, B., Wang, Y.: RZKPB: a privacy-preserving blockchain-based fair transaction method for sharing economy. In: TrustCom/BigDataSE, pp. 1164–1169 (2018)
14. Li, B., Wang, Y., Shi, P., Chen, H., Cheng, L.: FPPB: a fast and privacy-preserving method based on the permissioned blockchain for fair transactions in sharing economy. In: IEEE International Conference on TrustCom/BigDataSE, pp. 1368–1373 (2018)
15. López-Pintado, O., Dumas, M., García-Bañuelos, L., Weber, I.: Controlled flexibility in blockchain-based collaborative business processes. Inf. Syst. **104**, 101622 (2022)
16. López-Pintado, O., García-Bañuelos, L., Dumas, M., Weber, I., Ponomarev, A.: Caterpillar: a business process execution engine on the Ethereum blockchain. Softw. Pract. Exp. **49**(7), 1162–1193 (2019)
17. Madsen, M.F., Gaub, M., Høgnason, T., Kirkbro, M.E., Slaats, T., Debois, S.: Collaboration among adversaries: distributed workflow execution on a blockchain. In: FAB, pp. 8–15 (2018)
18. Mendling, J., Weber, I., Van Der Aalst, W., et al.: Blockchains for business process management - challenges and opportunities. ACM Trans. Manag. Inf. Syst. **9**(1), 4:1–4:16 (2018)
19. Mühlberger, R., et al.: Foundational oracle patterns: connecting blockchain to the off-chain world. In: Asatiani, A., et al. (eds.) BPM 2020. LNBIP, vol. 393, pp. 35–51. Springer, Cham (2020). https://doi.org/10.1007/978-3-030-58779-6_3
20. Nakamoto, S.: Bitcoin: a peer-to-peer electronic cash system (2008). https://bitcoin.org/bitcoin.pdf
21. Odelu, V., Das, A.K., Khan, M.K., Choo, K.K.R., Jo, M.: Expressive CP-ABE scheme for mobile devices in IoT satisfying constant-size keys and ciphertexts. IEEE Access **5**, 3273–3283 (2017)
22. Pournaghi, S., Bayat, M., Farjami, Y.: MedSBA: a novel and secure scheme to share medical data based on blockchain technology and attribute-based encryption. J. Ambient Intell. Human. Comput. **11**, 4613–4641 (2020)
23. Rahulamathavan, Y., Phan, R.C.W., Rajarajan, M., Misra, S., Kondoz, A.: Privacy-preserving blockchain based IoT ecosystem using attribute-based encryption. In: ANTS, pp. 1–6 (2017)
24. Rivest, R.L., Shamir, A., Adleman, L.M.: A method for obtaining digital signatures and public-key cryptosystems (reprint). Commun. ACM **26**(1), 96–99 (1983)

25. Sahai, A., Waters, B.: Fuzzy identity-based encryption. In: Cramer, R. (ed.) EURO-CRYPT 2005. LNCS, vol. 3494, pp. 457–473. Springer, Heidelberg (2005). https://doi.org/10.1007/11426639_27

26. Stiehle, F., Weber, I.: Blockchain for business process enactment: a taxonomy and systematic literature review. In: BPM Blockchain Forum, September 2022

27. Tran, A.B., Lu, Q., Weber, I.: Lorikeet: a model-driven engineering tool for blockchain-based business process execution and asset management. In: BPM Demos, pp. 56–60 (2018)

28. Wang, H., Song, Y.: Secure cloud-based EHR system using attribute-based cryptosystem and blockchain. J. Med. Syst. **42**(8) (2018). Article number: 152. https://doi.org/10.1007/s10916-018-0994-6

29. Weber, I., Xu, X., Riveret, R., Governatori, G., Ponomarev, A., Mendling, J.: Untrusted business process monitoring and execution using blockchain. In: La Rosa, M., Loos, P., Pastor, O. (eds.) BPM 2016. LNCS, vol. 9850, pp. 329–347. Springer, Cham (2016). https://doi.org/10.1007/978-3-319-45348-4_19

30. Wolter, C., Schaad, A.: Modeling of task-based authorization constraints in BPMN. In: Alonso, G., Dadam, P., Rosemann, M. (eds.) BPM 2007. LNCS, vol. 4714, pp. 64–79. Springer, Heidelberg (2007). https://doi.org/10.1007/978-3-540-75183-0_5

31. Wood, G.: Ethereum: a secure decentralised generalised transaction ledger (2014). https://ethereum.github.io/yellowpaper/paper.pdf

32. Xu, X., Weber, I., Staples, M.: Architecture for Blockchain Applications. Springer, Cham (2019). https://doi.org/10.1007/978-3-030-03035-3

33. Zhang, R., Xue, R., Liu, L.: Security and privacy on blockchain. ACM Comput. Surv. **52**(3), 1–34 (2019)

# Challenges and Opportunities of Blockchain for Auditable Processes in the Healthcare Sector

Walid Fdhila[1,2]([✉]), Nicholas Stifter[1,2], and Aljosha Judmayer[1,2]

[1] Secure Business Austria (SBA-Research), Vienna, Austria
{walid.fdhila,nicholas.stifter,aljosha.judmayer}@sba-research.org
[2] University of Vienna, Vienna, Austria

**Abstract.** Blockchain technologies (BT) promise to offer exciting research directions for improving various aspects of business processes, in particular in cross-organizational settings where participants do not fully trust each other. However, while blockchain may readily provide transparency and immutability for the processes recorded on a shared ledger, these very characteristics can be problematic in regard to privacy and data protection requirements. In this paper, we address the challenges and opportunities of using BT to secure distributed processes where participants may have an incentive to make false claims or subvert pre-agreed compliance rules in their private processes. Specifically, our analysis is based on a real-world use case, namely how BT can secure (privacy preserving) commitments to processing steps that facilitate federated machine learning (FL) in the healthcare sector. Thereby, an immutable audit trail is created that can be used to detect deviations in retrospect. Hereby, we place a particular focus on the management of patient consent for accessing their data in FL. Our approach draws inspiration from the domain of Self-Sovereign Identity (SSI) where BT is also relied upon to enable the creation and management of decentralized identifiers while focusing on data minimization. The results of our work are not constrained to the particular use case and can be applicable to other emerging research areas of BPM, such as federated process mining.

**Keywords:** Blockchain · Healthcare · Consent management · Business process

## 1 Introduction

There is a dichotomy between user privacy/data protection requirements and the construction of a blockchain as a transparent and verifiable immutable ledger of transactions. On the one hand, BT promise to offer compelling characteristics and properties that can be leveraged, e.g., tamper resistance, high reliability, openness, and distributed or even decentralized trust [18,26]. For instance, they can be used to realize global data sharing and data traceability systems, where these

ⓒ Springer Nature Switzerland AG 2022
A. Marrella et al. (Eds.): BPM 2022, LNBIP 459, pp. 68–83, 2022.
https://doi.org/10.1007/978-3-031-16168-1_5

advantages make it possible to build larger scale, higher quality, and auditable global decentralized data platforms. On the other hand, the aforementioned properties of BT also present fundamental challenges in respect to ensuring user privacy and confidentiality, as well as enabling the removal of undesirable content or otherwise deleting or changing the recorded transaction history [22]. In some application domains, such as the healthcare sector, the ability to both withhold and even delete data due to privacy and regulatory requirements, e.g. the General Data Protection Regulation (GDPR), constitutes a necessity for compliant systems. At first, it would appear that in such cases incorporating BT is not an ideal approach. However, with careful design considerations the advantages of BT can be leveraged while avoiding these privacy issues.

In this paper, we highlight the potential utilization of BT in cross organizational business processes with untrusted parties *where ensuring data privacy and compliance constitutes a necessity* by presenting and analyzing a real-world scenario from the healthcare domain. The use case deals with the management of patient consents in FL. It outlines how access control and audit logs can be implemented through BT in a setting where privacy/confidentiality requirements are high and how commitments in the audit log can be used as a deterrent for misbehavior, even if compliance can not be fully verified automatically. The solution not only showcases how BT may be employed for various aspects of BPM where similar trust-issues could arise, e.g., in federated process mining, it also illustrates how BT can be integrated into legacy systems and processes without full digitization. We hereby bridge an important theory-practice gap in regard to novel proposals leveraging BT and an integration into existing information systems. Further, we show how new paradigms and approaches to identity management in the form of SSI may also be leveraged for the particular use case.

The remainder of this paper is structured as follows: Sect. 2 contextualizes the research problem and covers related work. Section 3 presents a detailed description of our use case scenario and outlines both, the main requirements, as well as the associated threats, while Sect. 4 covers system design details. Finally, Sect. 5 highlights future research challenges and insights gained.

## 2    Background and Related Work

This section contextualizes the addressed research problem and highlights related work, as well as open challenges, in regard to employing BT in the context of BPM and FL with a focus on privacy and confidentiality.

### 2.1    Blockchain in Business Process Management

In the field of BPM, the compelling characteristics of BT have garnered interest, in particular in regard to supporting and securing *cross-organizational* processes where involved parties may not fully *trust* each other [7,12,18,26]. Other application areas may be the provision of (immutable) audit trails [1,4,25]. The properties of the underlying blockchain data structure as an immutable totally ordered log of transactions has also been used for *process mining* [8,14,15,20]. It is expected that BT will fundamentally shift how organizations manage their business processes within their network, thereby opening up new challenges and research directions [18]. One such challenge in BPM is the aforementioned dichotomy between privacy/confidentiality and transparency when integrating BT. In cross-organizational settings, where sensitive data is involved, it can be necessary for certain data and processes to remain private, rendering it difficult to verify if they were performed correctly. Distributed compliance checking is able to capture if processes deviate from pre-agreed rules [16], however some misbehavior, such as utilizing healthcare data in private processes for which no consent was given, can evade such detection as the public part of the process may still appear conformant. When relying on BT, the results of private processes and other data that is extraneous to the blockchain must be fed into the system by a so called *oracle* [5]. As the name implies, the oracle[1] may not necessarily be truthful or provide correct results, requiring a certain degree of trust. A tangible example for an oracle is a weather monitoring station that feeds temperature data to a smart contract. [5] investigate how to ensure data confidentiality during business process execution on blockchain even in the presence of an untrusted oracle. However, the solution they propose requires a trusted setup/intermediary and relies on homomorphic encryption schemes for more complex patterns, rendering a practically feasible application limited to specific use cases.

## 2.2 Applications of Blockchain Technology for FL in Healthcare

A pressing problem in training artificial intelligence (AI) and machine learning (ML) models in healthcare is that the required data, which is usually distributed over various hospitals that may even be located in different jurisdictions, needs to be globally accessible for the training phase, e.g. in a central cloud. This challenge is further compounded by the increasing legal (e.g., GDPR[2],) and technical demands on data providers. In light of enormous data leaks and controversy surrounding how third parties protect data, the public opinion, as well as the patient trust, demand secure ways of storing and processing their data.

Unlike existing approaches, where the ML algorithm runs on data aggregated from different healthcare providers, FL [17] could help address these legal and privacy issues by enabling an ML algorithm to be executed locally at each data provider's site, and the output trained models are subsequently collected and aggregated. This avoids insecure client-cloud and inter-cloud communication, and can ensure that all data remains within (legally and technically) the data provider infrastructures. However, because data is kept and processed locally, external parties cannot readily ascertain from the output results that the process was carried out correctly. In particular, this approach relies on the assumption that all actors involved in the FL are trusted and behave honestly according to the established protocol, meaning that the parties are required to provide correct results and rely on eligible and untampered training data.

---

[1] Referred to by Weber et al. in [26] as a *trigger*.

[2] Cf. Art. 17 GDPR Right to erasure https://gdpr-info.eu/art-17-gdpr/.

Experience has shown that such strong trust assumptions often do not hold in practice [24], and thus, a clear threat model that accounts for both privacy and security threats needs to be elaborated. For example, in the context of a study on COVID-19 vaccine efficiency, a healthcare data provider could attempt to bias the output results in order to render the results of a specific vaccine more favourable, or illicitly include data for which they do not have patient consent. Hence, some form of manual or automatic auditing process is required to verify the integrity of the results and that no data manipulation was carried out during the learning process. To address this problem, prior art has both, considered Byzantine resistant *aggregation* of ML outputs that rely on advanced cryptographic schemes such as MPC, e.g. [13], and proposed approaches that seek to render the process more accountable, e.g. through use of blockchain [19,21,27].

In the context of improving clinical research quality, in particular securing the auditing process, BT could offer desirable characteristics [3] while retaining the decentralization afforded from FL as well as remaining largely compatible with legacy system designs. However, as the application of BT in electronic health is still a young concept careful consideration and sufficient risk analysis must be conducted [22] to avoid security threats and flawed system architectures.

## 3   Use Case: Auditable Consent Management for Federated Machine Learning in Healthcare

This section describes a use case scenario from the healthcare domain, captures the main requirements, and outlines the associated threats. The use case is part of the European H2020 Featurecloud project[3], which provides a privacy preserving solution for FL of healthcare data.

### 3.1   Use Case Description, Roles and Workflow

In the healthcare sector, FL is a distributed and privacy-preserving learning process that involves several medical data providers who collaboratively train a model. Data of different providers are locally trained and the output models are aggregated either i) centrally by a coordinator, or ii) in a distributed manner, e.g., using cryptographic techniques such as secure multi-party computation (SMPC). Figure 1 depicts a simplified BPMN collaboration diagram that captures the main steps and necessary interaction flows for a FL study. In this example, a research entity (i.e., *the project coordinator*) wants to conduct a study on breast cancer survival. Therefore, gene expression data from patients are required to learn a model that predicts tumor recurrence. The project coordinator would select and invite participating hospitals (i.e., *a participant*) where each has a local project manager (LPM) assigned to the project. The invitation includes meta-data about the scope of the study and participation requirements (e.g., conditions on data). Upon acceptance to join the study, a participant

---

[3] https://featurecloud.eu/.

should perform a data discovery check to compute the amount of consented data relevant to the study, upon which its participation in the study is either confirmed or denied. If the conditions are met, the project coordinator sends a participation approval that includes a data analysis (ML) application that can be executed locally by the participants (i.e., data providers). The latter are also responsible for ensuring the pre-processing pipeline (e.g., format conversion, data standardisation). After collecting all ML output models, the project coordinator aggregates and eventually publishes the results.

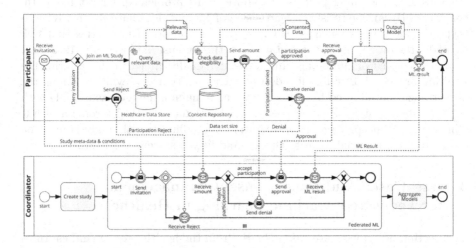

**Fig. 1.** Collaboration diagram: federated machine learning in healthcare

## 3.2   Requirement Analysis

Although, in principle, federating the ML process across participants solves particular privacy and legal issues encountered in centralized learning (as the data remains within its jurisdiction), in practice it raises additional fundamental challenges that need to be addressed. Particularly, in adversarial or cooperative settings, ensuring the integrity and compliance of the local training data, or the reliability of the participating actors can be challenging. Next, we will analyse some of the technical, data protection and privacy requirements for an auditable and compliant federated ML process in the healthcare sector (cf. Table 1).

For example, the legal requirement $R1$ reflects the text of Article 7 (3) of the GDPR, in which patients shall have the right to give, update or revoke consents for future use of their data: *"The data subject shall have the right to withdraw his or her consent at any time"*. Furthermore, during an audit, participants must be able to prove that they were entitled to use the data when the study was performed. This proof of consent ($R2$) is inline with the GDPR (see Article 7 (1)): *"Where processing is based on consent, the controller shall be able to demonstrate that the data subject has consented to processing of his or her personal data"*. Note that consent revocation means that the information

**Table 1.** Requirements

| ID | Type | Description |
|---|---|---|
| R1 | Legal | Revoking, modifying of consent must be possible |
| R2 | Legal | Proof of consent must be possible |
| R3 | Legal, privacy | Consent must be deleted if revoked |
| R4 | Technical, Procedural | Commitment to used data items (and the related consents) |
| R5 | Technical, Procedural | The coordinator must commit results collection & aggregation |
| R6 | Technical, Procedural | Random selection of sites/hospitals for audit |
| R7 | Technical, Procedural | Studies & commitments must be audited |
| R8 | Technical, Procedural | Auditors must commit the results of the audit |
| R9 | Privacy, Technical | The used commitments must not reveal the original input data (hiding) |
| R10 | Technical | Commitments must be cryptographically signed by respective parties |

that the patient has given consent should be deleted ($R3$). This deletion is not retroactive, i.e., any study which took place before the revocation will not become illegal on withdrawal. Consent can be either digital or a digitized signed paper form. Besides, they can be i) static, i.e., are given once for all future studies, limited with an expiry date, or (ii) dynamic, i.e., are given on the fly for each study. Note that consents are often locally managed by participants, but in an ideal scenario, may be managed by patients (cf. Sect. 4).

To ensure reliable audits, the tasks executed by the coordinator and each participant should be committed to the blockchain ($R4$–10). Commitments can be realized through cryptographic primitives, which allow one to commit to a chosen input value that is not revealed. In $R4$, the participant must commit to the inputs and outputs of the locally executed ML study, as well as the respective consents. These commitments might be checked by a competent authority, i.e., the *auditor*, against the actual data. In combination with the appropriate penalties in case of a wrong doing, this provides a credible threat that deviations are detected and thus probabilistically guarantees the integrity of the FL results and the process compliance with the aforementioned rules (e.g., GDPR).

### 3.3   Threat Model

In the following, a threat model is elaborated with a focus on threats that influence basic design decisions for securing FL processes and managing consents. The threat modelling follows a mixture of the threat modelling defined in the Microsoft SDL[4] and attack trees[5], where threats are modeled as attack trees with attack goals as roots and alternative ways to achieve that goal as tree branches. The full list of identified threats as well as mitigations can be found in [11].

As participants through their local project managers (LPMs) obviously have access to all patient data for technical reasons, there is no straightforward way to prevent them using this data in an unlawful or unauthorized manner for which the patient has not given its consent to. Therefore, it is out-of-scope of this

---

[4] https://www.microsoft.com/en-us/securityengineering/sdl/threatmodeling.

[5] https://www.schneier.com/academic/archives/1999/12/attack_trees.html.

**Table 2.** Classification of threats

| ID | Actor | Class | Threat description |
|----|-------|-------|--------------------|
| T1 | Participant | Data poisoning | Selects non consented data or data with expired/revoked consent |
| T2 | Participant | Data poisoning | Uses fake data as input to the ML algorithm |
| T3 | Participant | Data poisoning | Excludes eligible data |
| T4 | Participant | Tampering | Issues fake consents or manipulates their scope |
| T5 | Participant | Integrity | Deletes data necessary for audit |
| T6 | Participant | Integrity | Data loss due to hardware issues on-site |
| T7 | Participant | Integrity | Does not create an audit log entry |
| T8 | Coordinator | Privacy | Provides a malicious ML algorithm to Leak data |
| T9 | Coordinator | Maniplulation | Provides a malicious ML algorithm to manipulate the outcome |
| T10 | Coordinator | Tampering | Manipulates the model aggregation |
| T11 | Patient | Audit Integrity | Requests data deletion after a study |

paper to devise technical means that generally prevent the leakage of patient data in all possible ways, as this would require hospitals to give up at least some control over their IT infrastructure and thus might have other error-prone side effects. Instead we assume, that the participants (i.e., hospitals) do not want to intentionally leak patient data, but they might be willing to influence FL studies by including non consented data, fake data, or omit consented data. Moreover, they might want to learn about patient data from other hospitals. In relation to the IT security literature, the participants can thus be considered *covert adversaries* [2], that only act maliciously when the chance of not being detected is high. Therefore, it is important to identify wrongdoing or misbehaviour of the main actors during such a FL process. Based on the threat model, a secure design for trusted FL processes will ensure that only eligible and consented data is used. Table 2 provides a classification of possible threats grouped by actors.

In FL, only the output models are aggregated, which means that it is difficult for an external entity to tell whether the model was in fact trained based on consented, poisoned ($T2$) or non consented ($T1$) data. Because the internal processes (e.g., fetching and preparing data) are not visible to the outside, it is possible for the LPM to introduce fake data, thereby biasing the outcome of the study, e.g., to manipulate the efficiency of a specific COVID-19 vaccine. Inversely, an adversary may also exclude eligible input data ($T3$), e.g., data that negatively influences a recommendation for a specific drug. Similarly, a participant can issue fake consents or manipulate the scope of existing ones ($T4$).

In order to render the audit impractical, a participant may intentionally delete patient data ($T5$) or omit the creation of an audit entry ($T6$), both required for integrity and compliance checks on the conducted studies. Alternatively, due to the possibility of technical failures in the processes assigned to the LPM, it is possible that failures exhibit the same outcome as a malicious project manager would ($T6$). This is in particular the case for omission failures, e.g. data loss of digitized consent forms, but can also occur during the digitization process, e.g. flipping of digits in patient data, or erroneous OCR. Hence, even if we assume a strong system model where the LPM is completely trustworthy, it is necessary to contemplate the possibility of (random) technical failures and their potential impact to the correctness and security of the conducted ML studies.

With respect to the project coordinator, it is possible that the provided ML algorithm embodies malicious code that exfiltrates sensitive patient data (*T*8), e.g., leaking information about the input data in the result. Furthermore, the ML algorithm may be designed in a way that it will always yield the outcome wanted by the coordinator, regardless of the input (*T*9). Alternatively, the coordinator can manipulate the aggregation of the collected models (*T*10).

Finally, one of the legal rights of patients with respect to GDPR is to have their data deleted if requested. This, in general, does not represent a threat, but may influence audit results where the actual data is compared to the commitments (audit records). Therefore it becomes impossible to check whether the corresponding commitment stored for auditing purposes corresponds to the data used for a study. This may influence a possible replay of a study on the input data as some of the data were deleted. The deletion may be requested by the subjects for privacy reasons, or maliciously executed by LPMs.

# 4   System Design

Based on the requirements gathered in Sect. 3.2, and the threat model elaborated in Sect. 3.3, we propose a system design that uses BT to secure FL processes and improve their auditability. A key challenge for the design is to ensure accountability of the involved actors and prove that only legitimate data was used in the FL process, thereby enabling detect wrong doings such as the use of non consented data, or fake patient data, identities and consents. This will help improve the integrity of FL studies and tackle the data poisoning problem in the presence of Byzantine actors where data protection is an issue.

Another important challenge when dealing with consents and data lies in the difficulty for verifiers to prove that the latter were issued or belong to real patients, and not created by participants to bias a study. Therefore, the solution design should carefully consider how identities are linked to consents and data. So far, most healthcare systems relied on identities that are issued by central authorities (e.g., social security number), which if not managed correctly, may in retrospect become a correlation point across FL executions.

We opted for a design that enables a reliable *post auditing* process, and which supports different identity schemes. The design uses a consensus algorithm and a blockchain as an audit trail (for hashes with high min-entropy input, and signatures of involved parties). It also supports both digital and paper-based processes (e.g., handwritten or digital signatures on consent forms). This allows hospitals with different internal processes to participate in the FL.

Next, we first briefly outline the importance of a framework that deals with governance policies for establishing and maintaining a trusted blockchain network, and managing identities. Then, we discuss different identities schemes that may be used within the framework, with a particular focus on self-sovereign identity, i.e., an emerging approach that can give patients more control over their data and consents. Afterwards, we provide a verifiable approach for managing consents, which will enable auditors verify that i) consents are linked to real patient identities (cf. Sect. 4.1), and that ii) only consented data and most up-to-date consents are used for an FL study (cf. Sect. 4.2). Due to space restrictions, Sect. 4.3 only briefly discusses how to extend some activities in the FL process of Fig. 1 with constructs that render their auditability more trustworthy (e.g., using commitments on data inputs). Finally, we provide a short overview on the prototype implementation.

### 4.1 Governance Framework, DPKI and Identities

**Healthcare Trust Framework.** The trust framework (e.g. European actors from the healthcare sector, i.e. Health ministries) defines what issuers will issue what identity or credentials under which policies. This network of trusted authorities (TA) defines the governance rules for all stakeholders and enables legally binding relationships (e.g., onboarding policies, applicable regulations, governance rules, protocols). Members of the trust framework are also responsible for onboarding new members to FeatureCloud (e.g., hospitals, pharmaceutical companies, test labs, insurances) with specific access rights (e.g., writing to the ledger), and monitoring their compliance with the rules. All accredited members in this trust framework may act as both validator nodes and identity providers. Table 3 gives a summary of the stakeholders' roles in the system.

**DPKI and Identities.** Most approaches on healthcare processes (cf. Sect. 2) rely on the assumption that identities of all actors involved in such collaborative settings are well defined, and that there is an established public key infrastructure (PKI) that governs the issuance of digital certificates, and enables trustworthy communication and authentication. However, dealing with sensitive data such as in healthcare requires a particular level of protection against data misuse by providing mechanisms and means that improve their privacy and security. Thus, an elaborate design of an identity layer is crucial for trusted infrastructure that secures and links public keys to real identities. Recent developments in this area such as certificate transparency already embrace authenticated data structures as a means of identifying manipulation attempts. For example, electronic health certificates have increasingly been relied upon in an effort to contain the spread of COVID-19 infection. Hereby, the difficulty lies in striking a balance between ensuring an individual's privacy and the ability to verify that the claims made in the certificate are authentic and tied to a particular identity.

Unlike previous identities systems, which used to rely on centralized or federated architectures that have already proven their inefficiency and lack of security and privacy, a new identity approach, i.e. SSI [10,23], has emerged, which promises users control over their data, and ensures individuals are at the center of interactions. SSI often relies on blockchain technologies to record identity information or serve as the basis for a Decentralized public key infrastructures (DPKI) [6]. Such Blockchain-based DPKI infrastructure could, for instance, help provide more robust mechanisms for establishing (and revoking) digital identities that are used for aspects such as access control or rights management. SSI initiatives resulted in two key concepts i) verifiable credentials (VC), and ii) decentralized identifiers (DiDs) by the World Wide Web Consortium (W3C):

- *Decentralized Identifiers (DiDs)*: A DID is a globally unique identifier, cryptographically generated (not necessarily registered on a distributed ledger), which points to a DID document (e.g. a JSON-LD object) that specifies cryptographic keying material (e.g., public keys for authentication), verification methods essential for proving ownership of the DID and eventually service endpoints for trustworthy and persistent communication channels.

**Table 3.** Roles and responsibilities

| Coordinator | Participant (e.g., hospital) |
|---|---|
| • Initiates a FeatureCloud study | • Ensures local data security & privacy |
| • Selects study participants | • Performs data querying & preparation |
| • Provides ML algorithms | • Executes & manages FL studies |
| **Trusted Authorities** (e.g., Health ministry) | • Manages patient identities & consents |
| • Define policies & onboarding rules | **Patient** • Gives, updates & revokes consents |
| • Establish & governs the trust framework | |
| • Register & manage stakeholders Identities | **Auditor** • Audits studies |

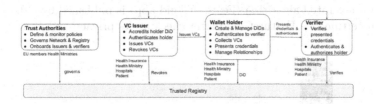

**Fig. 2.** SSI actors and responsibilities

- *Verifiable Credentials (VCs)*: VCs are tamper-evident identity attributes and assertions about a specific subject issued by an identity provider. In contrast to other types of digital credentials, a relying party (third party service) can check the validity of a VC without having to interact with the issuer (i.e. preventing correlation) (cf. Fig. 2). Note that the same stakeholder may play different roles, i.e., issuer, holder or verifier (e.g., a patient issues consents to hospital (holder), but also authenticates within a hospital (verifier)).

Unlike other alternatives such as x.509 certificates or qualified digital signatures, self-sovereign identities can be constructed in a way that is i) decentralized, ii) privacy-preserving (e.g., prevents linkability and supports zero-knowledge proofs), and iii) more secure (e.g., no single point of failure). In a report by the European Union Agency for Cybersecurity (ENISA) [9], an assessment of the potential of SSI technologies and other eID solutions for ensuing secure electronic identification and authentication was provided. It was acknowledged that

SSI provides an effective basis for digital identities, which protects the privacy of personal data, but also needs to co-exist/operate with established technologies such as X.509 PKI, OpenID Connect and other identification schemes.

Despite all the capabilities offered by SSI technologies, in practice, it is optimistic to assume that all stakeholders (e.g., hospitals, health insurances) will employ the same identity scheme. Therefore, in the context of FeatureCloud, we recommend using SSI solutions (e.g., DiDs, VCs), but we do not restrict stakeholders from adopting different methods as long as they securely identify and authenticate subjects and protect their privacy. Additionally, it is noteworthy that a binding identity is necessary to prevent malicious participants from biasing output ML models through poison attacks, i.e. using fake identities, consents or data. This means that identities need to be accredited by members of the trust framework or stewards. Furthermore, the identity model must avoid cross-correlating patient data across multiple participants.

### 4.2   Verifiable Consent Management Process

**Setup and Assumptions.** It is assumed that the software (e.g., a docker container) provided by the coordinator is trusted. This means that the software does not leak data and is designed to execute as reproducible as possible, i.e., ideally the execution is fully deterministic such that running it with the same inputs results in the exact same output. In the FeatureCloud AI store[6], the docker image of a federated algorithm (FeatureCloud app) is published by a developer, and then certified by FeatureCloud after checking its privacy and performance.

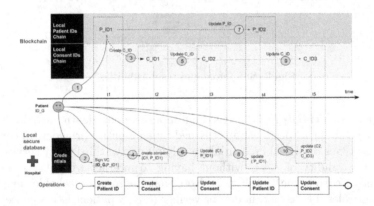

**Fig. 3.** Process example of consent commitments

In FeatureCloud, both paper and digital consents as well as digital and non-digital identities are supported to prevent discrimination. In the following, we focus solely on the use case where consents are digital or digitized by participants, and patients can use applications to cryptographically sign and manage

---

[6] https://featurecloud.ai/.

credentials. We also assume that public representations of cryptographic keying material generated by the patients was signed by a competent authority (e.g., ministry of health), and that there exist one or multiple trusted registries (not necessarily a blockchain) for revoking identities and credentials (cf. Fig. 2).

Based on the legal and technical requirements identified in Sect. 3, a permissioned blockchain can serve as a distributed and immutable audit trail. Access controls (read and write to the ledger) are also defined to restrict access to non-authorized subjects. A smart contract that is deployed and governed by the Trust framework, will be used by authorized entities for registering/updating and revoking local identifiers (e.g., patient identifier within a hospital). Separate smart contracts are also deployed for managing ML studies and consents respectively (i.e., define the rules/logic for commitments).

**Patient Identity Registration.** In order to avoid linkability of patient data across participants, each participant must accredit a separate local identifier $(PID_l)$ for each patient. First, the patient has to authenticate himself to the participant using a supported authentication mechanism. Then, for example if SSI is supported, the patient may create a new DiD and registers it within the participant, e.g., by issuing a Verifiable credential of the DiD signed with $PID_g$.

The local identifier (DiD) is then added by the participant to a trusted registry specific to identified patients. Note that using SSI, only the patient controls the DiD, and can easily rotate the keys (using CRUD operations), while the authority that issued/signed cannot link him to the new $PID_l$. This is also important because during an audit it proves for an auditor that this $PID_l$ corresponds to an actual patient and prevents generating fake patient/consent identifiers. A patient will use a $PID_l$ to issue or revoke consents to a participant.

**Consent Collection.** As described in Sect. 3.2, we distinguish two cases with regard to the time at which a consent was given: $i)$ upfront, where consents from patients are collected by hospitals upfront in a generic way that is not tailored towards a specific study, i.e., they are usually broader and might be eligible to multiple studies. Here, legal criteria regarding consents in the healthcare system have to be taken into account that might be different for each EU member state. $ii)$ on demand, where the hospital actively asks matching patients for consent to use their data in a specific study. In both cases, the consent is never stored in the blockchain or transmitted to another party despite the hospital.

*Give consent:* using the local identifier, the patient digitally signs and issues the consent to the participant. Ideally, with the right permissions defined in the consent management smart contract (given by the participant), the patient may use his application to commit the consent $(CID)$ to the blockchain (cf. Fig. 3). Alternatively, the participant commits the consent on his behalf. The consent itself is stored locally at the hospital site.

*Update/revoke consent:* the patient informs the hospital that he wants to update or revoke his consent. The updated/revoked consent is signed using $PID_l$, then issued to the participant. Ideally, the patient uses his wallet to update/revoke the previous commitment after permission is granted by the participant (using the CRUD operations in the smart contract). Operations on the same consent are linked, that an auditor can later verify the consent history to check if it was valid (e.g., expiration date) when the study was executed.

### 4.3  Federated ML Execution

**Execute the Federated Study and Commit to Input, Output and Used Consents (Participants).** Each participant executes the study software (e.g., docker container) with the required and consented input data. To facilitate auditability, the participants commit to the patient data used as input to the study, the eventual data transformation operations, the associated consents, and the output results. The commitments must be cryptographically hiding, signed and published on a secure bulletin board, e.g. BT, accessible to all involved parties.

**Aggregate all Submitted Results and Signal Successful Aggregation (Coordinator).** When all participants have submitted their trained models, i.e., partial results, the coordinator performs the aggregation of the submitted results. After all results have successfully been aggregated and the final output has been computed, the coordinator has to commit to the collected results, as well as to the outcome of the final aggregation. Again, ideally the study is deterministic, and thus the aggregation of the same partial inputs yields exactly the same overall result. In this case, the commitment can be a cryptographic hash of used inputs and outputs. Otherwise, other techniques such as fuzzy hashing, which enable checking hashes similarity may be employed.

**The Audit Is Performed (Auditors & Participants).** In this step, the on-site audits are performed by the auditors. To avoid bias, auditors are randomly selected and assigned to participants. The federated study is recalculated locally on hardware of the auditor, using the patient data provided by the selected participant. In case of a fully deterministic ML setup, the resulting output of the audit's execution should match the study result which was previously submitted and committed by the respective participant. The scope, state and associated signatures of the consents are checked. If the federated study was not designed to be completely deterministic, the entire committed-to input and output, has to be retained to allow for reproducibility during the audit. The auditor re-runs the federated study on his hardware and then determines if the previous output model is sufficiently close to his output (using the appropriate similarity techniques), provided the same input data and training parameters are used.

Although not yet considered in the design, by having the patients additionally committing to the data collected about them e.g., during hospital visits, it becomes possible for the auditor to detect if the participant has manipulated or excluded eligible data. Furthermore, by signing the underlying medical data (hash of the data) in their consents, patients ensure that they really had the described medical treatment/disease. This double checks that the data the participant can use in the studies is correct.

## 4.4 Implementation

This work has been implemented as part of the European H2020 Featurecloud project, which provides a privacy preserving platform for FL of healthcare data. Part of the framework is an AI Store[7] for FL as an all-in-one platform for biomedical research and other applications. The platform allows to run, develop and publish federated and privacy-preserving machine learning algorithms. All apps are stored as Docker images in a registry where images can be pushed or pulled in accordance with the access rights controlled through an authentication server. On the one hand, each participant needs to run a FeatureCloud controller, which manages the local execution of the ML application. On the other hand, a coordinator controller will be responsible of orchestrating the execution and instructing the participants' controllers to ensure a globally synchronous execution. The PoC is still being tested and improved and has not yet been integrated within the FeatureCloud platform[8]. A Hyperledger Fabric test net has been deployed with few nodes acting as the root of trust (governance framework). In the current version, only x.509 certificates and Idemix identities are supported for authentication as they are inherently supported by Fabric. We are still experimenting with both Hyperledger Indy and Aries to integrate issuance and verification of DIDs and verifiable credentials within Fabric. The implementation also relies on two smart contracts (chaincode) for managing consents and study related data commitments.

## 5 Insights, Applications, and Future Research Challenges

In this paper we have demonstrated how blockchain can be leveraged in cross-organizational business processes where there exist additional privacy requirements and constraints through a tangible use case. A key problem that we address is how trust in the correctness and adherence to compliance rules within private processes of participants can be increased, as the necessary data for complete verification is not publicly available. We rely on the compelling properties of blockchain to secure commitments to the execution of these private processes that can later be audited. In this regard, the utilization of BT does not *prevent* misbehavior, however it offers *detectability* and non-repudiation as a strong deterrent. Hereby, our solution bridges the current gap between research proposals where verification of private processes is made possible in a privacy preserving manner through MPC or by relying on trusted execution environments (TEEs), and real-world information systems where such designs are currently infeasible or even impossible to deploy. We believe that the herein presented approach presents a practical design pattern for a variety of real-world cross-organizational business processes. Further, our approach can also apply to other related research areas of BPM such as federated process mining. Future research

---

[7] https://featurecloud.ai/.

[8] The source code will be released on gitlab. The code is still in review as it may fall under a temporary NDA agreement.

challenges include how the aforementioned advanced cryptographic proof techniques can be effectively integrated into legacy information systems, as well as exploring the potential SSI has to offer in the context of BPM.

**Acknowledgments.** This research is based upon work partially supported by (1) SBA Research (SBA-K1); SBA Research is a COMET Center within the COMET – Competence Centers for Excellent Technologies Programme and funded by BMK, BMDW, and the federal state of Vienna. The COMET Programme is managed by FFG. (2) the European Union's Horizon 2020 research and innovation programme under grant agreement No 826078 (FeatureCloud) (3) the FFG ICT of the Future project 874019 dIdentity & dApps. (4) the FFG Industrial PhD project 878835 SmartDLP. (5) the Christian-Doppler-Laboratory for Security and Quality Improvement in the Production System Lifecycle; We would also like to thank Fenghong Zhang for her valuable contributions.

# References

1. Akhtar, A., Shafiq, B., Vaidya, J., Afzal, A., Shamail, S., Rana, O.: Blockchain based auditable access control for distributed business processes. In: International Conference on Distributed Computing Systems (ICDCS). pp. 12–22 (2020)
2. Aumann, Y., Lindell, Y.: Security against covert adversaries: efficient protocols for realistic adversaries. J. Cryptol. **23**(2), 281–343 (2010)
3. Benchoufi, M., Ravaud, P.: Blockchain technology for improving clinical research quality. Trials **18**(1), 1–5 (2017)
4. Bonyuet, D.: Overview and impact of blockchain on auditing. Int. J. Digit. Account. Res. **20**, 31–43 (2020)
5. Carminati, B., Rondanini, C., Ferrari, E.: Confidential business process execution on blockchain. In: International Conference on Web Services (ICWS), pp. 58–65 (2018)
6. Christopher, A., et al.: Decentralized public key infrastructure. https:// danubetech.com/download/dpki.pdf
7. Corradini, F., Marcelletti, A., Morichetta, A., Polini, A., Re, B., Tiezzi, F.: Engineering trustable and auditable choreography-based systems using blockchain. Trans. Manag. Inf. Syst. **13**(3), 1–53 (2022)
8. Duchmann, F., Koschmider, A.: Validation of smart contracts using process mining. In: ZEUS. CEUR Workshop Proceedings. vol. 2339, pp. 13–16 (2019)
9. Evgenia, N., Viktor, P., Marnix, D.: Leveraging the self-sovereign identity (SSI) concept to build trust. Tech. Rep. 10.2824/8646, The European Union Agency for Cybersecurity (ENISA) (2022)
10. Fdhila, W., Stifter, N., Kostal, K., Saglam, C., Sabadello, M.: Methods for decentralized identities: evaluation and insights. In: González Enríquez, J., Debois, S., Fettke, P., Plebani, P., van de Weerd, I., Weber, I. (eds.) BPM 2021. LNBIP, vol. 428, pp. 119–135. Springer, Cham (2021). https://doi.org/10.1007/978-3-030-85867-4_9
11. Fenghong, Z., Aljosha, J., Walid, F., Nicholas, S.: D6.2: "model for defining user rights in federated machine learning". Tech. rep., EU H2020 FeatureCloud (2021)
12. Garcia-Garcia, J.A., Sánchez-Gómez, N., Lizcano, D., Escalona, M.J., Wojdyński, T.: Using blockchain to improve collaborative business process management: systematic literature review. IEEE Access **8**, 142312–142336 (2020)

13. He, L., Karimireddy, S.P., Jaggi, M.: Secure byzantine-robust machine learning. arXiv preprint arXiv:2006.04747 (2020)
14. Hobeck, R., Klinkmüller, C., Bandara, H.M.N.D., Weber, I., van der Aalst, W.M.P.: Process mining on blockchain data: a case study of augur. In: Polyvyanyy, A., Wynn, M.T., Van Looy, A., Reichert, M. (eds.) BPM 2021. LNCS, vol. 12875, pp. 306–323. Springer, Cham (2021). https://doi.org/10.1007/978-3-030-85469-0_20
15. Klinkmüller, C., Ponomarev, A., Tran, A.B., Weber, I., Aalst, W.v.d.: Mining blockchain processes: extracting process mining data from blockchain applications. In: International Conference on Business Process Management, pp. 71–86 (2019)
16. Knuplesch, D., Reichert, M., Pryss, R., Fdhila, W., Rinderle-Ma, S.: Ensuring compliance of distributed and collaborative workflows. In: 9th IEEE International Conference on Collaborative Computing: Networking, Applications and Worksharing, pp. 133–142. IEEE (2013)
17. McMahan, B., Moore, E., Ramage, D., Hampson, S., y Arcas, B.A.: Communication-efficient learning of deep networks from decentralized data. In: Artificial Intelligence and Statistics, pp. 1273–1282. PMLR (2017)
18. Mendling, I., et al.: Blockchains for business process management-challenges and opportunities. ACM Trans. Manag. Inf. Syst. 9(1), 1–16 (2018)
19. Mugunthan, V., Rahman, R., Kagal, L.: Blockflow: an accountable and privacy-preserving solution for federated learning. arXiv preprint arXiv:2007.03856 (2020)
20. Mühlberger, R., Bachhofner, S., Di Ciccio, C., García-Bañuelos, L., López-Pintado, O.: Extracting event logs for process mining from data stored on the blockchain. In: Di Francescomarino, C., Dijkman, R., Zdun, U. (eds.) BPM 2019. LNBIP, vol. 362, pp. 690–703. Springer, Cham (2019). https://doi.org/10.1007/978-3-030-37453-2_55
21. Passerat-Palmbach, J., Farnan, T., Miller, R., Gross, M.S., Flannery, H.L., Gleim, B.: A blockchain-orchestrated federated learning architecture for healthcare consortia. arXiv preprint arXiv:1910.12603 (2019)
22. Ploder, C., Spiess, T., Bernsteiner, R., Dilger, T., Weichelt, R.: A risk analysis on blockchain technology usage for electronic health records. Cloud Comput. Data Sci. 20–35 (2021)
23. Preukschat, A., Reed, D.: Self-sovereign Identity. Manning Publications (2021)
24. Schüler, P., Buckley, B.: Re-Engineering Clinical Trials: Best Practices for Streamlining the Development Process. Academic Press (2014)
25. Snow, P., et al.: Business Processes Secured by Immutable Audit Trails on the Blockchain. Brave New Coin (2014)
26. Weber, I., Xu, X., Riveret, R., Governatori, G., Ponomarev, A., Mendling, J.: Untrusted business process monitoring and execution using blockchain. In: La Rosa, M., Loos, P., Pastor, O. (eds.) BPM 2016. LNCS, vol. 9850, pp. 329–347. Springer, Cham (2016). https://doi.org/10.1007/978-3-319-45348-4_19
27. Weng, J., Weng, J., Zhang, J., Li, M., Zhang, Y., Luo, W.: Deepchain: auditable and privacy-preserving deep learning with blockchain-based incentive. IEEE Trans. Depend. Secure Comput. 18(5), 2438–2455 (2021)

# Measuring the Effects of Confidants on Privacy in Smart Contracts

Julius Köpke(✉) ⓘ and Michael Nečemer

Department of Informatics Systems, Alpen-Adria-Universität Klagenfurt,
Universitätsstraße 65, 9020 Klagenfurt, Austria
julius.koepke@aau.at, michaelne@edu.aau.at
https://www.aau.at/isys/

**Abstract.** Blockchain Systems provide highly welcome properties such as immutability, observability, availability, and distribution for implementing smart contracts without the need for intermediaries. While the smart contract goals of observability and enforceability can easily be achieved on blockchains, the goal of privity is much harder to tackle. In the context of smart contracts on blockchains, privity aims in limiting the spread of knowledge to the participants with a contractual need-to-know. However, limiting access to data can limit the possible degrees of proactive enforcement of correct decisions and it can negatively impact their availability. Therefore, it can be required to find a proper balance between privity and enforceability or availability requirements. Designers may be forced to include additional participants (confidants) in the decision process only for the sake of enforceability or availability.

In this paper, we introduce measures for assessing the impact of confidants for decisions within smart contracts on privacy. We model smart contracts in form of inter-organizational business processes and provide modeling constructs for privity requirements and the inclusion of confidants.

**Keywords:** Smart contracts · Blockchain · Confidentiality · Privacy · Privity · Enforceability · Distributed oracles · Measure

## 1 Introduction

When developing smart contracts developers are often confronted with goal-conflicts [7,11]. Nick Szabo coined the major design objectives observability, enforceability, and privity for smart contracts in [11] already in the last millennium. Today, blockchain systems can natively support observability and enforceability: The actions of each participant can be tracked on an immutable shared ledger and smart contract code with access to all actions and their data can proactively enforce that only permissible actions can be performed. However, this strong degree of enforcement is limited if privity is taken into consideration. Privity of smart contracts on blockchains aims in limiting the spread of

© Springer Nature Switzerland AG 2022
A. Marrella et al. (Eds.): BPM 2022, LNBIP 459, pp. 84–99, 2022.
https://doi.org/10.1007/978-3-031-16168-1_6

knowledge to the participants with a contractual need to know [7]. This requires limiting access to data of the smart contract to only the interacting participants or even subsets thereof. However, this can have a negative impact on the degree of enforceability as the correctness of decisions over such data can only be verified by a subset of participants rather than the entire blockchain network. While there exist solutions such as non-interactive Zero Knowledge Proofs [3] that allow verifying the correctness of computations without access to hidden input data, this only solves one part of the problem. A decision that can only be taken by one participant leads to a single point of failure. This is especially problematic in low-trust environments. For enforcing the correct execution of the contract it can thus be required to introduce additional actors who can take the decision. Technically, there are various approaches where developers are facing the discussed problem such as distributed or decentralized oracles [2], endorsing peers on Hyperledger Fabric [1] - in particular in combination with private data collections [13] - or redundant oracle services using Zero-Knowledge Proofs [3]. All these techniques have in common that additional participants need to be added by designers to meet requirements on enforceability or availability. However, if these participants must access confidential input data this has negative impacts on privity. We generalize from these approaches here and suppose that when designing a smart contract, a developer might need to add additional actors to decisions. We assume that a model-based development approach is followed, and smart contracts are modeled in form of inter-organizational business processes. In this setting, our goal is to assist modelers at design time. Our main contribution in this paper is a measure for precisely assessing the impact of additional actors of decisions on privity. There are various application scenarios for such a measure. It can be used to assist users in selecting endorsing peers or members of a distributed oracle with a minimal penalty on privity. However, it can also be used for automatically finding optimal sets of additional actors with minimal costs on privity for given requirements on enforceability or availability.

The remainder of the paper is structured as follows: Sect. 2 discusses related work. In Sect. 3 we lay the ground for the measure by introducing our meta-model for inter-organizational business processes including privity constraints and additional actors for decisions. Section 4 introduces a measure for the impact on privity of a process model with only one decision with additional actors. Section 5 generalizes the measure to general processes. Section 6 concludes the paper.

## 2    Related Work

Blockchain Systems allow enforcing the correct execution of business processes via custom transactions (smart contract code). Various execution environments such as [10,12] for business processes on Blockchains emerged over the recent years. Early approaches focused on the correct control-flow, and later approaches also addressed the correctness of role bindings [8], and the correct computation of decisions [4,7]. In the absence of constraints on confidentiality/privity, decisions can be fully based on smart contract transactions and thus backed up by the blockchain network. In the presence of confidentiality constraints over decision inputs, decisions cannot be executed on-chain. In this case oracle [9] based solutions such as [5,7] can provide some degree of enforceability. However, since such an oracle should be distributed in low-trust environments, additional participants may be required to participate in the distributed oracle [2]. If this is either impossible, or enforcement of the correctness of the whole network is required, non-interactive Zero-Knowledge Proofs [3] can be used. In such a case, the deciding participant generates a proof of the correctness of the decision which can then potentially be verified by all participants of the network. However, while at first sight this resolves the goal-conflicts, it introduces a single point of failure. The single actor can intentionally or due to technical failure prevent a decision. A solution is to include alternative decision takers. This can again limit the degree of privity.

Over the years several patterns for blockchain applications [15] emerged. The work in [14] provides an approach for selecting blockchain patterns based on application requirements. While the approach provides a structure for the vast number of existing patterns and their applications, it still does not take into account interdependencies between different requirements and their supporting patterns.

Nick Szabo originally coined the term privity for smart contracts as: *"privity, the principle that knowledge and control over the contents and performance of a contract should be distributed among parties only as much as is necessary for the performance of that contract"* [11]. This definition formulates very strict confidentiality requirements and additionally restricts the control aspect. In [7] we sketched goal conflicts between privity and enforceability for BPM-based smart contracts on blockchains. For supporting trade-offs between privity and enforceability we introduced privity spheres expressing different levels of confidentiality for BPM-based smart contracts. Since in comparison to limiting read access for achieving confidentiality, limiting control to specific participants at runtime is not a major issue on blockchains, we did not focus on this aspect in [7]. According to Szabo the term privity also subsumes privacy of smart contracts. However, privacy in general addresses a much broader field including rights such as the right to be forgotten or user-consent required by privacy acts. It should be noted that Szabo generalized the term privity from the law. Therefore, his definition differs from its common meaning in contract law where privity requires that contracts cannot specify obligations of third parties. See [11] for a discussion on the root of the term privity in the scope of smart contracts.

In this paper, we refine and formalize privity spheres [7] and use them as a basis for inspecting the inter-play between privity requirements and additional actors of decisions and introduce a comprehensive measure. To the best of our knowledge, this is the first approach for precisely measuring the effects on privity relations imposed by confidants.

## 3    Process Meta Model

We model smart contracts in form of block-structured inter-organizational business process models [6] and repeat the essential definitions here. In the remainder of the paper we use the usual object-style dot notation to access components. E.g. for a tuple $t = (a, b)$, we write $t.a$ for accessing $t_a$.

**Definition 1 (Process Model).** A business process model is a tuple $P = (N, E, D, A)$ with a set of nodes $N$ connected by a set of directed edges $E$ with a set of data objects $D$ and a set of participants $A$ forming a directed acyclic graph. For each node $n$ we define $n.type \in \{activity, xor-split, xor-join, and-split, and-join, business-rule-task)\}$ to declare the node type. $n.name$ is the label of the node, $n.d^r \subseteq P.D$, the set of data objects read and $n.d^w \subseteq P.D$, the set of data objects written, and $n.a \in P.A$, the actor executing the node. Each edge $(m, n) \in P.E$ describes a precedence constraint between nodes $m \in P.N$ and node $n \in P.N$. There is one node without predecessor, called start node and one node without successor called stop node. Nodes of type $xor-split$ and $and-split$ have exactly 2 successors, nodes of type xor-join and and-join have exactly 2 predecessors. $DV \subset P.D$ is a set of Boolean decision variables. Each xor-split node $x$ is located immediately after a node of type $business-rule-task$. The business rule task $br$ for $x$ may read data objects and writes to a unique decision variable $dv$ ($br.d^w = \{dv\}$), which is also assigned to the xor-split node ($x.d^r = \{dv\}$). One of the outgoing edges of $x$ is adorned with $dv$, the other with $\neg dv$, indicating which path is chosen at runtime. This decision variable of a xor-block is not modified by any other node except the corresponding business rule task. The process model is full-blocked, i.e. each split node is associated with exactly 1 join node such that each path originating in the split node to the end node includes the associated join node. □

During the execution of a process instance, business rule tasks assign values to their decision variables. These values result in either following the *true* or *false* branch of the corresponding gateway.

**Definition 2 (Instance Type).** An instance type of a process $P^I$ is determined by an instantiation of decision variables $I = \{(d_1, v_1), ..., (d_n, v_n)\}$, where $d \in D$ and v is a Boolean value. $P^I$ is a sub-graph of $P$ where each Xor-split node has exactly one successor, the one that matches the value of the corresponding decision variable in $I$. We define $PI(P)$ as the set of all possible instantiations of decision variables. □

**Definition 3 (Origin).** The *origin* for a data object $d$ of a node $n$ in an instance type $P^I$, denoted $o(P^I, n, d)$, is defined as a node $m$ such that there exists a path $p$ in $P^I$ starting at $m$ and ending at $n$ and there is no step $m'$ writing to $d$ between $m$ and $n$ in $P^I$. □

The correctness criteria ensures that the Process Model is free of race-conditions of data objects.

**Definition 4 (Correct Process Model).** A process model $P = (N, E, D, A)$ is correct, iff for every instantiation $I \in PI(P)$ of decision variables, for each input data object $x \in D$ of each activity or business rule task $n \in N$: $o(P^I, n, x)$ exists and is unique. □

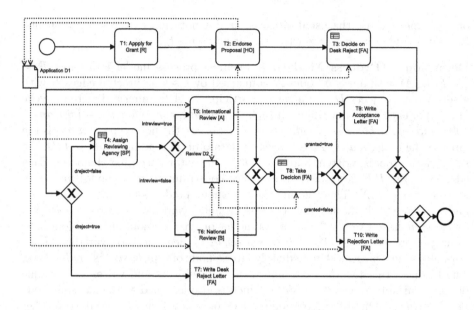

**Fig. 1.** Example collaboration between participants R, HO, FA, SP, A, B

## 3.1   Example Process

In Fig. 1 an example collaboration between a researcher $R$, a host organization $HO$, a funding agency $FA$, a specialist $SP$, an international reviewing agency $A$, and a national reviewing agency $B$ is shown. We denote the actor of some activity in square brackets in Fig. 1. In the text, we denote the actor in subscript. First, the researcher applies for a grant in $T1_R$. This task creates the data object $D1$ containing the research proposal and a formal application. Next, $D1$ is processed and updated by the host organization in $T2_{HO}$. Then the funding agency decides based on $D1$ if the processing is continued or a desk reject is done in $T3_{FA}$. Technically $T3_{FA}$, sets the Boolean decision variable *deskreject*. If no desk-reject is issued, a specialist $SP$ assigns either an international $A$ or a national reviewing agency $B$ in $T4_{SP}$ by setting the Boolean decision variable *intreview*. The review document $D2$ is then created either by $T5_A$ or $T6_B$ and the final decision is taken by the funding agency in $T8_{FA}$ based on the application and the reviews. Finally, either an acceptance or rejection letter is created by the funding agency in one of the tasks $T7_{FA}, T9_{FA}, T10_{FA}$.

## 3.2   Privity Spheres

We assume that smart contracts are represented in form of business process models. This also implies that the read and write sets of tasks in the process model implicitly define the privity relations between participants and data objects. We base our measure on privity spheres that we first sketched in [7]. We now refine and formalize the original definitions of privity spheres:

**Definition 5 (Private Sphere).** Let P be a process model. A participant is in the private sphere if she is an actor of the process: $PrivateSphere(P) = P.A$      □

*Example 1 (Private Sphere).* For the process model in Fig. 1 the private sphere is {R, HO, FA, SP, A, B}.      □

**Definition 6 (Static Sphere).** A participant $a$ of a process is a member of the static sphere of some data object $d$ if $a$ is an actor of any task accessing $d$ in $P$.

$$StaticSphere(P, d) = \{a | a \in P.A : n \in P.N \wedge n.a = a \wedge (d \in n.d^w \vee d \in n.d^r)\}$$

*Example 2 (Static Sphere).* For the process model in Fig. 1 there are two static spheres. The one for D1 is $\{R, HO, FA, SP, A, B\}$ and the one for $D2$ is $\{A, B, FA\}$.      □

**Definition 7 (Weak-Dynamic Sphere).** Let $d \in P.D$ be a data object, $w$ be a task writing to $d$. An actor $a$ is in the weak-dynamic sphere of $d$ for $w$, iff $a$ is the actor of $w$ or $a$ is an actor of some task reading $d$ where $w$ is a possible origin for $d$:

$$WeakDynamicSphere(P, d, w) = \{w.a\} \cup \{a | a \in P.A : n \in P.N \wedge n.a = a \wedge d \in$$
$$n.d^r \wedge \exists I \in PI(P) : o(P^I, n, d) = w\}$$      □

*Example 3 (Weak-Dynamic Sphere).* The weak-dynamic sphere for data object $D1$ of writer $T1$ in Fig. 1 is $\{R, HO\}$, since $R$ is the actor of the activity $T1$ and $HO$ executes task $T2$ that will read the value from $T1$. The weak-dynamic sphere for data object $D1$ of writer $T2$ is $\{HO, FA, SP, A, B\}$, since they can all execute activities reading the value from $T2$.      □

While the weak-dynamic sphere of a writer contains all participants, that can execute some task reading the written data value, the strong-dynamic sphere requires, that participants must certainly execute some tasks reading the data value.

**Definition 8 (Strong-Dynamic Sphere).** Let $d \in P.D$ be a data object, $w$ be a task writing to $d$, $r$ be a node. An actor $a$ is in the strong-dynamic sphere of $w$ for $d$ at node $r$, iff for every instance type where $w$ is the origin of $r$ for $d$, $a$ will execute some node reading the value of $d$ from $w$ or $a$ is the actor of $w$.

$$StrongDynamicSphere(d, w, r) = \{a | a \in P.A : \exists I \in PI(P) : o(P^I, r, d) =$$
$$w \wedge \forall I \in \{I | I \in PI(P) : o(P^I, r, d) = w\} \exists n \in P^I.N : (d \in n.d^r \wedge n.a =$$
$$a \wedge o(P^I, n, d) = w) \vee a = w.a\}$$      □

*Example 4 (Strong-Dynamic Sphere).* The strong-dynamic sphere of data object $D1$ of writer $T1$ at Position $T2$ in Fig. 1 is $\{R, HO\}$ like for the weak-dynamic case. For $D1$, writer $T2$ at Position $T3$ it is $\{HO, FA\}$. In contrast to the weak-dynamic sphere $SP$, $A$, $B$ are not included since it is not certain if they will execute some task reading $D1$. This will only be known, once the later decisions are actually taken.      □

In order to model privity requirements of data objects, our process meta-model offers the property *sphere* for each data object. The property may be assigned to any value of the sequence <private, static, weak-dynamic, strong-dynamic>. The value identifies the minimal sphere requirement for participants. A data object $d$ with sphere $d.sphere = x$ may only be read by participants who are members of Sphere $x$.

### 3.3 Modeling Additional Actors for Decisions

For enforcing the correctness of decisions or to guarantee their availability, modelers can specify additional participants who must be able to execute the decision. This is realized by extending each node $n$ of type *business-rule-task* with the additional attribute $n.addAct \subseteq P.A \setminus \{n.a\}$.

## 4  Measures for Simple Processes

We introduce measures for assessing the effect of additional actors on privity spheres. We first discuss, how the changes of spheres can be measured, when a process model contains only one business rule task with additional actors in this section. We will introduce measures for the general case in Sect. 5.

Given a process model with only one business rule task with additional actors, we want to measure the effect on the privity spheres of these additional actors. The general idea is to measure the effect by comparing the privity spheres of the process including the additional actors and the ones of the process without taking the additional actors into account. The introduced measures are based on counting the additional members of each sphere that were introduced by the additional actors. Since the input process model contains minimal sphere requirements of each data-object, only changes in the defined sphere and in less restrictive spheres than the one defined for the data object are counted. E.g., if a data object requires at least the weak-dynamic sphere, then changes to the static and weak-dynamic sphere are taken into account but changes to the strong-dynamic spheres of that data object are not taken into account. We now discuss separately, for each sphere, how implications of additional actors on the privity relations of a cooperation are assessed.

### 4.1  Measuring the Impact on the Static Sphere

We first define, how the static sphere changes due to the inclusion of additional actors.

**Definition 9 (Static Sphere with Additional Actors).** A participant $a$ of a process is a member of the static sphere of data object $d$ taking into account additional actors if $a$ is member of the static sphere of $d$ or $a$ is an additional actor of some business rule task with $d$ as input:

$$StaticSphere'(P, d) = StaticSphere(P, d) \cup \{a | a \in P.A : n \in P.N \wedge a \in$$
$$n.addAct \wedge d \in n.d^r)\} \qquad \square$$

We can now quantify the effect of additional actors on the static sphere of a data object $d$ as $|StaticSphere'(P,d)| - |StaticSphere(P,d)|$. This leads the way to calculate the overall impact on all static spheres of a process. As discussed earlier, the measure counts only sphere updates for data objects that have at least the static sphere requirement.

**Definition 10 (Measure for the Static Sphere).** We define the overall impact of additional actors on all static spheres as the sum of the impacts on the spheres of each data object as the measure $\alpha$:

$$\alpha(P) = \sum_{d \in P.D \wedge d.sphere \geq static} |StaticSphere'(P,d) \setminus StaticSphere(P,d)|$$

*Example 5 (Calculating Measure for the Static Sphere).* Considering the example process in Fig. 1 and assuming $D1$ and $D2$ require at least the static sphere: If we add the additional actor $FA$ to the business rule task $T4$, this has no impact on the static sphere of $D1$ since $FA$ is already member of the static sphere of $D1$. This results in a measure score of $\alpha(P) = 0$. If we add the additional actor $SP$ to $T8$, the static sphere of $D2$ changes from $\{A, B, FA\}$ to $\{A, B, FA, SP\}$ resulting in a measure score $\alpha(P) = 1$. □

### 4.2  Measuring the Impact on the Weak-Dynamic Sphere

While there is only one static sphere for each data object, there is one weak-dynamic sphere for each data object and each activity writing to that data object. We first define the weak-dynamic sphere with additional actors:

**Definition 11 (Weak-Dynamic Sphere with Additional Actors).** Let $d \in P.D$ be a data object, $w$ be a task writing to $d$. An actor $a$ is in the weak-dynamic sphere with additional actors of $d$ for $w$, iff $a$ is in the weak-dynamic sphere of $d$ and $w$ or $a$ is an additional actor of some business rule task reading $d$ where $w$ is a possible origin of $d$:

$$WeakDynamicSphere'(P,d,w) = WeakDynamicSphere(P,d,w) \cup \{a|a \in P.A :$$
$$n \in P.n \wedge a \in n.addAct \wedge d \in n.d^r \wedge \exists I \in PI(P) : o(P^I, n, d) = w\} \qquad □$$

We can now count the effect of additional readers on the weak-dynamic spheres of data object $d$ and writer $w$: $|WeakDynamicSphere'(P,d,w)| - |WeakDynamicSphere(P,d,w)|$.

However, which weak-dynamic spheres are actually instantiated during process execution is a dynamic phenomena and depends on the decisions taken. For calculating the overall impact on the weak-dynamic spheres of a process model this behavior should be taken into account. Therefore, we weight the measure of each weak-dynamic sphere based on the node writing the data object.

**Definition 12 (Weight of a Node).** Let $n$ be a node of in a process model $P$. The weight of $n$ is the fraction of instance types containing $n$ in relation to all instance types:

$$weight(n) = \frac{|\{I|I \in PI(P) : n \in P^I.N\}|}{|PI(P)|}$$

We can now define the overall impact of additional actors on all weak-dynamic spheres as the sum of the impacts on the spheres of each data object and writer, weighted by the weight of the writer:

**Definition 13 (Measure for the Weak-Dynamic Sphere).**

$$\beta(P) = \sum_{d \in P.D \wedge d.sphere \geq weak-dynamic} \sum_{w \in P.N \wedge d \in w.d^w} weight(w) * |\Delta_\beta(P,d,w)|$$

$$\Delta_\beta(P,d,w) = WeakDynamicSphere'(P,d,w) \setminus WeakDynamicSphere(P,d,w)$$

*Example 6 (Calculating Measure for the Weak-Dynamic Sphere).* Considering the example process in Fig. 1: Assuming $D2$ requires at least the weak-dynamic sphere: If we add $SP$ as an additional actor to $T8$, the weak-dynamic spheres of $T5_A$ and $T6_B$ change. In particular, without additional actors, the weak-dynamic sphere for $T5_A$ and $D2$ is $\{A, FA\}$. The one for $T6_B$ and $D2$ is $\{B, FA\}$. When $SP$ is added to $T8$, we get: $\{A, FA, SP\}$ for $T5_A$ and $\{B, FA, SP\}$ for $T6_B$. Since $T5_A$ and $T6_B$ are conditionally executed and nested in another xor-block, their individual weight is $\frac{1}{4}$. This results in a measure score of:

$$\frac{1}{4} * |\{SP\}| + \frac{1}{4} * |\{SP\}| = \frac{1}{2}$$

□

### 4.3   Measuring the Impact on the Strong-Dynamic Sphere

While there is one private sphere for a process model, there is one static sphere per data-object and one weak-dynamic sphere for each combination of data object and tasks writing to that data object. In case of the strong-dynamic sphere, there is one sphere for each combination of data object $d$, origin $w$ and node having $w$ as origin for that data object $r$. We first define the Strong-Dynamic Sphere with additional actors:

**Definition 14 (Strong-Dynamic Sphere with Additional Actors).** Let $d \in P.D$ be a data object, $w$ be a task writing to $d$, $r$ be a node. An actor $a$ is in the strong-dynamic sphere with additional actors of $w$ for $d$ at node $r$, iff for every instance type where $w$ is the origin of $r$ for $d$, $a$ will execute some node reading the value of $d$ from $w$ or act as an additional actor for $r$ or $a$ is the actor of $w$.

$$StrongDynamicSphere'(d,w,r) = \{a | a \in P.A : \exists I \in PI(P) : o(P^I, r, d) =$$
$$w \wedge \forall I \in \{I | I \in PI(P) : o(P^I, r, d) = w\} \exists n \in P^I.N : (d \in n.d^r \wedge (n.a = a \vee (a \in$$
$$n.addAct \wedge n = r)) \wedge o(P^I, n, d) = w) \vee a = w.a\}$$

□

We can now calculate the impact of additional actors for a single sphere for data object $d$, origin $w$ and business rule task $r$ as:

$$|StrongDynamicSphere'(d,w,r)| - |StrongDynamicSphere(d,w,r)|$$

We are now interested in measuring all changes to the strong-dynamic sphere for all data objects in a process model with exactly one business rule task with additional actors $r$. Since both, the origins and also the business rule task itself are conditional, we propose a weighted measure to reflect the proportion of instances for each sphere change. We introduce a second weight function $weight(w, r, d)$ that computes the fraction of instances types containing both $w$ and $r$ where $w$ is the origin of $d$ for $r$ and the instance types containing $w$. We can now provide the overall measure for the impact on the strong-dynamic sphere:

**Definition 15 (Measure for the Strong-Dynamic Sphere).**

$$\gamma_1(P, r) = \sum_{d \in ST(P)} \sum_{w \in P.N \wedge d \in w.d^w} weight(w) * weight(w, r, d) * |\Delta_\gamma(d, w, r)|$$

$$\Delta_\gamma(d, w, r) = StrongDynamicSphere'(d, w, r) \setminus StrongDynamicSphere(d, w, r)$$

$$ST(P) = \{d : d \in P.D \wedge d.sphere = strong - dynamic\} \qquad \square$$

*Example 7 (Calculating Measure for the Strong-Dynamic Sphere).* Considering the example process in Fig. 1: Assuming $D1$ requires at least the strong-dynamic sphere: If we assign $A$ and $B$ as additional actors for $T4_{SP}$, this will change the strong-dynamic sphere for data object $D1$, writer $T2$ at position $T4$ from $\{HO, FA, SP\}$ to $\{HO, FA, SP, A, B\}$. Since the last writer $T2$ is executed unconditionally, and $T4$ is nested in one xor-block, the overall $\gamma_1$ measure for this change is: $1 * \frac{1}{2} * |\{A, B\}| = 1$ $\qquad \square$

The overall impact of one business rule task with additional actors on privity is the combination of the impact on the static spheres, the weak-dynamic spheres and the strong-dynamic spheres. We assume that the weighting between the spheres is domain specific, we therefore allow to use custom weights for the impacts of each sphere. We use parameter $p_\alpha$ for weights of the impacts on the static sphere, $p_\beta$ for the for weights of the impacts on the weak-dynamic sphere and $p_\gamma$ for the weights of the impacts on the strong-dynamic sphere:

**Definition 16 (Measure for Privity Impacts for a process with one task with additional actors.).** Let $P$ be a process model, $r$ be the only task in $P.N$ where $r.addAct \neq \{\}$. $p_\alpha$, $p_\beta$, $p_\gamma$ are non-negative numeric parameters in $\mathbb{R}$ provided by the user. $MeasurePrivity1(P, r, p_\alpha, p_\beta, p_\gamma) = p_\alpha * \alpha(P) + p_\beta * \beta(P) + p_\gamma * \gamma_1(P, r)$ $\qquad \square$

## 5   Measures for General Processes

For process models with more than one task with additional actors, we follow the same principle as for process models with one task with additional actors. However, in this case, the measure needs to take into account that one task with additional actors $a$ can have consequences for another task with additional actors $b$. E.g., if the assignment of additional actors to $a$ includes some actor $A1$ who therefore gets a member of the static sphere for some data object $d$, then an additional assignment of $A1$ to $b$ should not result in an additional increase of the measure. We will now discuss the effects of multiple tasks with additional actors per sphere.

## 5.1   Measuring the Impact on Static- and Weak-Dynamic Spheres

The measure of the static sphere is solely based on counting members of two different sets: A set of actors with and a set of actors without taking additional actors into account. Since each member can only be added once to a set, we can also apply measure $\alpha(P)$ for processes with more than one task with additional actors. If multiple tasks with additional actors lead to the same sphere changes this will not have any influence on the size of the set and consequently it cannot have an influence on the measure.

Weak-Dynamic Sphere: While there exists one static sphere for each data object, there is one weak-dynamic sphere for each data object and writer. The $\beta$ measure also compares the size of each sphere with and without taking additional actors into account. The only difference is that the result is weighted by the writer of the sphere. We can therefore use the $\beta(P)$ measure for processes with more than one task with additional actors.

## 5.2   Measuring the Impact on the Strong-Dynamic Sphere

While effects of additional actors on the static and weak-dynamic spheres are always local to that sphere, this property does not hold for the strong-dynamic sphere. Including an additional actor for one task can have an impact on the spheres of other tasks. Additionally, combinations of tasks with additional actors can lead to effects on the spheres of other tasks with additional actors. However, a measure should count every effect only once.

*Example 8 (Inter-Dependencies between nodes with Additional Actors).* Figure 2 sketches a process model with the data objects $D1$ and $D2$. For improved visibility, we denote read and write operations in the task boxes with $R(d)$ and $W(d)$, rather than with data-flow edges. The process contains the tasks $T1_A$ writing to both data objects. Then data object $D1$ is conditionally updated by participant $B$ either in $T2_B$ or $T3_B$. Finally, $D2$ is conditionally updated by participant $C$ in $T4_C$. The process contains the business rule tasks $BR1_A$, $BR2_A$ and $BR3_A$ of participant $A$. Tasks without labels are only depicted for the sake of completeness. However, they are independent of Data objects $D1$ and $D2$ and therefore irrelevant to the discussion.

We now assume that participant $C$ should be assigned as an additional actor to the business rule task $BR3_A$. We also assume that Data object $D2$ is annotated with the strong-dynamic sphere. $D1$ is annotated with the *private* sphere. Therefore, only $D2$ needs to be taken into account for assessing the impact on the strong-dynamic spheres. We first show the strong-dynamic spheres of the data object $D2$ ignoring additional actors:

$$(T1, BR1, D2) = \{A\}, (T1, BR2, D2) = \{A\}, (T1, BR3, D2) = \{A\}$$

When taking into account additional actors, and assuming $C$ is assigned as the only additional actor of $BR3$ and no other nodes have additional actors the spheres for $D2$ change to:

$$(T1, BR1, D2) = \{A, C\}, (T1, BR2, D2) = \{A, C\}, (T1, BR3, D2) = \{A, C\}$$

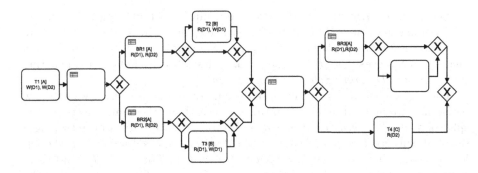

**Fig. 2.** Example process for assessing general processes

We can observe that the inclusion of one additional actor to one business rule task can have effects on the spheres of other business rule tasks. If we also assign $C$ as an additional actor for $BR1$ and $BR2$ we get exactly the same spheres for $D2$ as if $C$ was only assigned to $BR3$. Moreover, if we use $C$ as an additional actor for $BR1$ and $BR2$ but not for $BR3$ we will still get the exact same spheres as if only $C$ was added to $BR3$. Such implicit changes caused by inter-dependencies between business rule tasks must not lead to double-counting changes in a measure. Consequently, when assessing a solution where $C$ is assigned to $BR1$, $BR2$, and $BR3$, only the change of $(T1, BR3, D2)$ should be taken into account for Data object $D2$.                                                  □

To reflect this behavior, we propose to calculate the measure based on a minimal subset of tasks that lead to the same overall sphere updates as the inclusion of additional actors for every interdependent business rule task. In Example 8 for $D2$ there are the two minimal subsets $\{BR1, BR2\}$ and $\{BR3\}$ leading to the same impacts on the spheres as $\{BR1, BR2, BR3\}$. Our measure will return the measure for the subset $\{BR3\}$ as this leads to a $\gamma$ value of 0.5, instead of 0.5 + 0.5 = 1;

For defining the measure we first introduce the Strong-Dynamic Sphere with additional actors excluding a set of tasks.

**Definition 17 (Strong-Dynamic Sphere with additional actors excluding Tasks).** Let $d \in P.D$ be a data object, $w$ be a task writing to $d$, $r$ be a node, $TE$ be a set of excluded tasks. Actor $a$ is in the strong-dynamic sphere with additional actors excluding tasks $StrongDynamicSphere^*(d, w, r, TE, a)$, iff for every instance type where $w$ is the origin of $r$ for $d$, $a$ will execute some node $n$ or act as an additional actor for a node $n \notin TE$ where $n$ reads the value of $d$ from $w$ or $a$ $a$ is the actor of $w$                                          □

Business rule tasks with some additional actor $a$ can only have inter-dependencies if they can read the same data object from the same origin. Thus, for some actor $a$ and origin $w$, we can define the set of dependent business rule tasks as: $depT(a, w, d) = \{t | t \in P.N : t.type = business - rule - task \wedge a \in t.addAct \wedge \exists I \in PI(P) : o(P^I, t, d) = w\}$

Let $Spheres(a, w, d)$ be a function returning a set of tuples $SP$, where $SP$ contains one tuple $(t, w, d, S)$ for each element $t$ in $depT(a, w, d)$ where $S = StrongDynamicSphere^*(d, w, t, \{\}, a)$. $Spheres^*(a, w, d, TE)$ is defined in analogy to $Spheres(a, w, d)$ but the set of nodes $TE$ is excluded for calculating $S$: $S = StrongDynamicSphere^*(d, w, t, TE, a)$.

For a given set of dependent business rule tasks $depT(a, w, d)$, we can now define all minimal subsets of $depT(a, w, d)$ that have the same effects on the spheres as $depT(a, w, d)$ itself: Let $depTMin(a, w, d)$ be a function returning a set of all subsets $T$ of $depT(a, w, d)$ such that for every subset $T' \in T$ the following holds: $Spheres(a, w, d) = Spheres^*(a, w, d, depT(a, w, d) \setminus T')$ and there is no proper subset of $T'$ in $T$. Let $\gamma_{min}(a, w, d)$ be a function returning the minimal overall $\gamma$ measure of all elements $T'$ in $depTMin(a, w, d)$ in analogy to Definition 15.

We can now define the overall $\gamma$ value of a process as:

**Definition 18 (Measure for the Strong-Dynamic Sphere).**

$$\gamma(P) = \sum_{a \in P.A} \sum_{d \in P.D \wedge d.sphere=strong-dynamic} \sum_{w \in P.N \wedge d \in w.d^w} \gamma_{min}(a, w, d)$$

For general processes, we adopt the measure from Definition 16 as follows:

**Definition 19 (Measure for Privity Impacts for a process with additional actors).** Let $P$ be a process model, $p_\alpha$, $p_\beta$, $p_\gamma$ are non-negative numeric parameters in $\mathbb{R}$ provided by the user. The overall impact on privity of additional actors is: $MeasurePrivity(P, p_\alpha, p_\beta, p_\gamma) = p_\alpha * \alpha(P) + p_\beta * \beta(P) + p_\gamma * \gamma(P)$

### 5.3   Example Measure Calculation

We will now show the measure calculations using the business process model in Fig. 2, described in Example 8. In contrast to Example 8, we assume that Data objects $D1$ and $D2$ require the Strong-Dynamic Sphere.

**Impact on the Static Spheres.** Without any additional actors, we have the following static spheres: $StaticSphere(P, D1) = \{A, B\}$, $StaticSphere(P, D2) = \{A, C\}$. We now create a new process $P'$ derived from $P$ where $C$ acts as an additional actor for $BR1, BR2$ and $BR3$. This results in the spheres: $StaticSphere(P', D1) = \{A, B, C\}$, $StaticSphere(P', D2) = \{A, C\}$. This results in $\alpha(P') = |\{A, B, C\} \setminus \{A, B\}| + |\{A, C\} \setminus \{A, C\}| = 1$

**Impact on the Weak-Dynamic Spheres.** The calculation of the $\beta$ measure is shown in Table 1. The table contains one row for each writer of each data object. Column $WD()$ shows the members of the weak-dynamic sphere without $C$ as an additional actor for $BR1, BR2, BR3$. Column $WD'$ shows the spheres, when $C$ is used as an additional actor.

**Table 1.** $\beta$ measure with additional actor $C$ for $BR1$, $BR2$, $BR3$

| Sphere | WD() | WD'() | $\Delta$ | Weight(w) | Score |
|--------|------|-------|----------|-----------|-------|
| (T1,D1) | {A,B} | {A,B,C} | 1 | 1 | 1 |
| (T1,D2) | {A,C} | {A,C} | 0 | 1 | 0 |
| (T2,D1) | {A,B} | {A,B,C} | 1 | 0.25 | 0.25 |
| (T3,D1) | {A,B} | {A,B,C} | 1 | 0.25 | 0.25 |
| $\beta(P')$ | | | | | **1.5** |

**Impact on the Strong-Dynamic Spheres.** The calculations for the impact on the strong-dynamic spheres is shown in Table 2. The Column Sphere defines the current configuration for data object, writing task, and business rule task. Column $SD()$ shows the members of the sphere including the additional actors of the business rule tasks. Column $SD'()$ shows the spheres, without including additional actors. Sphere configurations that do not contribute to the measure using $\gamma_{min}()$ are marked with minus. This is the case for (D1,T1,BR3), and (D2,T1,BR1) and (D2,T1,BR2). The effects of (D1,T1,BR3) are implicitly achieved by the counted effects of (D1,T1,BR1) and (D1,T1,BR2). The effects of (D2,T1,BR1) and (D2,T1,BR2) are implied by the counted effects of (D2,T1,BR3).

**Table 2.** $\gamma$ measure of additional actor $C$ for $BR1$, $BR2$, $BR3$

| Sphere | SD() | SD'() | $\Delta$ | Weight(w) | Weight(w,r,d) | Score |
|--------|------|-------|----------|-----------|---------------|-------|
| (D1,T1,BR1) | {A,C} | {A} | 1 | 1 | 0.5 | **0.5** |
| (D1,T1,BR2) | {A,C} | {A} | 1 | 1 | 0.5 | **0.5** |
| (D1,T1,BR3) | - | - | - | - | - | - |
| (D1,T2,BR3) | {A,B,C} | {A,B} | 1 | 0.25 | 0.5 | **0.125** |
| (D1,T3,BR3) | {A,B,C} | {A,B} | 1 | 0.25 | 0.5 | **0.125** |
| (D2,T1,BR1) | - | - | - | - | - | - |
| (D2,T1,BR2) | - | - | - | - | - | - |
| (D2,T1,BR3) | {A,C} | {A} | 1 | 1 | 0.5 | **0.5** |
| $\gamma(P')$ | | | | | | **1.75** |

We can now calculate the overall effect based on the measures $\alpha$, $\beta$, $\gamma$ when using $C$ as an additional actor for $BR1$, $BR2$, $BR3$ as $MeasurePrivity$ $(P', 1, 1, 1) = 1 * 1 + 1 * 1.5 + 1 * 1.75$.

## 6 Conclusion and Future Work

A common pattern in applications targeting low-trust environments is the inclusion of additional actors for enforcing the correctness of decisions. Examples are distributed oracles or endorsing peers on permissioned blockchains. However, including additional participants can have negative impacts on privity. In this paper, we have defined measures for assessing the toll on privity that is paid for given sets of additional actors. The measures are based on counting changes in the privity relations of the collaboration. The resulting measure opens alleys for various applications such as: Interactively proposing optimal sets of additional actors to modelers, comparing alternative solutions, and detecting goal conflicts. Finally, the measure allows to automatically generate optimal sets of additional actors for given constraints on availability, enforceability, and privity. While the structure of the measure proposes that optimal sets of additional actors can be efficiently computed for the $\alpha$ and $\beta$ measure, the $\gamma$ measure is far more demanding due to the inter-dependencies of decisions. We are currently working on efficient, heuristic implementations for finding optimal sets of additional actors under given constraints on privity and enforceability or availability as well as user-defined positive and negative inclusion constraints on actors for decisions.

# References

1. Androulaki, E., Barger, A., et al.: Hyperledger fabric: a distributed operating system for permissioned blockchains. In: Proceedings of EuroSys 2018, pp. 1–15 (2018)
2. Basile, D., Goretti, V., Di Ciccio, C., Kirrane, S.: Enhancing blockchain-based processes with decentralized oracles. In: González Enríquez, J., Debois, S., Fettke, P., Plebani, P., van de Weerd, I., Weber, I. (eds.) BPM 2021. LNBIP, vol. 428, pp. 102–118. Springer, Cham (2021). https://doi.org/10.1007/978-3-030-85867-4_8
3. Blum, M., Feldman, P., Micali, S.: Non-Interactive Zero-Knowledge and Its Applications, pp. 329–349. Association for Computing Machinery, New York (2019)
4. Haarmann, S., Batoulis, K., Nikaj, A., Weske, M.: DMN decision execution on the ethereum blockchain. In: Krogstie, J., Reijers, H.A. (eds.) CAiSE 2018. LNCS, vol. 10816, pp. 327–341. Springer, Cham (2018). https://doi.org/10.1007/978-3-319-91563-0_20
5. Haarmann, S., Batoulis, K., Nikaj, A., Weske, M.: Executing collaborative decisions confidentially on blockchains. In: Di Ciccio, C., et al. (eds.) BPM 2019. LNBIP, vol. 361, pp. 119–135. Springer, Cham (2019). https://doi.org/10.1007/978-3-030-30429-4_9
6. Köpke, J., Franceschetti, M., Eder, J.: Optimizing data-flow implementations for inter-organizational processes. Distrib. Parallel Databases **37**(4), 651–695 (2018). https://doi.org/10.1007/s10619-018-7251-3
7. Köpke, J., Franceschetti, M., Eder, J.: Balancing privity and enforceability of BPM-based smart contracts on blockchains. In: Di Ciccio, C., et al. (eds.) BPM 2019. LNBIP, vol. 361, pp. 87–102. Springer, Cham (2019). https://doi.org/10.1007/978-3-030-30429-4_7
8. Mercenne, L., Brousmiche, K., Hamida, E.B.: Blockchain studio: a role-based business workflows management system. In: IEMCON 2018, pp. 1215–1220 (2018)
9. Mühlberger, R., et al.: Foundational oracle patterns: connecting blockchain to the off-chain world. In: Asatiani, A., et al. (eds.) BPM 2020. LNBIP, vol. 393, pp. 35–51. Springer, Cham (2020). https://doi.org/10.1007/978-3-030-58779-6_3

10. Pintado, O., García-Bañuelos, L., Dumas, M., Weber, I., Ponomarev, A.: Caterpillar: a business process execution engine on the ethereum blockchain. Softw. Pract. Exp. **49**(7), 1162–1193 (2019)
11. Szabo, N.: Formalizing and securing relationships on public networks. First Monday **9**(2) (1997)
12. Tran, A.B., Lu, Q., Weber, I.: Lorikeet: a model-driven engineering tool for blockchain-based business process execution and asset management. In: Proceedings of BPM 2018, pp. 56–60 (2018)
13. Wang, S., et al.: On private data collection of hyperledger fabric. In: ICDCS 1, pp. 819–829 (2021)
14. Xu, X., Dilum Bandara, H., Lu, Q., Weber, I., Bass, L., Zhu, L.: A decision model for choosing patterns in blockchain-based applications. In: 2021 IEEE 18th International Conference on Software Architecture (ICSA), pp. 47–57 (2021)
15. Xu, X., Pautasso, C., Zhu, L., Lu, Q., Weber, I.: A pattern collection for blockchain-based applications. In: Proceedings of EuroPLoP 2018, pp. 3:1–3:20 (2018)

# Threshold Signature
# for Privacy-Preserving Blockchain

Sara Ricci[1]([☒]) [iD], Petr Dzurenda[1] [iD], Raúl Casanova-Marqués[1,2] [iD],
and Petr Cika[1] [iD]

[1] FEEC, Department of Telecommunications, Brno University of Technology, Brno,
Czech Republic
{ricci,dzurenda,casanova,cika}@vut.cz
[2] Institute of New Imaging Technologies, Universitat Jaume I, Castellón, Spain
https://axe.vut.cz/

**Abstract.** Threshold signatures received renewed interest in recent
years due to their practical applicability to Blockchain technology. In
this article, we propose a novel $(n, t)$-threshold signature scheme suitable for increasing security and privacy in Blockchain technology. Our
scheme allows splitting a Blockchain wallet into multiple devices so that
a threshold of them is needed for signing. This increases the security of
the transactions, e.g., more devices need to be compromised to recover
the key and permits, and the privacy, e.g., the signing is made anonymously on behalf of the group of users sharing the Blockchain wallet.
Our experimental results show that the signing algorithm requires less
than 10 ms in the cases of 10 devices involved.

**Keywords:** Threshold signature · Multi-signature · Blockchain ·
Secret sharing · Paillier cryptosystem · Schnorr protocol

## 1 Introduction

Threshold signatures belong to distribute signature family where a threshold
number of participants have to cooperate to issue a signature that can be verified
by a single public key. Even if they have been studied for a long time [13,22],
these signatures received renewed interest in recent years due to their practical applicability to Blockchain technology and electronic transactions, including
cryptocurrencies such as Bitcoin [3].

A typical Blockchain consists of two parts: 1) a consensus mechanism for del-
egating the creation of new blocks including user transaction and 2) a signature
scheme for user transactions verification [24]. A standard transaction, namely a
single-signature transaction, involves only one private key, which is managed by
a single device. On the contrary, a multi-signature transaction involves at least
two keys that can be stored in different devices. This approach can bring several
new beneficial features [17], e.g., 1) `increased security`: splitting the wallet
keys between more user devices reduces the risk of compromising the wallet.

© Springer Nature Switzerland AG 2022
A. Marrella et al. (Eds.): BPM 2022, LNBIP 459, pp. 100–115, 2022.
https://doi.org/10.1007/978-3-031-16168-1_7

In fact, a malware is unlikely to infect them all. 2) `Joint accounts`: the transactions require the signatures of multiple users before the funds can be transferred, and 3) `wallet key backup`: if one key is lost in a "2-of-3" wallet, then the other two keys can be used to retrieve the wallet.

It is important to notice that Bitcoin already supports multi-signature transactions. These multi-signature wallets consist of several regular Bitcoin addresses. However, this approach has several privacy issues such as 1) several wallets of the same owner are publicly known, 2) the signing threshold and signer's identity are also known, and 3) the size of the transaction grows with the number of wallet owners and signers. Threshold signatures can help reduce the amount of data stored in the Blockchain and solve privacy issues by compressing the signatures together while keeping verification still possible.

In a $(n, t)$-threshold signature, $n$ parties can jointly generate a single public key from their $n$ private shares of the key. The key can be used to securely sign messages if and only if $t$ parties collaborate in the signing process. Moreover, no group of $t-1$ colluding parties should be able to recover the secret key. Threshold signatures are mainly based on RSA [12,33] and Elliptic Curve Digital Signature Algorithm (ECDSA) signatures [9,18,20,26,27]. In the ECDSA signatures group, we can split the signatures in the $(2, 2)$-threshold variant [7,14,25] and the more general $(n, t)$-threshold case [6,8,10,11,15,18,19,26]. Note that $(n, n)$-threshold signatures are a particular case of $(n, t)$-threshold ones. At the moment, the most efficient threshold signature schemes rely on pairing-based cryptography [4,5]. These schemes can perform signing operations in a single round among participants whereas the best non-pairing-based threshold schemes require multiple rounds of interaction during signing operations. These signatures are based on Schnorr's protocol. For instance, [21,34] require at least three rounds of communication during signing operations whereas FROST threshold signatures [23] needs only two rounds. Even if having more communication overload, these latter schemes guarantee robustness as a main feature, i.e., if any participant misbehaves, honest participants can detect this misbehavior, disqualify the misbehaving participant, and produce a signature as long as the threshold number of honest parties is achieved.

## 1.1 Contribution and Paper Structure

In this work, we proposed a novel $(n, t)$-threshold signature scheme suitable for increasing security and privacy in Blockchain technology. To do so, we provide a solution on how to securely split a Blockchain wallet between more devices. These devices can be held by a single user (i.e., it increases security, since more user devices are needed to sign Blockchain transactions) or by several collaborative users (i.e., it increases privacy, since the signing is made anonymously on behalf of the group of users sharing the Blockchain wallet). Our scheme is built on provable secure cryptographic primitives such as Schnorr signature [31], Pailler cryptosystem [29], and Shamir secret sharing scheme [32]. Furthermore, we implement our proposal with promising experimental results on a single board computer using ARM Cortex-A72 processor widely used in the internet of things environment.

The paper is organized as follows. Section 2 outlines the used notation and cryptographic primitives used in our proposal. Section 3 introduces our $(n, t)$-threshold signature scheme. Section 4 presents the security analysis of our proposal. Section 5 shows the possible deployment of threshold signatures to Blockchain technology. Section 6 presents our experimental results. In the last section, we conclude this work.

## 2    Cryptographic Preliminaries

In this section, we introduce used notation and cryptographic primitives. From now on, the symbol ":" means "such that", "$|x|$" is the bitlength of $x$, and "$||$" denotes the concatenation of two binary strings. We write $a \in_R A$ when $a$ is sampled uniformly at random from $A$. Let $\mathbb{G}$ be a additive cyclic group generated by elliptic curve $E$ over final field $\mathbb{F}_p$ and base point $g \in E(\mathbb{F}_p)$ of prime order $q_{EC}$, where $|q_{EC}| = \kappa$ and $\kappa$ is a security parameter. A secure hash function is denoted as $\mathcal{H} : \{0, 1\}^* \to \{0, 1\}^\kappa$, where $\kappa$ is a security parameter. We describe the Proof of Knowledge (PK) and the Signature of Knowledge (SK) protocols using the notation introduced by Camenisch and Stadler. In particular, the protocol for proving the knowledge of discrete logarithm of $c$ with respect to $g$ is denoted as $\text{PK}\{\alpha : c = g^\alpha\}$ and the protocol for proving the knowledge of discrete logarithm of $c$ with respect to $g$ and message $m$ is denoted as $\text{SK}\{\alpha : c = g^\alpha\}(m)$.

### 2.1    Schnorr Signature

Schnorr signature [31] is a digital signature scheme known for its simplicity, efficiency, and short signatures. It is based on the PK concept and it is frequently used in many cryptosystems, including privacy-enhancing schemes, such as group signatures, ring signatures, and attribute-based credentials. Using this scheme, the prover proves his/her statement on knowledge of a discrete logarithm (i.e., secret key $sk : pk = g^{sk}$) with respect to public parameters $\mathbb{G}, g, q, pk$. In contrast to the Schnorr identification protocol [31] which is a interactive 3-way protocol, the Schnorr signature scheme is non-interactive.

The scheme is depicted in Fig. 1. The prover commits a random number $r$, computes a challenge $e$ by using a secure hash function $\mathcal{H}$, and finally responds by the proof $z$ on the challenge $e$, secret key $sk$ and message $m$.

Furthermore, Schnorr signatures have linear characteristics as shown in [16, 28]. This linearity property of Schnorr signatures can be used to construct a multi-signature, see Eq. 1 describing case 2-of-2 multi-signature.

$$\begin{aligned}
pk &= pk_A * pk_B = g^{sk_A} * g^{sk_B} \\
c &= c_A * c_B = g^{r_A} * g^{r_B} \\
e &= \mathcal{H}(c||m) \\
z &= z_A + z_B = (r_A - e * sk_A) + (r_B - e * sk_B)
\end{aligned} \tag{1}$$

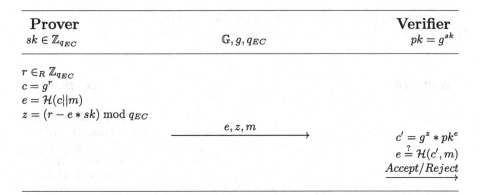

**Fig. 1.** Schnorr's signature of knowledge of discrete logarithm $SK\{sk : pk = g^{sk}\}(m)$.

The verification of 2-of-2 multi-signature is depicted in Eq. 2.

$$c' = g^z * pk^e = g^{(r_A - e*sk_A) + (r_B - e*sk_B)} * (g^{sk_A} * g^{sk_B})^e = g^{r_A} * g^{r_B} \quad (2)$$

## 2.2 Shamir Secret Sharing Scheme

Shamir secret share scheme [32] is a well-known $(n, t)$-threshold scheme where $n$ denotes the number of shares involved and $t$ the number of shares needed to reconstruct the secret $k$. This scheme is based on 1) unique polynomial property, i.e., there exists a unique $t-1$-th degree polynomial that passes through $t$ points in the plane, and 2) interpolation problem.

Let $k$ be the secret, where $k$ is in a field $\mathbb{F}_q$. The scheme involves a dealer who owns a secret and a set of $n$ parties. The dealer chooses $t - 1$ random elements $a_1, \ldots, a_{t-1}$ from $\mathbb{F}_q$ independently with uniform distribution and defines a polynomial $P(x) = k + \sum_{i=1}^{t-1} a_i x^i$. The share of party $j$ is the evaluation of the polynomial $\beta_j = P(\alpha_j)$, that is the pair $(\alpha_j, \beta_j)$. The secret can be recovered with the following formula:

$$k = P(0) = \sum_{j=1}^{t} \beta_j * \prod_{m=1, m \neq j}^{t} \frac{\alpha_m}{\alpha_m - \alpha_j}. \quad (3)$$

Shamir secret sharing scheme has both properties [2]: 1) **Correctness** the secret $k$ can be reconstructed by any authorized set of parties, and 2) **Perfect Privacy** every unauthorized set cannot learn anything about the secret (in the information theoretic sense) from their shares.

## 2.3 Paillier Cryptosystem

Paillier cryptosystem [29] is a probabilistic public-key algorithm for asymmetric encryption. This scheme runs in a RSA-modulo where $P, Q$ are two large primes of equal length. Figure 2 depicts its subroutines, namely `Keygen`, `Enc`, and `Dec` that states for key generation, encryption and decryption protocols, respectively.

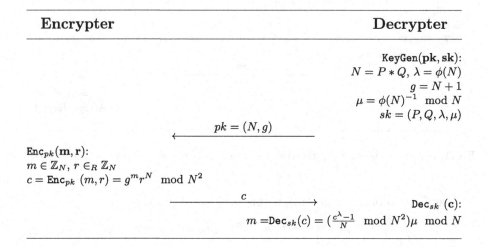

**Fig. 2.** Paillier cryptosystem.

Paillier scheme has additive homomorphic properties, i.e., allows doing additive operations on ciphertexts without corrupting the result. In particular, we will use the following two properties:

- $\text{Enc}_{pk}(m_1, r_1) * \text{Enc}_{pk}(m_2, r_2) = \text{Enc}_{pk}(m_1 + m_2, r)$,
- $\text{Enc}_{pk}(m_1, r)^{m_2} = \text{Enc}_{pk}(m_1 * m_2, r)$,

where $m_1$ and $m_2$ are two messages in $\mathbb{Z}_N$ and $r, r_1$, and $r_2$ are the random noises.

## 3 Proposed (n,t)-Threshold Scheme

In this section, we define the algorithms and entities of our (n,t)-threshold protocol. It employs two parties:

- **Signer**: jointly generates the signature in collaboration with other co-signers. The signer is typically a user which is accessing online services, such as cryptocurrency payments. The signing key (i.e., Blockchain Wallet secret key) is split and stored on different user devices such as smartphones, wearables, microcontrollers, and personal computers. In particular, a `Main Device` (MD) represents a user device with an activated signing mode. This mode is used by a signer when he/she wants to sign a Blockchain transaction. The main task of the MD is to initiate all events, i.e., communication with Blockchain, creation of user transactions, and generation of a signature on the transaction. To do so, the MD has to run the signing protocol jointly with the co-signers' devices. Moreover, a **Secondary Device** (SD) represents a user device with activated co-signing mode. This mode is used by co-signers when they want to join a signing process of a Blockchain transaction initiated by the MD.

- **Verifier**: is a Blockchain node receiving and validating the user transactions. Among others, it checks the validity of the signature in the transaction over the Blockchain Wallet public key. The Verifier can be represented by a powerful computer as well as a computationally less powerful device such as a single-board computer or microcontroller.

The (n,t)-threshold signature scheme consists of the following three algorithms: 1) **Setup**, 2) **Signing**, and 3) **Verifying** that are described in Sects. 3.1, 3.2, and 3.3, respectively. Our $(n, t)$-threshold signature scheme needs that $t$ out of $n$ authorized signers collaborate to generate the signature. Let $D_j$ be a signer's device performing the signature where $j = 1, \ldots, n$. Each $D_j$ owns its signing secret key $k_j$.

## 3.1  Setup Algorithm

Our scheme is based on Shamir protocol [32] and, therefore, requires that $n$ signers agree on a polynomial $f(x)$ of degree $t-1$ that has $sk = \sum k_i$ as constant term, please see Sect. 2 for more details. Note that $sk$ is the secret key used in the signature. We consider a polynomial with the following structure:

$$f(x) = (d_{t-1}^{(1)} + \cdots + d_{t-1}^{(n)})x^{t-1} + \cdots + (d_1^{(1)} + \cdots + d_1^{(n)})x + sk, \quad (4)$$

where $d_j^{(i)}$ is privately generated by $D_i$ for $i = 1, \ldots, n$ and $j = 1, \ldots, t-1$.

By using Paillier encryption [29], the polynomial $f(x)$ is evaluated in $n$ points without disclosing its coefficients, i.e., the polynomial is not "built" and shared among the devices but only evaluated. Accordingly, each device $D_j$ obtains the pair $(\alpha_j, f(\alpha_j))$ where $\alpha_j = j$ is publicly known and $f(\alpha_j)$ is kept secret. The **Setup** algorithm consists of two phases:

$(\Lambda_j, pk, pk_j, pk_{p,j}) \leftarrow \mathsf{ParGen} \leftarrow (n, t, \kappa)$: each device $D_j$ for $j = 1, \ldots, n$ does as follow and with respect of a security parameter $\kappa$:

- generate at random $d_1^{(j)}, \ldots, d_{t-1}^{(j)}$,
- generate the Paillier's key pair $(pk_{p,j}, sk_{p,j})$,
- generate at random $k_j$ in $\mathbb{Z}_{q_{EC}}$,
- compute $pk_j = g^{k_j}$.

The values $\Lambda_j = (k_j, d_1^{(j)}, \ldots, d_{t-1}^{(j)}, sk_{p,j})$ are privately stored in each device whereas $(pk_j, pk_{p,j})$ are made public. An agreed user compute the common key

$pk = \prod_{i=1}^{n} pk_i$ (i.e., Blockchain wallet public key). Note that the coefficients $d_i^{(j)}$ need to meet the following dis-equality:

$$|d_i^{(j)}| < (\frac{|N|}{n} - |q_{EC}|)\frac{2}{t(t-1)}, \tag{5}$$

for $i = 1, \ldots, t-1$ and $j = 1, \ldots, n$.

**Lemma 1.** *Equation 5 allows the polynomial $f(x)$ to not be modified by applying Paillier cryptosystem.*

*Proof.* Since we are going to encrypt $f(x)$ with Paillier cryptosystem, we need that its evaluation $f(\alpha_j)$ is smaller than Paillier modulus $N$ for all $j = 1, \ldots, n$. To do so, we need that $|f(\alpha_n)| = |(d_{t-1}^{(1)} + \cdots + d_{t-1}^{(n)})\alpha_n^{t-1} + \cdots + (d_1^{(1)} + \cdots + d_1^{(n)})\alpha_n + \sum_i k_i| < |N|$, where $\alpha_i < \alpha_2 < \cdots < \alpha_n$ by construction. Let $d$ be equal to $\max_{i,j} d_i^{(j)}$, then

$$|f(\alpha_n)| < n|d||\alpha_n|^{t-1} + \cdots + n|d||\alpha_n| + n|q_{EC}| =$$
$$n|d|((t-1) * (t-2) * \cdots * 1)|\alpha_n| + n|q_{EC}| =$$
$$n|d|\frac{t*(t-1)}{2}|\alpha_n| + n|q_{EC}| < |N|.$$

Therefore, $|d| < (\frac{|N|}{n} - |q_{EC}|)\frac{2}{t(t-1)}$.

$(\alpha_j, f(\alpha_j)) \leftarrow \texttt{PolyEval} \leftarrow (\Lambda_j, pk_{p,j})$: all devices engage in the following step for the computation of each $f(\alpha_j)$ for $j = 1, \ldots, n$. Let $h$ be equal to $j+1$:

1. $D_h$ generates random value $r_{j,h}$ and compute $\epsilon_{j,h} = Enc_{pk_{p,j}}(\alpha_j, r_h)$,
2. $D_h$ generates random value $v_{j,h}$ and compute

$$c_h = \epsilon_{j,h}^{\alpha_j^{t-2} * d_{t-1}^{(h)}} * \epsilon_{j,h}^{\alpha_j^{t-3} * d_{t-2}^{(h)}} * \cdots * \epsilon_{j,h}^{d_1^{(h)}} * Enc_{pk_{p,j}}(k_j, v_{j,h}),$$

3. if $h = j + 1$, then $D_h$ sends $c_h$ to $D_{h+1}$,
4. if $h \neq j$, then set $h = h + 1 \mod n$ and go to Step (1),
5. if $h = j$, then $D_j$ computes

$$f(\alpha_j) = Dec_{sk_{p,j}}(c_{j-1}) + d_{t-1}^{(j)}\alpha_j^{t-1} + \cdots + d_1^{(j)}\alpha_j + k_j.$$

The algorithm outputs for each device $D_j$ the pair $(\alpha_j, f(\alpha_j))$ where $\alpha_j = j$ is publicly known and $f(\alpha_j)$ is kept secret. In Fig. 3, PolyEval phase is sketched in the case of five devices where $D_1$ acts as MD. At the end of the process $D_1$ obtains the evaluation of the $f(x)$ in $\alpha_1$, i.e., $f(\alpha_1)$.

### 3.2 Signing Algorithm

During the Signing algorithm, $t$ out of $n$ devices need to collaborate to generate the signature $\sigma$. To do so, $t$ specific devices need to be agreed. Therefore, let $\mathcal{J}_t \subset \{1, \ldots, n\}$ be the set of $t$ indices such that $D_j$ with $j \in \mathcal{J}_t$ has been selected for signing. The Signing algorithm consists of two phases:

$(\{s_j\}_{j \in \mathcal{J}_t}) \leftarrow \texttt{SessionKeyGen} \leftarrow (\mathcal{J}_t, \alpha_j, f(\alpha_j))$: each $D_j$ with $j \in \mathcal{J}_t$ does as follow:

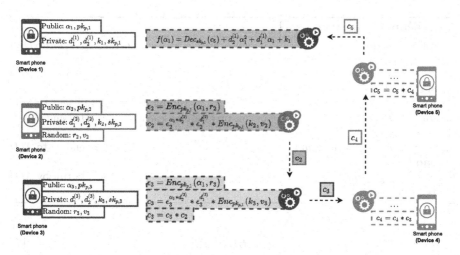

**Fig. 3.** `PolyEval` phase in `Set-up` algorithm of the proposed $(n, t)$−threshold signature.

– compute the session key:

$$s_j = f(\alpha_j) * \prod_{m \in \mathcal{J}_t \setminus \{j\}} \frac{\alpha_m}{\alpha_m - \alpha_j} \quad \text{mod } q_{EC},$$

– store $s_j$ privately.

Note that $s_j$ is the private session key of $D_j$ used to generate $\sigma_j$ and, therefore, $s_j$ is kept secret.

$(\sigma) \leftarrow$ `SignatureGen` $\leftarrow (\{s_j\}_{j \in \mathcal{J}_t}, m)$: at first, each device $D_j$ with $j \in \mathcal{J}_t$ commits to its random value $r_j$. At second, the commitments $c_j$ are sent to the MD device that aggregates them to receive the common commitment $c$. Then $c$ is sent back to SDs with the signing message $m$ (i.e., Blockchain transaction) so that each device $D_j$ generates its signature fragment $(z_j)$ on $m$. At last, the MD aggregates all signature fragments and outputs the Schnorr signature $\sigma = (e, z)$ on the Blockchain transaction which is sent to the Blockchain for validation, see Algorithm 1 for more details.

### 3.3 Verifying Algorithm

The `Verifying` algorithm allows the verifier (i.e., Blockchain node) to verify the signature (see Sect. 2.1 for more details).

$(0/1) \leftarrow$ `Verifying` $\leftarrow (pk, \sigma, m)$: this is run by the verifier that verifies if the signature is valid as follow:

$$c' = g^z * pk^e \tag{6}$$

---

**Algorithm 1.** SignatureGen($\{s_j\}_{j \in \mathcal{J}_t}, m$)

---
1: **for** $j \in \mathcal{J}_t$ **do:**                    ▷ run privately by each $D_j$ (i.e., $MD$ and $SDs$ )
2:    $r_j \in_R \mathbb{Z}_{q_{EC}}$
3:    $c_j = g^{r_j}$                          ▷ SDs send $c_j$ to $MD$
4: **end for**
5: $c = \prod_{j \in \mathcal{J}_t} c_j$                       ▷ run by $MD$, $c$ and $m$ sent to $SDs$
6: **for** $j \in \mathcal{J}_t$ **do:**                    ▷ run privately by each $D_j$ (i.e., $MD$ and $SDs$ )
7:    $e = \mathcal{H}(c\|m)$
8:    $z_j = r_j - e * s_j \bmod q_{EC}$              ▷ SDs send $z_j$ to $MD$
9: **end for**
10: $z \leftarrow \sum_{j \in \mathcal{J}_t} z_j \bmod q_{EC}$                 ▷ run by $MD$
11: **return** $\sigma = (e, z)$

---

$$e \overset{?}{=} \mathcal{H}(c'\|m). \qquad (7)$$

If Eq. 7 holds, the signature $\sigma$ is accepted and the algorithm returns true, false otherwise.

## 4    Security Analysis

In this section, we prove the security of our threshold signature scheme. The signature is based on provable secure cryptographic primitives, namely Schnorr signature [31], Pailler cryptosystem [29], and Shamir secret sharing scheme [32].

**Lemma 2.** *The proposed threshold signature is existentially* **unforgeable** *under chosen-message attacks in the random oracle model assuming that the discrete logarithm problem is hard.*

*Proof.* This is based on the fact that our proposal is built on the Schnorr signature and its unforgeability is proven in [30], Lemma 2.

**Lemma 3.** *The proposed threshold signature is* **sound and complete**, *i.e., valid signatures will be always verified correctly, and invalid ones will always fail verification.*

*Proof.* In order to prove the completeness of the signature, Eq. 7 has to hold. This happens if:

1. the sum of the sessions keys $s_i$ is equal to the private key $sk = \sum_{1=1}^{n} k_i$. This follows from Eq. 3, where $\beta_j = f(\alpha_j)$. In fact, $\sum_{j=1}^{t} s_i = \sum_{j=1}^{t} f(\alpha_j) * \prod_{m=1,m\neq j}^{t} \frac{\alpha_m}{\alpha_m - \alpha_j} = f(0) = sk$, where a re-labeling of the elements of $\mathcal{J}_t$ is applied;
2. the commitment $c'$ (i.e., Eq. 6) is correctly reconstructed

$$c' = g^z * pk^e = g^{\sum_{i=1}^{t}(r_i - e*s_i)}(g^{sk})^e = g^{\sum_{i=1}^{t} r_i - e*sk}(g^{sk})^e = g^{\sum_{i=1}^{t} r_i}$$
$$= \prod_{i=1}^{t} g^{r_i} = \prod_{i=1}^{t} c_i = c.$$

Soundness (sketch of proof by contradiction): if an unauthorized signer is able to produce at least two valid signatures $\sigma = (e, z)$ and $\sigma' = (e', z')$ without knowing $sk$, then Eq. 6 has to hold, i.e., both signatures pass the verification phase. Note that $1 = pk^{e-e'} * g^{z-z'}$ and, therefore, $pk = g^{\frac{z'-z}{e-e'}}$ where $sk = \frac{z'-z}{e-e'}$ , i.e., the unauthorized signer knows the secret key $sk$.

**Lemma 4.** *The proposed threshold signature is* **zero-knowledge**. *This means that there exists a simulator* $\mathbb{S}$ *that is able to efficiently generate a protocol transcript indistinguishable from a real protocol transcript without the knowledge of the private key sk.*

*Proof.* We prove the zero-knowledge property by constructing the zero-knowledge simulator $\mathbb{S}$. Let's assume that the simulator can program the random oracle $\mathcal{H}$ in a way that on inputs $\hat{c}$, $m$ outputs $\hat{e}$. Then, the simulator does as follows:

1. Randomly selects the response $\hat{z} \in_R \mathbb{Z}_q$.
2. Randomly selects the challenge $\hat{e} \in_R \mathbb{Z}_q$.
3. Computes the commitment $\hat{c} = pk^{\hat{e}} * g^{\hat{z}}$.

The simulator's output is computationally indistinguishable from the real protocol transcript, i.e., $(\hat{e}, \hat{z}) \cong_c (e, z)$, because all pairs are selected randomly and uniformly from the same sets.

**Lemma 5.** *The proposed threshold signature provides both secret sharing properties, i.e.,* **correctness and perfect privacy**.

*Proof.* Note that the secret $sk$ can be recovered with either the secret keys $k_i$ or the session keys $s_i$. Therefore, correctness and perfect privacy properties need to be proven in both cases. For $k_i$, the proof is straightforward and follows the Shamir secret sharing scheme ones [2]. **Correctness:** each session key $s_i$ secrecy relies on $f(x)$ one. In fact, if one knows $f(x)$, then can evaluate $f(x)$ in all $\alpha_i$ and reconstruct $s_i$. Thanks to PolyEval algorithm, any signer knows only partial values of each coefficient of $f(x)$. Accordingly, $f(x)$ can be reconstructed only knowing all $d_j^{(i)}$. **Perfect privacy:** the modulus and Paillier encryption prevent to have information on the coefficient $d_j^{(i)}$ of $f(x)$ and, therefore, to reconstruct the session keys $s_i$, where $sk = \sum_i s_i$.

# 5 Deployment of $(n, t)$-Threshold Scheme to the Blockchain

Two main use case scenarios for $(n, t)$-threshold scheme deployment to the Blockchain are depicted in Fig. 4. On one hand, the Multi-device wallet scenario aims at higher protection of the Blockchain wallet. The user owns a Blockchain wallet of which a secret key is split between his/her several wearable devices. In this case, we use a $(5, 2)$-threshold signature where two of five devices have to collaborate to sign a transaction by the Blockchain wallet secret key. On the other

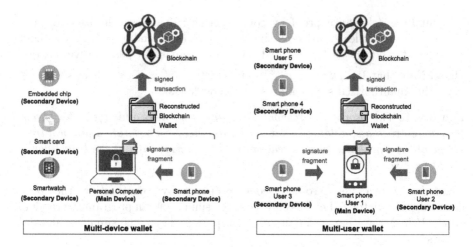

**Fig. 4.** Practical use cases of our $(n, t)$-threshold signatures in the Blockchain.

hand, the Multi-user wallet scenario focuses on stronger privacy protection of users and sharing property of the Blockchain wallet between several users. The users own a common Blockchain wallet of which a secret key is split between all of them. Here, we consider a $(5, 3)$-threshold signature, where three of five users have to collaborate to reconstruct the Blockchain wallet, i.e., sign a transaction by the Blockchain wallet. The signature on the transaction is verifiable by the Blockchain wallet public key. However, no one is able to track back the signers, since, the signature is generated anonymously on behalf of the group of users sharing the wallet. In both scenarios, the verifier or even eavesdropper will learn nothing about the signers, i.e., the number of users/devices sharing the Blockchain wallet and the required threshold for reconstructing the wallet. In fact, there is only one group Blockchain wallet public key and several fragments of the Blockchain wallet secret key distributed between several users/devices.

## 6   Experimental Results

In this section, we provide experimental results of the proposed $(n, t)$-threshold scheme. We assess the execution time for all deployed algorithms (i.e., Setup, Signing, and Verifying) independently, as well as the overall execution time. We used one single Raspberry Pi 4 Model B with 4 GB of RAM to represent all system entities, i.e., Signer' MD and SDs devices, and the Verifier's device. The testing application is written in C programming language and uses several external cryptographic libraries. The cryptographic core follows the key length recommendations defined by the National Institute of Standards and Technology (NIST) [1] for 112-bit security strength. We use Paillier cryptosystem with modulus size $|N| = 2048$ bits, where $N = P * Q$ and $|P| = |Q| = 1024$ bits primes and Shamir protocol with elements of 256-bit length sizes. Both protocols were

**Table 1.** Benchmarks in ms of the Setup, Signing, and Veryfing algorithms for (5,3)-threshold scheme.

| Setup algorithm | | Signing algorithm | | |
|---|---|---|---|---|
| ParGen | PolyEval | SessionKeyGen | SignatureGen $(\sigma_j)$ | *Total* $(\sigma)$ |
| 273.7 | 570.9 | 0.02 | 8.57 | 8.59 |
| Veryfing algorithm | | *Total* | | |
| Verifying | | **870.44** | | |
| 16.95 | | | | |

implemented by using the GMP library . Furthermore, we use the micro-ecc library to implement the Schnorr signature over elliptic curve. Namely, we use standardized elliptic curve secp256r1 where $|p| = |q| = 256$ bits. Finally, we utilized OpenSSL library to perform SHA-256 hash algorithm.

We follow the environment model depicted in Fig. 3, with a total of five devices, of which three were required to perform the signature. Each device is simulated by a separated application thread. Table 1 shows the benchmarks of the Setup algorithm, the benchmarks of the Signing algorithm, and the benchmarks of the Veryfing algorithm. In addition, we provide the whole protocol execution timings. Note that the benchmarks shown for ParGen algorithm relate to the execution time on each device (owing to running in parallel), whilst the polynomial evaluation refers to the overall execution time across all five devices. Regarding the Signing algorithm, the session key generation is performed in parallel on each device whereas the times for signature generation are divided into two parts: 1) the partial signature on each device and 2) the joint signature. Additionally, we run several experiments with varying numbers of devices and signers. Figure 5 depicts the execution timings of the protocol for $n = 5$ and $t = \{2, 3, 4\}$, whereas Fig. 6 illustrates the speed for $n = 10$ and $t = \{2, \cdots, 9\}$. Since signature verification is consistent across all devices, we simply included

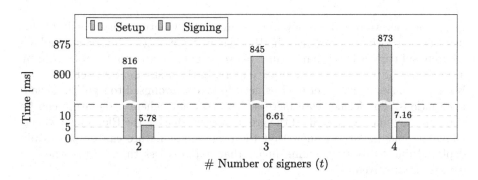

**Fig. 5.** Speed comparison of the threshold scheme for $n = 5$ and $t = \{2, 3, 4\}$.

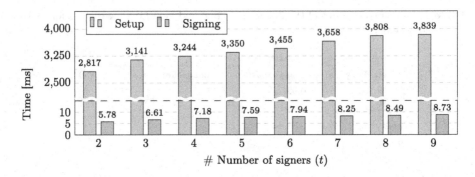

**Fig. 6.** Speed comparison of the threshold scheme for $n = 10$ and $t = \{2, \cdots, 9\}$.

the Setup and Signing algorithms. The most costly component of the scheme is the Setup algorithm, which requires around 1 s for $(5, 4)$ setting and around 4 s for $(10, 9)$ setting. Fortunately, this procedure only has to be performed once. On the other side, the Signing algorithm requires less than 10 ms in all settings.

Since our benchmarks do not consider communication overhead, we compute how much data need to be sent and evaluate how long it would take via Ethernet and Bluetooth connections. During the PolyEval phase, the protocol requires to transfer 2,048 B between all devices in one round (i.e., $|c_i| = 512$ B, 5 devices deployed, i.e., $4 * 512 = 2,048$ B). We consider parallel processing of all 5 rounds of the PolyEval phase. Using Transmission Control Protocol (TCP), the communication latency is negligible. However, the Bluetooth requires ca. 3 s. The SignatureGen phase requires to transfer 160 B (i.e., $|c_j| = 64$ B, $|c| = 64$ B, and $|s_j| = 32$ B) between MD and one SD. Also in this case, we consider parallel communication processing with all signing SDs. TCP communication latency is again negligible whereas the Bluetooth communication takes ca. 3 s.

## 7  Conclusion

In this article, we propose a novel $(n, t)$-threshold signature scheme suitable for increasing security and privacy in Blockchain technology. Our scheme allows securely splitting a Blockchain wallet between more devices that can be held by a single user or by several collaborative users. In the first case, the user's security is increased, since more user devices need to be compromised to sign Blockchain transactions. The former case increases user privacy, where a signature can be anonymously made on behalf of the group of users sharing the Blockchain wallet. Our experimental results show that the proposed signature can be practically deployed due to its fast signing phase that requires less than 10 ms when 10 devices are involved.

**Acknowledgements.** Research described in this paper was financed by the Technology Agency of the Czech Republic 'DELTA 2 Programme' under grant TM02000036. The authors gratefully acknowledge funding from European Union's Horizon 2020

Research and Innovation programme under the Marie Skłodowska Curie grant agreement No. 813278 (A-WEAR: A network for dynamic wearable applications with privacy constraints, http://www.a-wear.eu/).

# References

1. Barker, E.: Recommendation for key management part 1: general (revision 5). NIST Spec. Publ. Part 1 **800**(57), 1–171 (2020)
2. Beimel, A.: Secret-sharing schemes: a survey. In: Chee, Y.M., et al. (eds.) IWCC 2011. LNCS, vol. 6639, pp. 11–46. Springer, Heidelberg (2011). https://doi.org/10. 1007/978-3-642-20901-7_2
3. BitcoinCore: Technology roadmap - schnorr signatures and signature aggregation (2017). https://bitcoincore.org/en/2017/03/23/schnorr-signature-aggregation/
4. Boneh, D., Drijvers, M., Neven, G.: Compact multi-signatures for smaller blockchains. In: Peyrin, T., Galbraith, S. (eds.) ASIACRYPT 2018. LNCS, vol. 11273, pp. 435–464. Springer, Cham (2018). https://doi.org/10.1007/978-3-030-03329-3_15
5. Boneh, D., Lynn, B., Shacham, H.: Short signatures from the Weil pairing. J. Cryptol. **17**(4), 297–319 (2004). https://doi.org/10.1007/s00145-004-0314-9
6. Canetti, R., Gennaro, R., Goldfeder, S., Makriyannis, N., Peled, U.: UC non-interactive, proactive, threshold ECDSA with identifiable aborts. In: Proceedings of the 2020 ACM SIGSAC Conference on Computer and Communications Security, pp. 1769–1787 (2020)
7. Castagnos, G., Catalano, D., Laguillaumie, F., Savasta, F., Tucker, I.: Two-party ECDSA from hash proof systems and efficient instantiations. In: Boldyreva, A., Micciancio, D. (eds.) CRYPTO 2019. LNCS, vol. 11694, pp. 191–221. Springer, Cham (2019). https://doi.org/10.1007/978-3-030-26954-8_7
8. Castagnos, G., Catalano, D., Laguillaumie, F., Savasta, F., Tucker, I.: Bandwidth-efficient threshold EC-DSA. In: Kiayias, A., Kohlweiss, M., Wallden, P., Zikas, V. (eds.) PKC 2020. LNCS, vol. 12111, pp. 266–296. Springer, Cham (2020). https:// doi.org/10.1007/978-3-030-45388-6_10
9. Cogliati, B., et al.: Provable security of (tweakable) block ciphers based on substitution-permutation networks. In: Shacham, H., Boldyreva, A. (eds.) CRYPTO 2018. LNCS, vol. 10991, pp. 722–753. Springer, Cham (2018). https:// doi.org/10.1007/978-3-319-96884-1_24
10. Dalskov, A., Orlandi, C., Keller, M., Shrishak, K., Shulman, H.: Securing DNSSEC keys via threshold ECDSA from generic MPC. In: Chen, L., Li, N., Liang, K., Schneider, S. (eds.) ESORICS 2020. LNCS, vol. 12309, pp. 654–673. Springer, Cham (2020). https://doi.org/10.1007/978-3-030-59013-0_32
11. Damgård, I., Jakobsen, T.P., Nielsen, J.B., Pagter, J.I., Østergaard, M.B.: Fast threshold ECDSA with honest majority. In: Galdi, C., Kolesnikov, V. (eds.) SCN 2020. LNCS, vol. 12238, pp. 382–400. Springer, Cham (2020). https://doi.org/10. 1007/978-3-030-57990-6_19
12. Damgård, I., Koprowski, M.: Practical threshold RSA signatures without a trusted dealer. In: Pfitzmann, B. (ed.) EUROCRYPT 2001. LNCS, vol. 2045, pp. 152–165. Springer, Heidelberg (2001). https://doi.org/10.1007/3-540-44987-6_10
13. Desmedt, Y., Frankel, Y.: Threshold cryptosystems. In: Brassard, G. (ed.) CRYPTO 1989. LNCS, vol. 435, pp. 307–315. Springer, New York (1990). https:// doi.org/10.1007/0-387-34805-0_28

14. Doerner, J., Kondi, Y., Lee, E., Shelat, A.: Secure two-party threshold ECDSA from ECDSA assumptions. In: 2018 IEEE Symposium on Security and Privacy (SP), pp. 980–997. IEEE (2018)
15. Doerner, J., Kondi, Y., Lee, E., Shelat, A.: Threshold ECDSA from ECDSA assumptions: the multiparty case. In: 2019 IEEE Symposium on Security and Privacy (SP), pp. 1051–1066. IEEE (2019)
16. Dzurenda, P., Ricci, S., Casanova-Marqués, R., Hajny, J., Cika, P.: Secret sharing-based authenticated key agreement protocol. In: The 16th International Conference on Availability, Reliability and Security, pp. 1–10 (2021)
17. Freemanlaw: Cryptocurrency transactions: Multi-signature arrangements explained. https://freemanlaw.com/cryptocurrency-transactions-multi-signature-arrangements-explained/
18. Gennaro, R., Goldfeder, S.: Fast multiparty threshold ECDSA with fast trustless setup. In: Proceedings of the 2018 ACM SIGSAC Conference on Computer and Communications Security, pp. 1179–1194 (2018)
19. Gennaro, R., Goldfeder, S., Narayanan, A.: Threshold-optimal DSA/ECDSA signatures and an application to bitcoin wallet security. In: Manulis, M., Sadeghi, A.-R., Schneider, S. (eds.) ACNS 2016. LNCS, vol. 9696, pp. 156–174. Springer, Cham (2016). https://doi.org/10.1007/978-3-319-39555-5_9
20. Gennaro, R., Jarecki, S., Krawczyk, H., Rabin, T.: Robust threshold DSS signatures. In: Maurer, U. (ed.) EUROCRYPT 1996. LNCS, vol. 1070, pp. 354–371. Springer, Heidelberg (1996). https://doi.org/10.1007/3-540-68339-9_31
21. Gennaro, R., Jarecki, S., Krawczyk, H., Rabin, T.: Secure applications of Pedersen's distributed key generation protocol. In: Joye, M. (ed.) CT-RSA 2003. LNCS, vol. 2612, pp. 373–390. Springer, Heidelberg (2003). https://doi.org/10.1007/3-540-36563-X_26
22. Itakura, K.: A public-key cryptosystem suitable for digital multisignatures. NEC J. Res. Dev. **71**, 1–8 (1983)
23. Komlo, C., Goldberg, I.: FROST: flexible round-optimized schnorr threshold signatures. In: Dunkelman, O., Jacobson, Jr., M.J., O'Flynn, C. (eds.) SAC 2020. LNCS, vol. 12804, pp. 34–65. Springer, Cham (2021). https://doi.org/10.1007/978-3-030-81652-0_2
24. Li, C.Y., Chen, X.B., Chen, Y.L., Hou, Y.Y., Li, J.: A new lattice-based signature scheme in post-quantum blockchain network. IEEE Access **7**, 2026–2033 (2018)
25. Lindell, Y.: Fast secure two-party ECDSA signing. In: Katz, J., Shacham, H. (eds.) CRYPTO 2017. LNCS, vol. 10402, pp. 613–644. Springer, Cham (2017). https://doi.org/10.1007/978-3-319-63715-0_21
26. Lindell, Y., Nof, A.: Fast secure multiparty ECDSA with practical distributed key generation and applications to cryptocurrency custody. In: Proceedings of the 2018 ACM SIGSAC Conference on Computer and Communications Security, pp. 1837–1854 (2018)
27. MacKenzie, P., Reiter, M.K.: Two-party generation of DSA signatures. Int. J. Inf. Secur. **2**(3), 218–239 (2004)
28. Okamoto, T.: Provably secure and practical identification schemes and corresponding signature schemes. In: Brickell, E.F. (ed.) CRYPTO 1992. LNCS, vol. 740, pp. 31–53. Springer, Heidelberg (1993). https://doi.org/10.1007/3-540-48071-4_3
29. Paillier, P.: Public-key cryptosystems based on composite degree residuosity classes. In: Stern, J. (ed.) EUROCRYPT 1999. LNCS, vol. 1592, pp. 223–238. Springer, Heidelberg (1999). https://doi.org/10.1007/3-540-48910-X_16
30. Pointcheval, D., Stern, J.: Security arguments for digital signatures and blind signatures. J. Cryptol. **13**(3), 361–396 (2000)

31. Schnorr, C.P.: Efficient identification and signatures for smart cards. In: Brassard, G. (ed.) CRYPTO 1989. LNCS, vol. 435, pp. 239–252. Springer, New York (1990). https://doi.org/10.1007/0-387-34805-0_22

32. Shamir, A.: How to share a secret. Commun. ACM **22**(11), 612–613 (1979)

33. Shoup, V.: Practical threshold signatures. In: Preneel, B. (ed.) EUROCRYPT 2000. LNCS, vol. 1807, pp. 207–220. Springer, Heidelberg (2000). https://doi.org/10.1007/3-540-45539-6_15

34. Stinson, D.R., Strobl, R.: Provably secure distributed Schnorr signatures and a (t, n) threshold scheme for implicit certificates. In: Varadharajan, V., Mu, Y. (eds.) ACISP 2001. LNCS, vol. 2119, pp. 417–434. Springer, Heidelberg (2001). https://doi.org/10.1007/3-540-47719-5_33

# Robotic Process Automation (RPA) Forum

# Preface

## Robotic Process Automation (RPA) Forum

Robotic process automation (RPA) is an emerging technology in the field of business process management (BPM) that enables the office automation of intensive repetitive tasks. In essence, it relates to software agents called software robots that mimic how humans use computer applications when performing rule-based and well-structured tasks in a business process. Examples of tasks that software robots perform include data transfer between applications, automated email query processing, and collation of payroll data from different sources.

RPA is drastically more than just technological innovation. It enables a digital task force and, what is more important, a control mechanism over it. The objective for RPA also extends beyond cutting costs; it directly addresses the digital transformation of companies by creating new value, improving the quality of services, reducing and controlling task times, and improving work satisfaction by freeing employees from repetitive and tedious tasks. Moreover, RPA has a generative capacity when combined with technologies such as optical character recognition (OCR), machine learning (ML), and artificial intelligence (AI), among others, creating a new breed of "smart" automation tools.

The capabilities and opportunities of RPA challenge a broad set of research communities. Computer scientists are attracted to its various technical aspects, while economists study the impact of RPA on labor and organizational effectiveness and engineers are enabled to connect different data sources, improve the quality of the data, and accelerate data analysis. Another question is how RPA fits within a corporate program of digital innovation. Finally, RPA has social implications since it may reduce work opportunities for those people who are carrying out simple, manual work.

The RPA Forum aims to bring together researchers from the above communities to discuss challenges, opportunities, and new ideas that relate to RPA and its application to business processes in private and public sectors. It is a unique setting where technical, business-oriented, and human-centered perspectives come together. Given the growing adoption in the RPA context of natural language inputs to the development of software robots, this year, a keynote speech by Tathagata Chakraborti (IBM Research) on "From Natural Language to Workflows: Towards Emergent Intelligence in Robotic Process Automation" enriched the program of the forum and is included in these proceedings as invited paper. In addition, the forum attracted 16 international submissions on different topics, including human-in-the-loop and conversational approaches, novel platforms and governance models for RPA, and analysis of the socio-human implications of RPA. All submissions were reviewed by three Program Committee members or their sub-reviewers and the best nine papers were finally accepted. We believe that the accepted papers provide a novel mix of conceptual and technical contributions that are of interest for the RPA community.

Flechsig et al. show the results of design science research conducted to identify concrete requirements and conceptual design of a holistic BPM-RPA platform based on business process modeling notation. The practicability of the approach is substantiated by evaluation interviews and a prototypical implementation, which outline interesting research directions on this topic.

Modliński et al. explore the logic behind unsuccessful RPA implementation, resulting in the so-called "re-manualization" phenomenon, which pushes workers taking over robotized tasks to perform them manually again. Relying on interviews, group discussions with managers experienced in RPA, and secondary data analysis, the authors found four types of "cause and effect" narratives that reflect reasons for re-manualization to occur.

Borghoff and Plattfaut investigate the design of RPA-specific governance models in practice, identifying organizational internal context factors that may influence the implementation mechanisms of a lightweight-specific model. The research is built through a qualitative approach based on in-depth interviews with practitioners, consultants, and a vendor representative.

Harmoko et al. present a systematic mapping study to review the literature that studies the impact of RPA on humans, identifying negative and positive influences and implications. The results of the study show that the positive implications of RPA projects for humans outweigh the negative ones. In contrast, human influence in RPA projects is dominated by negative inputs.

Agostinelli et al. propose a human-in-the-loop approach to support human experts in the identification of allowed routine segments from a user interface (UI) log. Their approach can be leveraged by segmentation techniques that are able to discover from scratch the structure of the routines under analysis recorded in a UI log, with the aim to increase the quality of discovered routine segments.

Yaevi et al. present an exploratory approach to the problem of next-best-skill recommendation in human-robot collaboration. The paper highlights the key characteristics of the problem, examines existing approaches, and calls out challenges for implementing a concrete solution. An implementation architecture that can serve as an integrated recommendation strategy is finally proposed.

Rybinski and Schüler examine literature process discovery algorithms to investigate the feasibility of their application in the context of RPA. This study points out that the process models generated by the discovery algorithms tend to neglect the data perspective, which is crucial to convert such models into flowcharts amenable to automation. To mitigate this, a proof-of-concept implementation is provided to translate the discovered process models into RPA flowcharts.

In their work, Rizk et al. argue that the next generation of software robots must leverage AI to learn from user interactions and generalize to unseen settings. To achieve this, the authors first assess the current state of the art on this topic. Then, they identify some key research challenges at the intersection of AI, RPA, and interactive task learning that must be addressed to realize the vision of software robots that continually learn new automation solutions from UI.

Finally, Průcha and Skrbek investigate the possibility of employing applications' API during automation via RPA, which could help to solve some well-known stability issues of software robots.

We wish to thank all those who contribute to making the 2022 edition of the RPA Forum a success: the authors who submitted papers, the members of the Program

Committee who carefully reviewed the submissions, and the speakers who presented their work at the forum. We also express our gratitude to the BPM 2022 chairs and organizers for their support in preparing the RPA Forum.

September 2022                                                    Andrea Marrella
                                                                 Bernhard Axmann

# Organization

## Program Chairs

Andrea Marrella                    Sapienza University of Rome, Italy
Bernhard Axmann                    Technische Hochschule Ingolstadt, Germany

## Program Committee

Simone Agostinelli                 Sapienza University of Rome, Italy
Aleksandre Asatiani                University of Gothenburg, Sweden
Tathagata Chakraborti              IBM Research, USA
Adela del Río-Ortega               University of Seville, Spain
Carmelo Del Valle                  University of Seville, Spain
José González Enríquez             University of Seville, Spain
Peter Fettke                       German Research Center for AI (DFKI)
                                   and Saarland University, Germany
Christian Flechsig                 Technische Universität Dresden, Germany
Lukas-Valentin Herm                Universität Würzburg, Germany
Hannu Jaakkola                     University of Tampere, Finland
Christian Janiesch                 TU Dortmund University, Germany
Andrés Jiménez Ramírez             University of Seville, Spain
Volodymyr Leno                     University of Melbourne, Australia
Fabrizio Maria Maggi               University of Bolzano, Italy
Massimo Mecella                    Sapienza University of Rome, Italy
Tommi Mikkonen                     University of Helsinki, Finland
Hajo A. Reijers                    Utrecht University, The Netherlands
Yara Rizk                          IBM Research, USA
Rehan Syed                         Queensland University of Technology,
                                   Australia
Maximilian Völker                  Hasso Plattner Institut, Germany
Inge van de Weerd                  Utrecht University, The Netherlands
Jonas Wanner                       Universität Würzburg, Germany
Moe Wynn                           Queensland University of Technology,
                                   Australia

# From Natural Language to Workflows: Towards Emergent Intelligence in Robotic Process Automation

Tathagata Chakraborti, Yara Rizk[✉], Vatche Isahagian, Burak Aksar, and Francesco Fuggitti

IBM Research AI, Cambridge, MA, USA
`yara.rizk@ibm.com`

**Abstract.** RPA technologies allow the automation of repeated processes through indirect or direct instruction from the end-user. While declarative authoring techniques provide a powerful tool to scale up process complexity with RPA elements, often such techniques are difficult to use without expertise. In this work, we will explore systems (in the context of web service composition and goal-oriented conversational agents) that both consumers and developers can interact with, in natural language, to compose RPA elements that demonstrate emergent intelligence as a composition of smaller units of automation. We will also discuss the overhead in authoring such systems, and potential learning opportunities in reducing said overhead. Finally, we will explore issues of explainability for the developer and transparency of dynamic compositions for the consumer in dealing with such systems with aggregated automation.

**Keywords:** Web service composition · Automated planning · Natural language processing · Process automation

## 1 Constructing Flows for Automation

A vast number of automation tools require the user to construct flows that embody some form of automation or business process. This user is not the end user but rather the developer or administrator of that process. Such applications range from goal-oriented conversational agents such as Dialogflow[1] or Watson Assistant[2], data processing flows such as AutoAI/ML[3,4], and web service composition such as in App Connect[5] or Zapier[6]. The examples are many and diverse, but eventually they take the form of a flow or decision tree. They model a mixture of actions that determine or sense user intent and a composition of one or

---

[1] https://developers.google.com/learn/pathways/chatbots-dialogflow.
[2] https://www.ibm.com/products/watson-assistant.
[3] https://aws.amazon.com/machine-learning/automl/.
[4] https://cloud.google.com/automl.
[5] https://www.ibm.com/cloud/app-connect.
[6] https://zapier.com/how-it-works.

© Springer Nature Switzerland AG 2022
A. Marrella et al. (Eds.): BPM 2022, LNBIP 459, pp. 123–137, 2022.
https://doi.org/10.1007/978-3-031-16168-1_8

more units of automation and manual processes, realized as steps in that flow, that are able to satisfy the requirements of that intent.

This is, of course, not news to the business process management community, which has decades of work to show for representing and reasoning about such processes, including sophisticated specification languages like BPMN[7] as well as workflow construction tools and languages, both commercial and academic, such as the likes of YAWL [49], FLOWer [1], DECLARE [51], and others. One natural outcome of this is that the ability of process administrators to write sophisticated constraints has significantly grown; however, this has come at the cost of a much higher barrier to entry in terms of the expected expertise of users who are able to write such specifications.

This limitation is not due to a lack of subject matter expertise on the user's part, but rather due to how different the modeling paradigms are. For example, a declarative modeling approach offers an exponential increase in the complexity of specification to the complexity of the realized model, but requires a completely different way of thinking from imperative modeling – not unlike the thought process that goes into the basics of programming. Increased sophistication of the specification language further increases the required programming knowledge of the user. As such, the full capabilities of these advanced modeling tools remain out of reach for non-expert users who have no training or experience in declarative modeling and programming [34]. Recent advances in natural language processing offer an intriguing way out of this conundrum.

## 1.1   Natural Language to Flows

One of the most fascinating outcomes of recent advances in natural language processing is the emergence of generic language models that can be instantiated for specific domains to perform non-trivial information processing using only an interface to natural language instruction from the user [4]. In this paper, we explore to what extent a natural language interface may be leveraged for a workflow construction task, where compilations to more formal languages (that are more expressive and unambiguous) are hidden away from the user, thereby lowering the barrier of entry and expertise required for using such systems.

In rudimentary form, this form of interaction can manifest in constructing pipelines in Bash syntax (command line interface) through natural language [2]; or composing web services through IFTTT (if-this-then-that) instructions[8]. In general, this no-code/low-code approach [25] applies to the space of programming through natural language, such as in [29].[9] There are also recent examples of workflow construction starting from natural language/document or multi-document [14,20,21,42]. However, these flow construction systems do not feature

---

[7] https://www.omg.org/spec/BPMN/2.0/.

[8] https://ifttt.com.

[9] Interestingly, although not strictly code, programming languages can be used as an intermediate representation in the task of converting natural language instruction to workflow representations. This allows the use of off-the-shelf language models trained on generic code, for which there exists plenty of data [22].

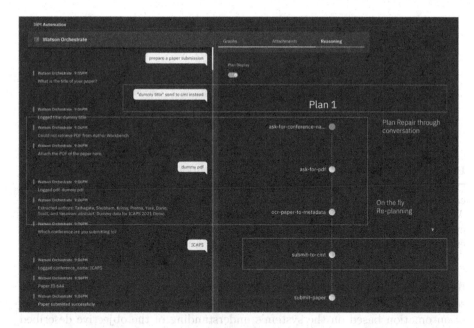

**Fig. 1.** An illustration of the user completing a workflow for submitting a paper to a conference through conversation (example uses dummy data). The user begins by declaring their intent of preparing a paper submission and the system interprets that intent as a sequence of steps it can do internally to service the request – acquiring the title of the paper, looking it up in the internal company database, gathering the metadata, and final submission. This fails when the metadata is not found and the system re-plans internally to ask for the PDF instead and uses automated extraction (OCR) to acquire the same information. Evidently, the internal machinations are hidden from the end user conversing with the system – we will revisit this in Sect. 3.1.

an active user experience. In fact, among them, only Doc2Dial[10] [20, 21] provides a user-agent dialogue flow, though with an associated grounding-document helping with user utterances semantic parsing – the documents, and not user specifications, remain the source of knowledge on the workflow.

In this paper, we will focus on Watson Orchestrate[11] and App Connect as two canonical examples of natural language to flow construction driven actively by the user – one online and the other offline. While both offer embodiments of the natural language to workflow use case, we will see in Sect. 1.2 just how disparate that task can become in practice. However, a detailed exploration of both will reveal some underlying challenges that apply in general before we can realize the full promise of a natural language interface to workflow management.

**Watson Orchestrate (WO).** This is a multi-agent conversational system [36], where users can describe their objectives in natural language for the system

---

[10] https://doc2dial.github.io/.

[11] https://www.ibm.com/cloud/automation/watson-orchestrate.

**Fig. 2.** Here, the user provides a description of a workflow in natural language and the system strings together a sequence of steps in response. Each steps is an API call it can make corresponding to entities identified in the text – each entity is a triplet of an API or "connector" (e.g., Gmail), a business object the connector operates on (e.g., an email) and the mode of operation (e.g., create, delete, retrieve, etc.). Contrary to the example in Fig. 1, this flow is under construction and is not going to execute yet.

to executes a sequence of automation units to complete this objective. These sequences or flows may be dynamically composed from RPA bots or other units of automation based on the system's understanding of the objective described in natural language. WO consists of one or more components that leverage AI technology to transform the input (natural language utterance) to the output sequence (from an automated planner). Figure 1 offers an example.

**App Connect (AC).** In this interface to the App Connect tool, the user types in a brief description of their desired workflow in natural language, and the system responds with the corresponding flow using the components available to it. These components are usually API calls to units of automation available in its catalog. This construction happens through a mixture of natural language parsing, data support from a knowledge graph, and an automated planner as a constraint solver [5]. An example is provided in Fig. 2.

## 1.2   Run Time Versus Design Time Considerations

It should be evident from the examples above that, while both WO and AC drive a natural language to workflow engine in the background, the user experience is quite different. In particular, the former oversaw the construction and execution of a dynamic flow at run time while the latter produced a static construction to be executed later, as the final artifact of the interaction. Some of these differences are outlined in detail in Table 1. In the design of a natural language interface to workflow construction, these differences become critical in determining what the user experience is going to look like. However, there is a common theme in all of them: each system is producing an artifact that embodies a process, described in natural language, and this artifact is composed of units of automation none of which were specifically designed to support that particular process. In a throwback to [6], we refer to this as *emergent intelligence*.

**Table 1.** Online versus offline automated workflow construction.

| Dimension | Online | Offline |
|---|---|---|
| Example | Goal-oriented conversational agents: chatbots that operate on some underlying business process e.g. customer support | Web service composition – a business process as a composition of several API calls |
| Run time | Needs a dynamic composition tool that can recompute new flows on the fly, e.g., a classical planner [44]. This mode thus incurs some reasoning complexity at run time for planning and replanning | Possible to compute process with all outcomes statically e.g. using a non-deterministic planner [32]. Here, the run time does not incur cost of reasoning but merely involves determining which of all the planned for outcome has occurred |
| Execution | The workflow is executed as it is constructed | The workflow is being constructed for later execution |
| Extensibility | Relatively simpler to extend knowledge of the system e.g. add a new RPA in the catalog, since the flows are constructed anew at run time | Static construction and online knowledge discovery do not square well since the point of a static composition is that it is fit for purpose at run time |
| User role | The user interacting with the flow is part of the unfolding process, e.g., in Fig. 1 the user interacting with the system is submitting a paper | The user is creating and deploying the flow for an end user (who may or may not be themselves). For example, in this case, the user will define how a paper submission process should work when someone wants the system to do it |
| User experience | Since this involves execution at run time, the system must decide among alternatives. Its priority is to achieve common grounds to avoid unintended mistakes, and adjust appropriately based on user feedback | The goal of the system here is to make sure that the user understands the all the possibilities of the constructed workflow, and iteratively refine it until it meets their desired objective, before they deploy it [43] |
| Natural Language Interface | Multi-turn conversation | A mixture of generative search (open world information retrieval), as in Fig. 2, and limited conversation |

## 2   Emergent Intelligence

Three salient themes have emerged so far: 1) the user interaction in natural language leads to the automated construction of a workflow; 2) this interaction can happen online i.e. during execution where the user is a stakeholder in the flow or offline where the end goal of the user is to construct a static flow; and 3) the elements in the flow exist by themselves as independent entities, i.e., they were not built to be participants in the constructed flow. We posit that through a mixture of automated composition, natural language instruction, and generalizable units of automation, it is possible to achieve intelligence of a system that was not specifically designed for but rather emerged from the collection.

### 2.1   Units of Automation

So far in the discussion, we have kept the details of the "units of automation" in the composed workflows open-ended. These can take different forms depending on the application: they can be individual tasks in a task management tool, they can be web services or APIs that can be called to perform a task, or in general, they can be some piece of code, callable by the automated construction, that achieves a unit of work. Keeping them independent of any task description offers up two kinds of advantages in terms of scaling up:

- **Development overhead:** A distributed architecture, such as in [36], allows developers of individual units of automation a.k.a. "skills" to not worry about the rest of the system, thereby allowing scale-up of complexity by distributing the development and maintenance cost of individual components of a system – with the added side-effect of fostering a thriving developer ecosystem. Such aggregated assistant architecture has become increasingly common of late, both in personal assistants like Alexa[12] as well as in enterprise systems like Automation Anywhere[13], UiPath[14], etc.
- **Declarative design:** It allows a system to tap into declarative programming paradigms such as [19,33] to achieve exponential scale-up from the complexity of the specification to the complexity of the composed flow, i.e., a much smaller specification for the same size of a flow or a much more complex flow for the same size of a specification [12,32]. Indeed, the natural language specification in the sample interactions in Sect. 1.1 get converted to declarative form internally, for this purpose.

RPAs are prime candidates to form the units of automation in this paradigm. This is because RPAs present a specification task as mixture of instruction and learning that is ideal for automation through composition. RPA vendors have lately begun to infuse AI capabilities into their existing pipelines by leveraging machine learning (ML) and deep learning (DL) models. These capabilities can help to overcome the constraints of rule-based specifications and take advantage of existing AI models that represent cognitive abilities such as perceiving and reasoning [28]. Tasks that can be conducted with these abilities include process identification, image recognition, (process) prediction, natural language processing, chatbot functionality, etc. [13,24,46]. However, there is limited literature on how to generalize RPAs to be reusable and composable [37,46].

## 2.2  Automated Composition is Key

Ultimately, the ability to scale up to complex processes, the ability to construct flows from abstract instruction such as in natural language, and the ability to adjust and adapt to changing task and domain descriptions, all point to the requirement of automated composition. A key technology underlying all these composition examples is automated planning, as a vehicle for constructing agile workflows from declarative composition [30]. The exact flavor of planning varies between applications: for example, in Table 1, we mention classical planning as a way to model dynamic workflows at run time, through a mechanism of planning and re-planning [44], and non-deterministic planning [12,32,43] as means to construct the offline static workflow modeling all outcomes for the user to inspect and deploy. An interesting intermediate is the top-k variant [27], used in [5], which does not compute a full offline policy but rather produces $k$ alternative

---

[12] https://developer.amazon.com/alexa/alexa-skills-kit.

[13] https://botstore.automationanywhere.com/.

[14] https://marketplace.uipath.com/listings.

compositions within a quality bound – this can be useful both at offline composition time as a substitute for the policy visualization, as well as at execution time to establish possible outcomes and alternatives with the end user.

As previously mentioned, a clear advantage of automated planning in business process management is that of using declarative specification languages seamlessly, allowing flexibility in task representation. This is possible thanks to the Planning Domain Definition Language (PDDL) [31], which is the standard language to specify automated planning problems. Although, in this paper, we focus on automated planning specifically, PDDL is not the only declarative formalism that has been successfully employed in the business process management community. For instance, DECLARE specifications can encode several common process behavioral patterns based on Linear-time Temporal Logic on finite traces ($\text{LTL}_f$) [18]. The idea there is to discard explicit ad-hoc representations of processes control flow in favor of the specification of a set of $\text{LTL}_f$ formulas defining allowed finite executions, i.e., process behaviors.[15] Using this, one can use DECLARE to constrain a process behavior to a set of predefined rules.

While leveraging declarative specifications, planning techniques play a critical role to effectively solving Business Process related problems as, e.g., it is the case for conformance checking [17]. In fact, there is an entire line of work studying and implementing synergies among Artificial Intelligence, Formal Methods and Business Process Management (see [15] for an overview of such synergies), including compilations of $\text{LTL}_f$ into PDDL, as, e.g., [9,10,16,47].

## 2.3  Current Deficiencies

One of the major challenges towards automated composition is in figuring out how data flows between components – this is as old and hard a problem as any since components designed independently are not readily composable because reusable objects are simply not defined in the same terms, and more importantly even if they were identified to be reusable, it is hard to make a reliable transformation during execution time between data produced by one component that might be required by another. For the example in Fig. 1, this relation has been specified manually. There has been decades of work [23, 26] on using additional knowledge, e.g., an ontology, to make this transformation – however, this makes the already difficult specification overhead even worse. In the following, we will explore two active areas of research in reducing that overhead: 1) **Explainability** of composed models so that the user can trust the automation; and 2) **Model Acquisition** so that the user does not have to provide any additional specification or knowledge to make the automated composition possible. The success of automated composition very much hinges on the ability to lower the specification overhead – the use of natural language instruction, explainability of composed models, and learning of composition artifacts are all geared towards lowering this overhead as much as possible.

---

[15] Interestingly, LTL provides a natural pathway for integration with language-based interfaces. Recent attempts have been made to get a formal description out of a natural language description (see [7] for a survey, and [40] for a BPM-related application).

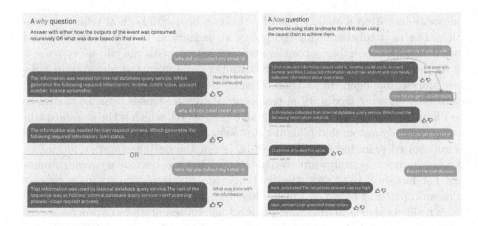

**Fig. 3.** Exploration of the flows composed on the fly in terms of a set of how- and why-questions [44]. The answers to why-questions can be forward looking e.g. what parts of the flow a certain step enables (top left) as well as backward looking e.g. why that step exists in the flow (bottom left); while the how-questions give summary insight into everything that was done (top right). These process summaries are generated using landmarks – i.e. things that must be true to achieve the process goals – that the user can iteratively drill down into more details through further questions.

# 3   Explainability

The explainability requirement of automated composition is a natural outcome of moving towards a more abstract specification, since the user is no longer in charge of constructing a flow manually and is thus relieved of a certain level of control. This spans the entire experience – the explainability question applies to the final composed flow (Sects. 3.1 and 3.2) to how the system enables the natural language instruction (Sect. 3.3) to get to that flow.

## 3.1   Transparency of Emergent Intelligence

One closely intertwined term with explainability is *interpretability* [11]. According to Rudin et al. [39], explainable ML aims to provide post-hoc explanations for existing black-box models (i.e., models that are incomprehensible to humans or are proprietary), whereas interpretable ML focuses on developing models that are intrinsically interpretable. The latter becomes critical in the context of automatically composed flows, where the first point of loss of control is when users lose sight of the composed flow. This is a problem especially in the online mode where plans may be produced, partially executed and discarded, and it is unclear to an end user what process was followed by the system in the background. For a standalone RPA, this is not a problem; for a composition of RPAs and other units of automation, especially when sourced independently and from third parties, this is at best a desired feature to have in order to establish common grounds,

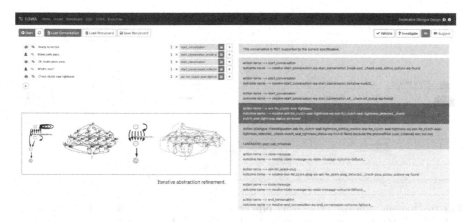

**Fig. 4.** Illustration of iterative refinement of a complex flow constructed offline – the system in [43] allows the user to fix a problem in the simplest version or "abstraction" [45] of the specification that contains the same problem and then test in the full model (the left inset shows this repeated process of debugging in minimal abstractions and testing in maximal form). The system further allows the user to outline individual process instances that are failing and explains those failures in terms of unsatisfied process landmarks that must be fixed in the specification.

and at least a necessary feature for conforming to GDPR requirements to establish provenance of data flow among the components within the system. Figure 3 illustrates this in the context of online composition in WO.

Another problematic piece may be the decision-making process of the RPAs themselves, many of which leverage black-box ML models. While the benefits of these high-performing models are well-known, production use is contingent on solving major issues in the decision making process. Black-box models are frequently cryptic since they can have incredibly intricate and interdependent relationships, which directly impacts the decision making logic. Furthermore, black-box ML models might undertake multiple data changes before achieving classification, making them difficult to comprehend. Having a composition of RPAs and other units of automation only increase the number of downsides, including making it difficult to debug predictions, eroding user trust, and lowering the system's overall utility. Figure 3 (bottom right) illustrates how process explanations from the composition can be composed with explanations from individual black-box components like RPAs e.g. why a loan application is denied along with the process that went into filing and determining that outcome.

### 3.2 Imperative Consequences of Declarative Specification

While iterative exploration of decisions made by the system is useful at run time when the user is interacting with one single realized instantiation of the underlying policy, this approach does not scale at the time of constructing highly complex processes offline. As we discussed before, modeling processes declaratively

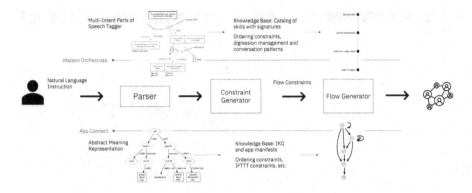

**Fig. 5.** Domain agnostic processing pipelines for natural language to flow generation in WO and AC, facilitating interpretable behavior of the systems end to end. Explanations can be provided by surfacing contributions from individual components of the system. For example, for AC, the system can say *"[IKG] The loan agent processes loan applications and [AMR] I detected loan intent in what you said. [Planner] The banking agent provides credit score to the loan agent."*; while for WO, it can say *"[Catalog] The loan agent processes loan applications and [MI-POS] I detected loan intent in what you said. [Planner] The banking agent provides credit score to the loan agent."*.

requires a shift in mindset of process administrators who have been describing imperative process elements for a long time. The goal of explainability in the offline composition task is thus to ensure that the user understands the *imperative consequences of their declarative specification*. Through a mixture of a novel model abstraction technique [45] and domain landmarks [35], Fig. 4 shows how this was achieved in the context of very large process-oriented conversational agents [32,43] composed automatically from declarative elements.

### 3.3   Natural Language is Noisy

While the previous two cases for explainability were in support of making the composed flows more accessible to the user, the use of natural language to further simplify the specification introduces an additional dimension of explainability – how the system ended up with the constraints that support the flow from the instruction. This is not so much of an explanation of the flow but rather an explanation of how the system itself works. Natural language utterances inputted to these systems by users are notoriously unstructured and diverse, often times grammatically incorrect and ambiguous. Hence, systems like WO and AC, with their multiple stages of AI models, may produce an outcome that was not expected by users or aligned with their intentions – this is not because of the system's knowledge being faulty but rather the system's interpretation of the user's input and/or the mapping of that information to the system's knowledge of the world was incorrect. Any error in one of the stages could lead to a cascade of decisions that lead to a different outcome. Hence, providing insights into how the system works can help the user modify their input.

Interestingly, although the user experience for both WO and AC are very different outside of the input (natural language) and the output (workflow), the underlying technology to make that transformation is quite similar and can be easily swapped in and out for alternative technologies. As shown in Fig. 5, while each of the systems have a component to 1) interpret the user utterance; 2) match detected items to its knowledge; and 3) enforce detected items and constraints in the final flow; the realizations of those components can vary wildly. WO uses a combination of an abstract meaning representation (AMR) [3] and latent link discovery and matching using graph neural networks in a knowledge graph [41] for the first two stages, WO may chose to use a completely different parser-to-constraint engine, focused on multi-intent classification and other conversation-oriented aspects. The processing pipeline can also be agnostic to the kind of flow – i.e. a sequence or a decision tree, generated eventually (with respect to offline versus online modeling requirements in Table 1.2). The individual explanations from these components can be pooled together by templating [38], as described in Fig. 5, for a holistic view into the system.

## 4   Model Acquisition

While natural language instruction, and explainability of composed flows, reduce the specification overhead, it introduces requirements of new domain artifacts that are not required for manual workflow construction. As we discussed previously, automated composition requires additional knowledge on composable elements that are produced and composed by individual components – e.g. knowledge on how to interpret and disambiguate natural language instruction, how to transform one kind of data to another that is semantically equivalent or contained and thus eligible for reuse, and so on. The additional knowledge requirement defeats the purpose of the progress made in reducing the original specification overhead. Thus, in the final section, we explore how generalizable domain artifacts required for automated composition can be either actively taught via instruction or passively learned from observation.

### 4.1   Learning from Instructions

To compose RPA elements into a flow, systems can learn from end users through natural language interactions where the users provide instructions on how to compose these flows. Learning from instructions allows an interactive and intuitive method that reduces the learning curving for end users. There is still a learning curve though since natural language processing technology has not matured enough to fully parse and comprehend users' instructions. Therefore, users need to constrain their phrases to ones that are within the scope of the system's processing abilities. Furthermore, end users (who do not have any formal training in programming) will most likely provide instructions for the happy path. Exception paths may only be thought of when errors occur. Therefore, systems like WO and AC should learn iteratively and, as errors pop up, users should be able to provide instructions on how to handle such exceptions [50].

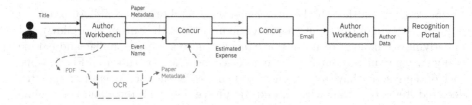

**Fig. 6.** The user here uses the AuthorWorkbench skill for two different purposes, once to fetch the paper data for use as justification for a travel approval and once to fetch their publication record to submit an application for internal recognition. There are two tasks unfolding here simultaneously – as we discuss in Sect. 4.2, an effective model learning technique needs to establish prosody in this observed data. Furthermore, in relation to learning from direct instruction, the learned model should be able to represent the latent requirements of a task, and not the observed steps themselves, so that it can generalize to new executions when the current one fails or a new task is presented that is doable as a composition of available elements, as shown through an alternative completion of the task using an OCR RPA.

## 4.2   Learning from Observing

Another way for the system to acquire knowledge is through observing a user passively and learning one or more models that explain the observed behavior. In general, this is referred to as the task of process mining [48]. Specifically as it relates to automated composition, there is decades worth of work aimed at learning planning models from observed data [8]. Unfortunately, those techniques have looked at learning representations for planning in isolation, focusing on learning a viable model given a set of plan traces, with little regard for 1) how those traces are obtained and 2) the properties of the system that is going to consume the learned model in the end.

This throws up several challenges: a system such as WO that wants to be self-learning through repeated usage, i.e. learn annotations for its skills over time so as to facilitate automated composition, has no non-determinism and noise in the classical sense to deal with – instead it has components that are deterministic and incompletely specified. This means that learning a model in the traditional sense, with a single mapping from inputs to outputs, in insufficient. In planning parlance, the system needs to learn *conditional effects*. However, in existing model learning literature there is no notion of a task (or tasks are required as input e.g. for hierarchical task networks) and thus the number of effects of actions conditioned on tasks (that the system can reuse) will blow up exponentially, unless the system is able to acquire both of them effectively through a joint learning problem. Such a situation is illustrated in Fig. 6.

**Work in Progress: Generalization and Prosody.** We thus observe two disparate forces in model acquisition through instruction versus through observing – the former offers up precise task-oriented instruction but is hard to generalize beyond a particular task, while the latter has the ability to generalize from many observations without the user having to provide additional information intentionally, but at the same time has to pick apart the nuances of an observed sequence of events in terms of reusable components the system can compose later into new workflows. Similar to the purpose of prosody in songwriting, the joint learning problem as described above requires modeling of agent rationality in order to learn models that can effectively orchestration composable elements to service a diversity of user intent. While we identify this gap in the state-of-the-art in acquisition of planning models in [8], we hope to report on how this manifests in WO and AC in the near future.

## 5   Conclusion: Bigger Picture, Bigger RPAs

RPAs have been successful in addressing the initial gaps in business automation. Their application and influence can be further enhanced by composing RPA components into "bigger" RPAs to address complex processes. Until recently that composition was restricted to expert users. In this work, by relying on recent advances in natural language processing and planning, we bridge that gap by enabling non-experts to compose flows from natural languages. We explore two systems that both consumers and developers can interact with, in natural language, to compose flows from RPA elements, and highlight key features, benefits and gaps that need to be addressed to unleash the full potential of this approach.

## References

1. Van der Aalst, W.M., Weske, M., Grünbauer, D.: Case handling: a new paradigm for business process support. Data Knowl. Eng. **53**, 129–162 (2005)
2. Agarwal, M., et al.: Neurips 2020 NLC2CMD competition: translating natural language to bash commands. In: NeurIPS 2020 Competition and Demonstration Track. PMLR (2021)
3. Astudillo, R.F., Ballesteros, M., Naseem, T., Blodgett, A., Florian, R.: Transition-Based Parsing with Stack-Transformers. arXiv:2010.10669 (2020)
4. Bommasani, R., et al.: On the opportunities and risks of foundation models. arXiv:2108.07258 (2021)
5. Brachman, M., et al.: A Goal-driven natural language interface for creating application integration workflows. In: AAAI Demonstration Track (2022)
6. Brooks, R.A.: A robot that walks; emergent behaviors from a carefully evolved network. Neural Comput. **1**(2), 253–262 (1989)
7. Brunello, A., Montanari, A., Reynolds, M.: Synthesis of LTL formulas from natural language texts: state of the art and research directions. In: TIME (2019)
8. Callanan, E., Venezia, R.D., Armstrong, V., Paredes, A., Chakraborti, T., Muise, C.: MACQ: a holistic view of model acquisition techniques. In: ICAPS Workshop on Knowledge Engineering for Planning and Scheduling (KEPS) (2022)
9. Camacho, A., McIlraith, S.A.: Strong fully observable non-deterministic planning with LTL and LTLf goals. In: IJCAI (2019)
10. Camacho, A., Triantafillou, E., Muise, C.J., Baier, J.A., McIlraith, S.A.: Non-deterministic planning with temporally extended goals: LTL over finite and infinite traces. In: AAAI (2017)

11. Carvalho, D.V., Pereira, E.M., Cardoso, J.S.: Machine learning interpretability: a survey on methods and metrics. Electronics **8**, 832 (2019)
12. Chakraborti, T., Agarwal, S., Khazaeni, Y., Rizk, Y., Isahagian, V.: D3BA: a tool for optimizing business processes using non-deterministic planning. In: BPM Workshop on AI4BPM (2020)
13. Chakraborti, T., et al.: From robotic process automation to intelligent process automation: emerging trends. In: BPM RPA Forum (2020)
14. Chambers, A.J., et al.: Automated business process discovery from unstructured natural-language documents. In: BPM (2020)
15. De Giacomo, G.: Artificial Intelligence-based Declarative Process Synthesis for BPM (2021). (Invited Talk)
16. De Giacomo, G., Fuggitti, F.: FOND4LTLf: FOND planning for LTLf/PLTLf goals as a service. In: ICAPS, Demo Track (2021)
17. De Giacomo, G., Maggi, F., Marrella, A., Patrizi, F.: On the disruptive effectiveness of automated planning for LTLf-based trace alignment. In: AAAI (2017)
18. De Giacomo, G., Vardi, M.Y.: Linear temporal logic and linear dynamic logic on finite traces. In: IJCAI (2013)
19. van Der Aalst, W.M., Pesic, M., Schonenberg, H.: Declarative workflows: balancing between flexibility and support. Comput. Sci. Res. Dev. **23**, 99–113 (2009)
20. Feng, S., Patel, S.S., Wan, H., Joshi, S.: MultiDoc2Dial: modeling dialogues grounded in multiple documents. In: EMNLP (2021)
21. Feng, S., Wan, H., Gunasekara, R.C., Patel, S.S., Joshi, S., Lastras, L.A.: doc2dial: a goal-oriented document-grounded dialogue dataset. In: EMNLP (2020)
22. Groth, P., Gil, Y.: Analyzing the gap between workflows and their natural language descriptions. In: IEEE World Congress on Services (2009)
23. Hepp, M., Leymann, F., Domingue, J., Wahler, A., Fensel, D.: Semantic business process management: a vision towards using semantic web services for business process management. In: ICEBE (2005)
24. Herm, L.V., et al.: A consolidated framework for implementing robotic process automation projects. In: BPM (2020)
25. Hirzel, M.: Low-code programming models. arXiv:2205.02282 (2022)
26. Hoffmann, J., Bertoli, P., Pistore, M.: Web service composition as planning. Revisited. In: Between Background Theories and Initial State Uncertainty, AAAI (2007)
27. Katz, M., Sohrabi, S., Udrea, O.: Top-quality planning: finding practically useful sets of best plans. In: AAAI (2020)
28. Le Clair, C., UiPath, A.A., Prism, B.: The Forrester Wave™: Robotic Process Automation, Q2 2018. Forrester Research (2018)
29. Li, Y., et al.: Competition-level code generation with alphacode. arXiv:2203.07814 (2022)
30. Marrella, A., Chakraborti, T.: Applications of automated planning for business process management. In: Polyvyanyy, A., Wynn, M.T., Van Looy, A., Reichert, M. (eds.) BPM 2021. LNCS, vol. 12875, pp. 30–36. Springer, Cham (2021). https://doi.org/10.1007/978-3-030-85469-0_4
31. McDermott, D., et al.: PDDL - The Planning Domain Definition Language. DCS TR-1165 (1998)
32. Muise, C., et al.: Planning for goal-oriented dialogue systems. In: arXiv:1910.08137 (2020)
33. Pesic, M., Schonenberg, M., Sidorova, N., van der Aalst, W.M.: Constraint-based workflow models: change made easy. In: OTM Confederated International Conferences "On the Move to Meaningful Internet Systems" (2007)

34. Reijers, H.A., Slaats, T., Stahl, C.: Declarative modeling - an academic dream or the future for BPM? In: BPM (2013)
35. Richter, S., Helmert, M., Westphal, M.: Landmarks revisited. In: AAAI (2008)
36. Rizk, Y., et al.: A unified conversational assistant framework for business process automation. In: Workshop on Intelligent Process Automation (2020)
37. Rizk, Y., Chakraborti, T., Isahagian, V., Khazaeni, Y.: Towards end-to-end business process automation. In: Robotic Process Automation (2021)
38. Rosenthal, S., Selvaraj, S.P., Veloso, M.M.: Verbalization: narration of autonomous robot experience. In: IJCAI (2016)
39. Rudin, C.: Stop explaining black box machine learning models for high stakes decisions and use interpretable models instead. Nat. Mach. Intell. **1**, 206–215 (2019)
40. Sànchez-Ferreres, J., Burattin, A., Carmona, J., Montali, M., Padró, L.: Formal reasoning on natural language descriptions of processes. In: BPM (2019)
41. Sheikh, N., Qin, X., Reinwald, B., Miksovic, C., Gschwind, T., Scotton, P.: Knowledge graph embedding using graph convolutional networks with relation-aware attention. arXiv:2102.07200 (2021)
42. Shing, L., et al.: Extracting workflows from natural language documents: a first step. In: BPM (2018)
43. Sreedharan, S., Chakraborti, T., Muise, C., Khazaeni, Y., Kambhampati, S.: D3WA+ - a case study of XAIP in a model acquisition task for dialogue planning. In: ICAPS (2020)
44. Sreedharan, S., Chakraborti, T., Rizk, Y., Khazaeni, Y.: Explainable composition of aggregated assistants. In: ICAPS Workshop on Explainable AI Planning (2020)
45. Sreedharan, S., Srivastava, S., Kambhampati, S.: Hierarchical expertise level modeling for user specific contrastive explanations. In: IJCAI (2018)
46. Syed, R., et al.: Robotic process automation: contemporary themes and challenges. Comput. Ind. **115**, 103162 (2020)
47. Torres, J., Baier, J.A.: Polynomial-time reformulations of LTL temporally extended goals into final-state goals. In: IJCAI, pp. 1696–1703. AAAI Press (2015)
48. Van Der Aalst, W.: Process mining. Commun. ACM. **55**, 76–83 (2012)
49. Van Der Aalst, W.M., Ter Hofstede, A.H.: YAWL: yet another workflow language. Inf. Syst. **30**, 245–275 (2005)
50. Venkateswaran, P., Muthusamy, V., Rizk, Y., Isahagian, V.: Towards continual learning in interactive digital assistants for process automation. In: IJCAI 2022 International Workshop on Process Management in the AI era (2022)
51. Westergaard, M., Maggi, F.M.: DECLARE: a tool suite for declarative workflow modeling and enactment. BPM Demonstr. Track. **820**, 1–5 (2011)

# Towards an Integrated Platform for Business Process Management Systems and Robotic Process Automation

Christian Flechsig[1](✉)[iD], Maximilian Völker[2][iD], Christian Egger[1],
and Mathias Weske[2][iD]

[1] Technische Universität Dresden, Dresden, Germany
{christian.flechsig,christian.egger}@tu-dresden.de
[2] Hasso Plattner Institute, University of Potsdam, Potsdam, Germany
{maximilian.voelker,mathias.weske}@hpi.de

**Abstract.** Business Process Management Systems (BPMS) and Robotic Process Automation (RPA) gain increasing relevance in digital transformation to enact and automate business processes. Recent discussions in academia and practice indicate a promising yet challenging integration of both technologies based on a standardized language for consistent process modeling and orchestration, like the Business Process Model and Notation (BPMN). However, scientific literature lacks profound approaches. Guided by Design Science Research, this empirical study substantiates the current debate with a scientifically grounded concept for integrating BPMS and RPA. Resting upon data from 20 expert interviews, we present the requirements and conceptual design of a holistic BPMS-RPA platform based on BPMN. The practicability of our approach is substantiated by five evaluation interviews and a prototypical implementation. Finally, we outline directions for further research and organizational practice.

**Keywords:** RPA · BPMS · BPMN · Integration · Design science research

## 1 Introduction

Digital transformation and external disruptions like the Covid-19 pandemic are increasingly impacting Business Process Management (BPM) [17]. Along this line, conventional Business Process Management Systems (BPMS), which are designed to support the end-to-end process definition, enactment, and automation of business processes [3], have several shortcomings [15]. For example, traditional process automation through deeply ingrained and inflexible business processes can hardly keep up with today's fast-changing environment [17]. To this end, recent articles indicate the promising integration of BPM, specifically BPMS, with the emerging Robotic Process Automation (RPA) technology [4,8,9]. RPA is an umbrella term that merges robotics and business process automation and aims to automate repetitive, standardized, and rule-based

© Springer Nature Switzerland AG 2022
A. Marrella et al. (Eds.): BPM 2022, LNBIP 459, pp. 138–153, 2022.
https://doi.org/10.1007/978-3-031-16168-1_9

tasks based on digital input, such as collecting, preprocessing, and transferring data [5, 21, 22].

A recent debate at the *BPM Expert Forum* [16] corroborated the practical and scientific relevance of an integrated BPMS-RPA platform and indicated potential benefits, particularly regarding a unified language for process modeling, such as the Business Process Model and Notation (BPMN). Additionally, several articles highlight potential integration synergies like improved process monitoring and error handling [8], process optimization [4], and human-bot collaboration [2]. However, profound academic approaches to examining concrete designs are scarce to date. Therefore, we contribute to the recent practical and scientific discourse by investigating the following research question:

*How can BPMS and RPA be holistically integrated based on a consistent notation for process modeling and orchestration?*

Grounded in Design Science Research (DSR) [13] and based on empirical data from 20 expert interviews, our contribution is threefold: First, we provide eight requirements for an integrated BPMS-RPA platform, indicating the suitability of BPMN as an underlying notation. Second, we propose an initial conceptual design of such a holistic platform based on a multi-layer process visualization approach and exemplify several components with a publicly available prototype. Finally, we contribute to the scarce literature by providing directions for further research and organizational practice on integrating BPMS and RPA.

This paper is structured as follows: Sect. 2 provides the relevant background while Sect. 3 introduces related work. Our applied research methodology is described in Sect. 4. The requirements and design of the proposed BPMS-RPA platform are presented in Sect. 5, followed by insights into the evaluation process and the prototypical implementation in Sect. 6. We conclude with a discussion in Sect. 7, covering our study's theoretical and practical implications, limitations, and recommendations for future research.

## 2   Background

In the following, we detail the basic terms and concepts relevant to this paper, i.e., BPMS, BPMN, and RPA.

### 2.1   Business Process Management Systems

BPM is considered a holistic concept to control and improve business processes, including identification, discovery, analysis, design, implementation, and monitoring. To this end, BPMS enact various aspects of BPM, such as the modeling, analysis, and execution of business processes [3, 15]. BPMS typically consist of a *Process Modeler* to define and configure the process models, which are stored in a *Process Model Repository* and deployed to a *Process Engine* for coordinated execution [27]. Besides, diverse application programming interfaces (APIs) are provided to integrate external software, e.g., for process analysis and monitoring [3].

BPMS usually build on BPMN, which constitutes the de-facto industry standard for process modeling and orchestration [24, 27]. This standardized and readily understandable graphical notation allows for visualizing and implementing business processes with varying complexity levels. BPMN provides multiple modeling elements to describe the process behavior, interrelations, and involved stakeholders and translates this information into BPMS execution languages [3, 12]. Generally, *activities* pose core elements of BPMN process diagrams and can be distinguished into *tasks* and *sub-processes*. The latter represent compound activities that can be subdivided into finer levels of detail, whereas tasks constitute atomic activities already at the lowest level of process detail [12, 24].

Considering the different needs of process model stakeholders (e.g., managers, process owners, business analysts, or programmers), van Nuffel and de Backer [24] propose a multi-abstraction layered approach with defined relationships to model and structure business processes. Their framework includes five levels with descending degrees of abstraction: *process map, process variant, elementary process, activity,* and *task.* While the first two levels are more abstract and of organizational nature, the latter three levels target specific business processes and their parts, for which the authors propose the use of BPMN [24]. The *elementary process level* shows abstract activities, inputs and outputs, and actor roles of single business processes. The subordinate *activity level* describes a specific part of the business process without revealing irrelevant aspects to a particular stakeholder. Finally, the elementary *task level* decomposes each activity into its atomic tasks by providing all available details at the lowest level of granularity [24].

## 2.2 Robotic Process Automation

RPA employs so-called bots that represent single software licenses and operate on the user interfaces of existing applications and IT systems, mimicking human behavior [21, 26]. Unlike BPMS, RPA does not require extensive programming, as RPA bots are developed using low-code or no-code approaches and are configured via graphical user interfaces [11, 25]. Therefore, RPA is also referred to as "lightweight IT", focusing on agility and speed, whereas the more complex BPMS are considered "heavyweight IT", emphasizing security and reliability issues [14]. RPA is perceived to be cheaper, easier, and faster to introduce, configure, and maintain than BPMS initiatives since RPA does not (profoundly) change the IT architecture and thus entails only a fraction of the implementation costs and efforts [4, 21, 26]. Although RPA systems are less sophisticated and extensive than BPMS, they consist of similar components, like a *Process Modeler, Model Repository,* and *Orchestrator* [8]. Furthermore, RPA distinguishes between attended and unattended bots. Unattended bots are executed autonomously, e.g., on virtual machines, and are suitable for end-to-end automation of standardized and straightforward tasks with a limited scope. Attended bots, in turn, usually run on local desktops and require human input and interaction since they are triggered by business users to perform specific tasks of a process [5, 21]. During operation, both bot types follow specified procedures consisting of detailed work instructions, which we refer to as *RPA flows* or *RPA sequences* further on.

# 3    Related Work

Despite increasing research and practical dissemination, neither RPA nor BPM(S) have exploited their full potential yet [5, 15, 18]. Whereas BPM(S) rather focuses on the more complex and abstract *process level*, RPA tackles the atomic *task level* of a business process [11] and is thus regarded a complement for BPM(S), indicating the beneficial integration [4, 8, 16].

Therefore, Flechsig et al. [4] propose a high-level framework that combines the BPM and RPA life cycles to realize synergy effects, e.g., the prior optimization of the *as-is* process model to improve the RPA sequence, which is then subject to the BPMS monitoring and control. Similarly, König et al. [8] introduce an RPA-aware BPM life cycle to link both approaches and present a prototype that provides an API between BPMS and RPA systems for tandem use. The authors conclude that the BPMS can facilitate the upscaling of RPA and its capabilities for exception handling and managing automation on the process level [8], particularly if RPA is intended as a long-term solution [7]. More practically oriented, Romao et al. [18] present preliminary results of a BPMS-RPA integration project in the banking industry. However, the concrete platform design and the task orchestration between BPMS, RPA, and human operators remain an open issue [2, 18]. Along this line, Ludacka et al. [9] outline a related initiative at *Deutsche Bahn Group* and illustrate the interplay between the BPMS and RPA bots but do not elaborate on how to accomplish the conceptual and technical integration. Besides, the presented approaches neglect the impact of BPMN for standardized process modeling and orchestration within integrated systems, as emphasized by academia and industry [15, 16].

Although BPMN is widely used in BPMS [3], the notation itself is rarely applied to model bots since most RPA providers maintain individual languages [25]. Consequently, BPMN has not yet been studied regarding its inclusion in an integrated BPMS-RPA platform. However, recent articles [6, 8, 18] indicate the technological feasibility and beneficiary of such an approach, which would address several RPA issues (e.g., upscaling, monitoring) and facilitate comprehensive and consistent process modeling and orchestration familiar to business users [6, 16].

To tackle these open issues and supplement the existing methodological work with requirements and a technical concept of an integrated BPMS-RPA platform, we conducted an empirical study that is described in the following.

# 4    Research Methodology

We employed DSR as our methodological foundation to iteratively develop an artifact that addresses the presented research problem. The applied procedure followed the widely accepted activities proposed by Peffers et al. [13], i.e., *problem identification, definition of objectives, design and development, demonstration, evaluation,* and *communication.* We aim to develop a concept for an integrated BPMS-RPA platform based on a standardized and comprehensive notation that

**Table 1.** Overview of the participating organizations and informants

| Org. | Scope | Revenue (EUR M) | Informant's role and relevant work experience (years) | Duration (min) (Iteration 1 \| 2) |
|------|-------|-----------------|-------------------------------------------------------|-----------------------------------|
| A | BPMS/RPA (Consulting) | 100–1000 | 1. Senior Project Manager (15) | 49 \| 54 |
|   |   |   | 2. Head of IT Consulting (16) | 44 |
|   |   |   | 3. Senior RF Engineer (21) | 53 |
| B | **BPMS**a/RPA | 100–1000 | Sales Leader (21) | 45 \| 65 |
| C | **BPMS/RPA** | 10–100 | Senior Solution Consultant (28) | 50 \| 71 |
| D | **BPMS/RPA** | 10–100 | Senior Solution Consultant (4) | 40 |
| E | **BPMS/RPA** | 10–100 | 1. Sales Leader (16) | 25 |
|   |   |   | 2. Presales Consultant (2) |   |
| F | **BPMS/RPA** | 10–100 | Automation Engineer (2) | 45 |
| G | **BPMS/RPA** | <10 | Key Account Manager (22) | 49 |
| H | **BPMS/RPA** | <10 | Senior IT Consultant (24) | 42 |
| I | BPMS/**RPA**b | <10 | 1. Chief Executive Officer (23) | 70 , 60 , 72 \| 65 |
|   |   |   | 2. Chief Solution Architect (23) |   |
| J | RPA | 100–1000 | Business Developer (1) | 40 |
| K | RPA | 100–1000 | Senior Solution Consultant (27) | 49 \| 53 |

a**BPMS**: interface to partnered RPA solutions; b**RPA**: self-developed RPA solution

allows for unified process modeling and orchestration. In this vein, the exploratory nature of our study reflects our research question as we seek new insights and intend to provide implications for further research and organizational practice. The design process comprised three iterations, each resulting in several adjustments: (1) building the conceptual design based on the requirements derived from the interview study and related literature; (2) discussion and evaluation with six experts from the first iteration; (3) revision of the concept and prototypical exemplification of several components to demonstrate the practicability.

Our empirical inquiry followed the principles of case study research proposed by Runeson et al. [19], who consider expert interviews as essential data sources for software engineering, particularly when applying DSR. We selected the participating organizations based on theoretical sampling and paid attention that they differ in their scope (i.e., providing BPMS and/or RPA platforms and services) and size (i.e., revenues) to increase external validity. We ensured construct validity by conducting 20 semi-structured interviews with 15 experts of various functions and hierarchical roles (see Table 1). The average duration was 51 min, with one or two informants participating per interview. The organizations' names and revenues are anonymized for confidentiality reasons [19].

The applied interview guideline included open-ended questions related to five parts: (1) information on the experts' background, understanding of BPMS and RPA, and related experience; (2) feasibility of using a standardized notation for process modeling and orchestration for BPMS and RPA; (3) discussion of suitable integration approaches; (4) implications and application areas of an integrated BPMS-RPA platform; (5) challenges to be addressed by further research.

The procedures for data gathering and analysis were performed collectively by three authors to reduce bias and increase validity and reliability, including rich supplementary data for triangulation, e.g., websites, white papers, internal presentations, and software demonstrators [19]. The interviews were transcribed and coded with the software *MAXQDA 2022* following the guidelines for systematic qualitative content analysis [10]. We ensured internal validity by applying a hybrid inductive-deductive approach, i.e., we generalized emerging patterns through a combined within-case and cross-case analysis and assigned the coding elements to main categories deduced from related literature and sub-categories developed inductively [10,19]. The results of our study are presented next.

## 5 Towards an Integrated BPMS-RPA Platform

During the interviews, we recognized that many organizations employ separate BPMS and RPA systems, even though they acknowledged that the growing number of operational RPA bots necessitates sophisticated and standardized orchestration to align the execution of business processes and RPA flows. However, adjusting and connecting both systems requires considerable effort since there is not yet a common standard interface. That recurrent problem corroborates the need for a novel approach that enables consistent process modeling and orchestration while treating RPA bots as "first-class citizens", i.e., deeply embedded into processes. In this section, we derive the requirements and propose an initial conceptual design for such an integrated BPMS-RPA platform.

### 5.1 Requirements Engineering

The synthesized findings from the interview study and related literature yielded eight requirements of an integrated BPMS-RPA platform: four referring to organizational aspects (O1–O4) and four addressing technical issues (T1–T4). The requirements are described in the following and substantiated with representative quotes from the interviews [10] in a supplementary documentation.[1]

**O1 BPM Maturity.** Integrating BPMS and RPA systems requires a certain degree of BPM maturity, i.e., organizational maturity and respective process capabilities. In this vein, *maturity* relates to the extent and interplay of process modeling, process deployment, process optimization, process management, organizational culture, and organizational structure to enhance business process performance. In contrast, *capabilities* refer to the competencies necessary to achieve the intended process results [23]. Various experts [Org. C, G, H, I] indicated the poor BPM maturity of many organizations and corroborated related work [20] by emphasizing the importance of a well-prepared IT architecture and process landscape, know-how building, and strategic implementation for an integrated BPMS-RPA platform.

**O2 Mindset.** The participants also reported that the lacking understanding of integrated process automation impedes the deployment of respective initiatives

---

[1] https://github.com/bptlab/holistic-process-platform/raw/main/quotes.pdf.

[Org. B, C, H, I]. While many organizations have been triggered by the automation hype around RPA and focus on bot development, they tend to neglect the more expensive yet essential BPMS projects. Although the functionalities of both technologies are increasingly converging [7], it was emphasized that organizations should not follow the RPA-centric approach [2] by considering RPA as a replacement for BPMS but rather a beneficial complement [Org. A, C, G, H]. Along this line, introducing an integrated platform requires top-level management support to release necessary budgets and promote change management that facilitates user acceptance, familiarity with new procedures, knowledge sharing, and collaboration between the IT and business departments [7,9].

**O3 Economic Efficiency.** Profitability is essential for integrated platforms since the acquisition, deployment, operation, and maintenance require high efforts, particularly for small and medium-sized enterprises, which may not necessarily need a comprehensive (BPM) system for business process execution. Besides, multiple RPA bots must be employed to justify their incorporation since a limited number can also be managed manually and more cost-effectively [Org. A, G, H]. The more bots and tools involved, the higher the costs for licensing, operation, and orchestration. Therefore, an integrated platform must show a reasonable return on investment [16] and a favorable cost-benefit ratio [Org. A, D, E].

**O4 Integrated Organizational Structure.** The interviewees also reported on different departments being responsible for BPMS and RPA initiatives. However, an integrated platform requires integrated organizational structures, i.e., "bringing the two worlds together and overcoming the silo thinking" [Org. D]. Therefore, a consolidated "Center of Excellence" (CoE) centralizes the necessary competencies, responsibilities, technical capabilities, and human resources for operation and control [Org. A, C, D, E]. The CoE should also reflect the organizational and IT strategy and implement appropriate governance structures [5].

**T1 BPMS Fundament.** When scaling up RPA initiatives, many organizations recognize the need for a central BPMS platform to automate, control, and monitor processes holistically, i.e., "end-to-end" [Org. D, E, I]. BPMS usually include sophisticated procedures for process documentation, analysis, and orchestration and provide interfaces to integrate external applications (e.g., process mining tools), facilitating the further adoption of RPA [Org. A, C, G]. In that sense, BPMS could identify suitable routines for RPA, standardize process modeling and task orchestration, launch and monitor RPA bots to detect bottlenecks and exceptions, and drive comprehensive process optimization [1,4,6,8,9]. The BPMS fundament with a complementary RPA integration constitutes the most frequently mentioned requirement and reflects the BPM-centric approach [2].

**T2 Concerted Task Orchestration.** While RPA systems lack large-scale process orchestration, focusing on the management and alignment of bots and tasks [8], current BPMS are often restricted regarding their functionalities for human-bot collaboration [18,20]. Therefore, academics and practitioners emphasize the need for seamless coordination and collaboration between the BPMS, RPA bots, and human operators. To this end, an integrated platform should be based on transparent interfaces, profound decision logic, and predefined rules for requests [Org. A, B, G, I], providing functionalities for synchronous and asynchronous human-bot collaboration [2].

**T3 Consistent Process Modeling.** The need for uniform and comprehensive process documentation was consistently mentioned in the interviews and related work [4,6,8,18]. In this vein, multiple experts recommended using the standardized BPMN 2.0 notation as it is already applied for process modeling and execution by most BPMS [Org. C, D, E, G]. Besides, many RPA design languages are inspired or rest upon BPMN elements [Org. I, K]. Therefore, an integrated BPMS-RPA platform based on the BPMN 2.0 notation would facilitate consistent and comprehensive process documentation and automation. The holistic approach could enable the standardized design of BPMS and RPA workflows, automatically generate related flowcharts [1], and foster the upscaling of RPA bots and their integration into the BPMS [Org. D, H, I].

**T4 Limited Complexity.** Several articles [6,18] and interview participants stressed the necessity of limited complexity regarding technical implementation (i.e., preferably low-code or no-code programming) and process model representation. Adequate process visualization can be facilitated through a layered approach, enabling all stakeholders to illustrate the process model and relevant activities in the required level of detail [24]. However, some interviewees highlighted potential difficulties when linking the rather technical RPA bot configuration with the graphical BPMN 2.0 notation due to the different contexts and objectives, resulting in too complex and hardly readable process models. In this context, the experts reported on necessary BPMN extensions to adequately depict human-bot collaboration, complex decision-making, data extraction from user interfaces, and the use of artificial intelligence. Besides, status-affected objects, loop constructs, and function calls must be embedded [Org. A, C, H, I, K].

Emphasizing the high efforts for integrating the different BPMS and RPA technologies, several experts recommended the development of an entirely new platform resting upon a holistic and consistent approach for process modeling, orchestration, and automation [Org. B, C, K]. As indicated by expert E1, such an environment would allow specifying both BPMS routines and RPA bots based on a consistent notation: *"You need a uniform BPMS and RPA system. As long as you partner with an external RPA provider, it's difficult to implement an integrated platform, as the different RPA tools have their own notation."*

Although we noticed that some organizations are pioneering holistic approaches, no respective concept yet exists in the academic literature. Therefore, we tackle this research gap by proposing an initial conceptual design of an integrated BPMS-RPA platform based on BPMN in the following.

## 5.2  Conceptual Design

This section presents the final version of our artifact and explains how it addresses the revealed (technical) requirements. The changes made during the evaluation process are detailed in Sect. 6.1. We propose a holistic platform for

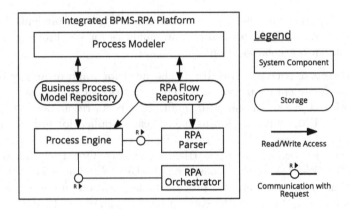

**Fig. 1.** High-level architecture of the integrated BPMS-RPA platform as FMC-Diagram

modeling, orchestrating, and executing business processes and RPA flows while capturing and considering their interplay and relation. Corresponding to the requirement T1, the platform's architecture (Fig. 1) mainly builds on the components of a traditional BPMS (cf. [27]), in particular on the *Process Modeler* and *Process Engine*. However, the components are functionally extended and supplemented with additional RPA-related elements. For example, we introduce an additional *RPA Flow Repository*, responsible for storing the definitions of end-to-end automated RPA workflows. Besides, the *Process Engine* is supported by an *RPA Orchestrator* and a *Parser* for RPA flows similar to those of stand-alone RPA tools. The various components are described below.

**Process Modeler.** Contributing to T3, the *Process Modeler* builds on BPMN to enable the seamless creation and visualization of business processes and RPA flows, which are stored in the respective repositories. The *Business Process Model Repository* is adapted from traditional BPMS and contains business process definitions that can be enriched with RPA functionality. Sequences in the *RPA Flow Repository* are end-to-end automated reoccurring activities applicable to various business processes of an organization, e.g., logging in to specific software or retrieving certain information. As shown in Fig. 2, these rather generic workflows are solely composed of BPMN tasks representing atomic RPA operations that need to be performed and are substantiated by an underlying technical RPA configuration. Therefore, the *RPA Flow Repository* employs flowcharts of automated sequences specified through the process or task variables rather than a "farm" of predefined bots tailored to concrete tasks [2]. These variables allow for flexible configuration of the automated activities and their reuse for different contexts and users, e.g., by dynamically requesting login data from a central credential store. Furthermore, it can be defined whether and how the RPA flow should be executed in an attended or unattended manner.

**Fig. 2.** Sample sequence of the RPA flow repository

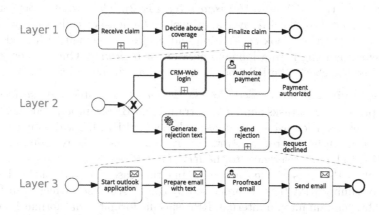

**Fig. 3.** Applied layered approach, exemplified through a sample business process

Although flows of the RPA repository usually pose stand-alone sequences, their integration into the superordinate business processes orchestrated by the BPMS is challenging due to the different scope and level of abstraction. While BPMS mainly include relatively high-level workflows that take additional knowledge on how they are performed, RPA is used to automate individual tasks within a business process and requires detailed technical instructions. To avoid extensive process models and reduce complexity (cf. T4), we adapt the layered approach of van Nuffel and de Backer [24] (cf. Sect. 2.1), which is technically implemented in the *Process Modeler* using the sub-process elements defined in BPMN 2.0. Therefore, our concept decomposes business processes into three layers with descending levels of abstraction. The first layer represents the *process level* and includes the main activities and interactions. The second layer poses the *activity level* and details the main activities. In accordance with van Nuffel and de Backer [24], this layer can be defined recursively for highly complex processes, i.e., the activities can be defined in more detail while remaining on a relatively abstract level. Finally, the third layer describes the *task level* and comprises atomic instructions on how to perform a particular activity.

The process modeler offers two approaches for integrating RPA functionality into business processes. On the one hand, RPA workflows specific to a particular business process can be directly defined in the third layer, as its atomic task level matches the granularity of RPA with its atomic work instructions. On the other hand, generic RPA sequences defined in the *RPA Flow Repository* can be referenced in the second layer using BPMN call activities. Since they are defined in a central instance, they can be reused quickly and need to be adjusted for updates only once. In contrast, specifying RPA operations directly in the process is helpful for tailored automation and allows for synchronous human-bot collaboration (cf. T2), i.e., RPA tasks and human tasks can be defined alternately. The two approaches are illustrated in Fig. 3. In the given example, the call activity "CRM-Web login" in the second layer references the respective end-to-end sequence from the *RPA Flow Repository* (cf. Fig. 2). The activity "Send rejection" is detailed in the third layer and involves human-bot collaboration. RPA is used to start the outlook application and prepare the email with predefined input before handing it to a human operator for reviewing. Once the employee has finished the check, the RPA bot is triggered again to send the email.

**Process Engine.** The proposed *Process Engine* orchestrates and executes the defined processes and tasks similar to a traditional BPMS process engine. Since the layers are modeled using sub-process elements of BPMN, it does not require any modification in this regard. However, its functionality is extended for RPA integration to handle references to the *RPA Flow Repository* on the second layer and specific RPA tasks on the third layer. In either case, the (part of the) BPMN model containing RPA tasks is processed by the *RPA Parser*, which transforms the BPMN diagram into an internal, RPA-specific format. That format is subsequently handed to the *RPA Orchestrator* for executing the RPA tasks within the appropriate environment as specified by the respective process variables. When a process on the third layer also includes non-RPA tasks, the *RPA Orchestrator* pauses the bot execution until the intervening non-RPA tasks have been completed and the *Process Engine* signals to proceed. In both cases, the *RPA Orchestrator* distributes pending RPA tasks to appropriate RPA bots considering the individual capabilities and task variables, e.g., to account for the configured mode (i.e., attended or unattended) and required software installations.

# 6   Evaluation and Demonstration

Following the DSR methodology, our concept was evaluated through further expert interviews and partially implemented in a prototypical demonstrator.

## 6.1   Follow-up Interviews

The evaluation rests upon five follow-up interviews with experts from Org. A, B, C, I, and K (cf. Table 1). We had lively yet constructive discussions and received positive feedback on our concept, entailing valuable proposals for modification. For example, we initially intended to generate RPA scripts from the RPA workflows and third-layer tasks directly after the modeling. Due to the experts' feedback, we changed this behavior to a more dynamic approach, i.e., the process models containing RPA definitions are both parsed and interpreted during the execution. That allows for short-term modifications and facilitates the detailed monitoring of the RPA execution. Consequently, the former *RPA Script Generator* component was replaced by an *RPA Parser* (cf. Fig. 1). Furthermore, we refrained from advocating bots configured for specific RPA workflows. Instead, we propose a "pool" of RPA sequences with predefined yet adjustable process or task variables due to its practicability and increasing relevance.

Generally, the study participants attested to the applicability of BPMN as the underlying notation to visualize business processes and their interrelated activities and tasks regarding the proposed layers. However, during the evaluation, several experts emphasized that the current BPMN 2.0 standard needs to be enhanced to adequately depict RPA sequences as intended with our concept. To this end, we discussed introducing a fourth layer that translates RPA tasks into code and represents technical aspects, while the superordinate layers show the business logic. However, this would increase the complexity of the process model and proposed platform, contrary to our objectives and derived requirements (cf. T4). Besides, it would involve profound (technical) amendments to BPMN 2.0, which does not fit the notation's original purpose [12]. Therefore, we opt for three layers and propose to configure the chosen RPA functionality using task attributes. For visual guidance, we now indicate the automated application in the process model by a respective icon in the task. Finally, the experts emphasized that activities require precise and descriptive labels due to the nested layers.

## 6.2 Prototypical Implementation

We prototypically implemented the extended *Process Modeler*, *Business Process Model Repository*, and *RPA Flow Repository* to provide an initial demonstration of our concept.[2] Based on our explanations in Sect. 5.2, the *Process Modeler* enables users to model RPA flows and layered business processes using BPMN. RPA tasks on the third layer or in the *RPA Flow Repository* can directly be configured through task variables in the user interface by choosing the appropriate RPA operation for execution. This configuration via the *Process Modeler* is based on the open-source *Robot Framework*[3], which could be integrated by an extended *Process Engine* and *RPA Orchestrator* to perform the RPA tasks. It is important to note that the amount and complexity of the automated workflows in both repositories will likely increase over time. Therefore, the *Process Modeler* enables direct navigation between the different layers and RPA flows via *subprocess activities* (referencing subordinate layers) and *call activities* (referencing sequences of the *RPA Flow Repository*) to handle the increasing complexity. Consequently, processes and their subordinate activities and tasks on the different layers can be navigated and explored intuitively.

## 7    Discussion and Conclusion

This empirical article rests upon the DSR methodology and examines how BPMS and RPA can be holistically integrated based on a consistent notation for process modeling and orchestration. To this end, the conducted interview study and

---

[2] The open-source prototype can be found here: https://github.com/bptlab/holistic-process-platform.

[3] https://robotframework.org/ (accessed: 11.04.2022).

related literature yielded four organizational and four technical requirements. We also developed an initial conceptual design of an integrated BPMS-RPA platform, reflecting the technical requirements. The proposed concept was evaluated by multiple experts and partially implemented through a publicly available prototype, indicating its feasibility and practicability. In the following, we discuss the implications of our study and provide recommendations for further research.

## 7.1   Theoretical and Practical Implications

Our concept amalgamates the functionalities of BPMS and RPA systems in a uniform platform that rests upon comprehensive and consistent process modeling, orchestration, and automation with BPMN, contributing to the recent practical and scientific discourse [4,6,16,18]. As highly recommended by the interviewed experts, the BPMS constitutes the central instance for process enactment and is complemented by RPA (cf. T1), thus enabling single-source and end-to-end process automation and preparing the ground for holistic process monitoring and exception handling. The consistent use of BPMN throughout the platform in line with the three abstraction layers allows for the integrated visualization of both the abstract business logic and specific atomic RPA tasks (cf. T3, T4). It also simplifies task orchestration between the BPMS, RPA bots, and human operators (cf. T2). Besides, the varying informative value and complexity at the different layers contribute to the various purposes of each process stakeholder and could foster coordination and understanding. Additionally, the concept enables a direct collaboration of RPA bots and humans on the task level, coordinated by the BPMS. As a result, the *Process Engine* can provide detailed execution information, while error handling could be facilitated, and the subsequent process analyses become more comprehensive. Hence, the need to evaluate which business logic has to be implemented in the BPMS and which in the RPA system, as is often the case with current solutions, is eliminated with our approach.

   In contrast to related work considering separated BPMS and RPA systems [8], we opt for in-depth integration of RPA into BPMS to increase process transparency and model consistency while preventing RPA "black boxes" and data silos that usually result from separated solutions. Consequently, our study corroborates the harmonization of the BPM and RPA life cycles [4,8] with a profound concept on how to realize that integration technically. In this vein, a bridging approach [8] is suitable for organizations that maintain deeply ingrained legacy BPMS and/or RPA systems for which immediate integration would cause excessive efforts. These organizations should prioritize increasing their BPM maturity (cf. O1) and fostering a mindset for holistic process automation (cf. O2) to enable the step-by-step migration towards an integrated BPMS-RPA platform. Depending on the needs and economic efficiency (cf. O3), individual RPA sequences can be gradually incorporated into the BPMS. However, an integrated BPMS-RPA platform also entails integrated organizational structures, i.e., a consolidated *Center of Excellence* pooling all necessary resources and capabilities (cf. O4).

## 7.2  Limitations and Recommendations for Further Research

This research is not without limitations. Even though our empirical inquiry rigorously followed DSR and case study principles to provide meaningful insights based on a representative sample, it is restricted regarding the number of examined organizations. Despite the interviewed experts attesting to our concept's practicability, it still lacks a real-world implementation and validation beyond the prototype. Besides, the presented requirements mainly rest upon the interview study due to the scarce scientific literature. Therefore, future research can conduct more comprehensive studies to confirm, complement, or disprove our findings.

As this work is intended to provide initial insights and starting points for an integrated BPMS-RPA platform, several aspects require further elaboration and should be addressed by future research. For example, concrete and holistic orchestration and governance mechanisms should be established to manage the two repositories [2,18], specify the configuration and upscaling of automated workflows [1,11], and control the input-output transfer across the layers and between the BPMS, RPA bots, and human operators. Besides, appropriate regulations for system security and self-government should be examined.

Although we did not explicitly discuss a rework of BPMN, a purposeful extension of the standard could be beneficial, particularly concerning the visualization of RPA-specific information. A process recorder could automatically generate BPMN models, which are then subject to verification, harmonization, and configuration before shifting them to the repository, e.g., to be executed by RPA bots. Targeted task mining approaches can also be incorporated. Finally, we examined non-intelligent BPMS and RPA systems. Therefore, future research needs to investigate their increasing integration with artificial intelligence and other automation technologies towards "hyperautomation" [7], as various functionalities converge, like process modeling, orchestration, or monitoring.

**Acknowledgments.** We thank Mika Göckel and Alexander Steiner for the constructive discussions that contributed to the improvement of this article.

## References

1. Agostinelli, S., Marrella, A., Mecella, M.: Towards Intelligent Robotic Process Automation for BPMers (2020). https://arxiv.org/pdf/2001.00804
2. Cabello Ruiz, R., Jiménez-Ramírez, A., Escalona Cuaresma, M.J., González Enríquez, J.: Hybridizing humans and robots: an RPA horizon envisaged from the trenches. Comput. Ind. **138**, 103615 (2022)
3. Dumas, M., La Rosa, M., Mendling, J., Reijers, H.A.: Fundamentals of Business Process Management, 2nd edn. Springer, Heidelberg (2018). https://doi.org/10.1007/978-3-662-56509-4
4. Flechsig, C., Lohmer, J., Lasch, R.: Realizing the full potential of robotic process automation through a combination with BPM. In: Bierwirth, C., Kirschstein, T., Sackmann, D. (eds.) Logistics Management. LNL, pp. 104–119. Springer, Cham (2019). https://doi.org/10.1007/978-3-030-29821-0_8

5. Hofmann, P., Samp, C., Urbach, N.: Robotic process automation. Electron. Mark. **30**(1), 99–106 (2019). https://doi.org/10.1007/s12525-019-00365-8
6. Hüller, L., Jenß, K.E., Speh, S., Woelki, D., Völker, M., Weske, M.: Ark automate - an open-source platform for robotic process automation. In: Proceedings of the Best Dissertation Award, Doctoral Consortium, and Demonstration & Resources Track at BPM 2021, pp. 126–130. CEUR-WS.org (2021)
7. Jiménez-Ramírez, A.: Humans, processes and robots: a journey to hyperautomation. In: González Enríquez, J., Debois, S., Fettke, P., Plebani, P., van de Weerd, I., Weber, I. (eds.) BPM 2021. LNBIP, vol. 428, pp. 3–6. Springer, Cham (2021). https://doi.org/10.1007/978-3-030-85867-4_1
8. König, M., Bein, L., Nikaj, A., Weske, M.: Integrating robotic process automation into business process management. In: Asatiani, A., et al. (eds.) BPM 2020. LNBIP, vol. 393, pp. 132–146. Springer, Cham (2020). https://doi.org/10.1007/978-3-030-58779-6_9
9. Ludacka, F., Duell, J., Waibel, P.: Digital transformation of global accounting at deutsche bahn group: the case of the TIM BPM suite. In: vom Brocke, J., Mendling, J., Rosemann, M. (eds.) Business Process Management Cases Vol. 2, pp. 57–68. Springer, Heidelberg (2021). https://doi.org/10.1007/978-3-662-63047-1_5
10. Mayring, P.: Qualitative content analysis: theoretical foundation, basic procedures and software solution. SSOAR (2014)
11. Mendling, J., Decker, G., Hull, R., Reijers, H.A., Weber, I.: How do machine learning, robotic process automation, and blockchains affect the human factor in business process management? Commun. Assoc. Inf. Syst. **43**, 297–320 (2018)
12. Object Management Group: Business Process Model and Notation (BPMN): Version 2.0.2 (2013). https://www.omg.org/spec/BPMN/2.0.2/PDF
13. Peffers, K., Tuunanen, T., Rothenberger, M.A., Chatterjee, S.: A design science research methodology for information systems research. J. Manag. Inf. Syst. **24**(3), 45–77 (2007)
14. Penttinen, E., Kasslin, H., Asatiani, A.: How to choose between robotic process automation and back-end system automation? In: ECIS 2018. AIS (2018)
15. Reijers, H.A.: Business process management: the evolution of a discipline. Comput. Ind. **126**, 103404 (2021)
16. Rinderle-Ma, S., Rücker, B.: 3rd BPM Expert Forum: Will RPA make BPMSs redundant? (2021). https://bpm-conference.org/bpma/expert-forum/3
17. Röglinger, M., et al.: Exogenous shocks and business process management. Bus. Inf. Syst. Eng. (2022). https://doi.org/10.1007/s12599-021-00740-w
18. Romao, M., Costa, J., Costa, C.J.: Robotic process automation: a case study in the banking industry. In: CISTI 2019. IEEE (2019)
19. Runeson, P., Höst, M., Rainer, A., Regnell, B.: Case Study Research in Software Engineering: Guidelines and Examples, 1st edn. Wiley, Hoboken (2012)
20. Simek, D., Sperka, R.: How robot/human orchestration can help in an HR department: a case study from a pilot implementation. Organizacija **52**(3), 204–217 (2019)
21. Syed, R., et al.: Robotic process automation: contemporary themes and challenges. Comput. Ind. **115**, 103162 (2020)
22. van der Aalst, W.M.P., Bichler, M., Heinzl, A.: Robotic process automation. Bus. Inf. Syst. Eng. **60**(4), 269–272 (2018)
23. van Looy, A., de Backer, M., Poels, G.: Defining business process maturity. A journey towards excellence. Total Qual. Manag. Bus. Excell. **22**(11), 1119–1137 (2011)
24. van Nuffel, D., de Backer, M.: Multi-abstraction layered business process modeling. Comput. Ind. **63**(2), 131–147 (2012)

25. Völker, M., Siegert, S., Weske, M.: Adding decision management to robotic process automation. In: González Enríquez, J., Debois, S., Fettke, P., Plebani, P., van de Weerd, I., Weber, I. (eds.) BPM 2021. LNBIP, vol. 428, pp. 23–37. Springer, Cham (2021). https://doi.org/10.1007/978-3-030-85867-4_3

26. Wanner, J., Hofmann, A., Fischer, M., Imgrund, F., Janiesch, C., Geyer-Klingeberg, J.: Process selection in RPA projects - towards a quantifiable method of decision making. In: ICIS 2019 Proceedings. AIS (2019)

27. Weske, M.: Business Process Management: Concepts, Languages, Architectures, 3rd edn. Springer, Heidelberg (2019)

# Rolling Back to Manual Work: An Exploratory Research on Robotic Process Re-Manualization

Artur Modliński[1]([✉]), Damian Kedziora[2], Andrés Jiménez Ramírez[3], and Adela del-Río-Ortega[4]

[1] University of Lodz, Lodz, Poland
artur.modlinski@uni.lodz.pl
[2] LUT University, Lappeenranta, Finland
[3] Universidad de Sevilla, Seville, Spain
ajramirez@us.es
[4] I3US Institute, SCORE Lab, Universidad de Sevilla, Seville, Spain
adeladerio@us.es

**Abstract.** Robotic process automation (RPA) is a technology that is presented as a universal tool that solves major problems of modern businesses. It aims to reduce costs, improve quality and create customer value. However, the business reality differs from this aspiration. After interviews with managers, we found that implementation of robots does not always lead to the assumed effect and some robots are subsequently withdrawn from companies. In consequence, people take over robotized tasks to perform them manually again, and in practice, replace back robots—what we call 're-manualization'. Unfortunately, companies do not seem to be aware of this possibility until they experience it on their own, to the best of our knowledge, no previous research described or analysed this phenomenon so far. This lack of awareness, however, may pose risks and even be harmful for organizations. In this paper, we present an exploratory study. We used individual interviews, group discussions with managers experienced in RPA, and secondary data analysis to elaborate on the re-manualization phenomenon. As a result, we found four types of 'cause and effect' narrations that reflect reasons for this to occur: (1) overenthusiasm for RPA, (2) low awareness and fear of robots, (3) legal or supply change and (4) code faults.

**Keywords:** Robotic process automation · RPA · Software robot · Investment · Information systems · Work manualization

This research is part of the projects PID2019-105455GB-C31 and RTI2018-101204-B-C22 funded by MCIN/AEI/ 10.13039/501100011033/and ERDF A way of making Europe; grant P18-FR-2895 funded by Junta de Andalucía/FEDER, UE; and grant US-1381595 (US/JUNTA/FEDER, UE).

A. Marrella et al. (Eds.): BPM 2022, LNBIP 459, pp. 154–169, 2022.
https://doi.org/10.1007/978-3-031-16168-1_10

# 1    Introduction

Robotic process automation (RPA) is an emerging technology in the business sector. Syed et al. define RPA as: *technology that comprises software agents called bots—or software robots—that mimic the manual path taken by a human through a range of computer applications when performing certain tasks in a business process* [1].

In the digital era, companies compete aggressively on price and efficiency [2]. It requires adapting so that solutions (1) improve their overall performance, (2) bring value to the customer, and (3) reduce both operational expenses and lead time [3]. All of this is possible due to the concept of 'intelligent competitive advantage' which is based on three elements: business analytics and intelligence [4], modular software development [5], as well as big data and cloud computing [6]. RPA embraces all these elements and may be adapted into existing information systems (IS), providing fast reimbursement [7]. For this reason, RPA has become one of the most popular technologies for delivering customer value [8]. Moreover, it brings several benefits to modern business: cost savings [9], increase in efficiency [10], value co-creation [8], quality improvement, work facilitation [11], increase in production, stable and accurate performance [2], and fast increase in RoI (Return on Investment) [9].

While robots are promoted as universal tools that mainly bring success to companies, our field observations suggest otherwise, that often this is rather an oneiric narration that does not have much in common with business reality. Moreover, sometimes robots need to be withdrawn and the related processes are taken over by a human workforce to execute them manually again. We call this the process *re-manualization* phenomenon. There is some recognition of the challenge of RPA readiness in a company, which somehow suggests the possibility of RPA not being appropriate in certain circumstances [12]. However, we found that the literature is scarce on unsuccessful implementation of robots and the reasons behind it. Therefore, the objective of our investigation was to discover what happens when robots do not work in accordance with a company's expectations. In particular, our research question is:

RQ: What are the reasons for RPA withdrawal in a company?

To answer this question, we performed an exploratory study involving three companies in Poland. The results suggest that process re-manualization occurs when (1) people are too enthusiastic about RPA and do not understand in which circumstances it works best, (2) employees' fear of software, (3) the internal procedures or supply are changed and the company is not able to adjust the robot accordingly, and (4) code faults exist and there is no one capable to repair it. In addition, we elaborate on cause and effect sequences of these four reasons.

The remainder of this paper is built as follows. Section 2 describes the main concepts of RPA and its advantages for business. In this section, we cite reports suggesting that RPA is sometimes withdrawn from the companies but a substantial research gap exists regarding why it happens. Building on this lacuna, in Sect. 3 we show the design of the exploratory study that help address the research question. Section 4 presents the original findings. Section 5 describes our contribution to both theory and practice as well as the limitations of the work. Finally, Sect. 6 concludes and set the future research lines.

## 2   Research Background

The initial application of RPA was limited to repetitive and error-prone processes, based on simple logics that added little value to business [13]. In the past few years, software for robots has been enhanced by technologies linked to artificial intelligence, such as process mining, sophisticated computing algorithms, data analytics, machine learning, natural language processing and optical character recognition [11,14]. All of these have helped robots perform more complex tasks [15]. As a result, they are already capable of handling payroll tasks, recruitment processes, accounting operations, inventory management, invoicing, reporting, software update, and data migration, among others. Though robots were first primarily used within IT-companies, they are now commonly adopted in banks, telecommunications, energy industries [16], judiciary processes [17] and outsourcing companies [3]. Moreover, it is foreseen that further dynamic adoption of RPA in other market areas will occur as cognitive RPA continues to develop. This technology will help to perform tasks that demand cognitive abilities, which so far have been perceived to be reserved only for humans. It is expected that such RPA will enable robots to 'see and read' unstructured text, learn, detect anomalies, forecast, and make decisions [1].

RPA helps to improve work accuracy and reduce complicated tasks [18]. It also facilitates data collection and processing [19] and helps to reduce the effort employees put into repetitive and simple tasks [2]. Consequently, time previously spent on routine and wearisome tasks can be saved and allocated elsewhere, and employees can focus on value-adding activities, resulting in innovative business solutions, services, and products [20].

Over the past decade, employee attitudes toward robots have been changing, which encourages companies to adapt RPA in their organizational space. Trust in robotic performance is systematically growing, as evidenced by the report from Oracle [21], in which 64% of respondents declared they would trust robots more than their own managers and 82% of employees said that robots are able to perform certain tasks better than humans. Wright et al. (2017) conducted research on 400 executives around the world and found that 53% use RPA in their companies, which has helped them improve compliance, quality/accuracy, productivity, Everest Research Group notes that not only large companies invest in RPA, but small- and medium-sized companies as well [20]. These findings across various types of companies confirm that the growing trend to invest in robots will shape the future of business in the coming years.

Although RPA offers numerous advantages for business, it also has limitations and creates challenges for managers. First of all, robots fail due to the lack of designing, executing, analytic tools or IT and business knowledge [22]. Secondly, there are still not enough experts who are able to design or redesign robots to perform their tasks optimally [22]. Consequently, companies delay RPA's implementation or modification for too long. Moreover, such investments demand financial resources that not every organization is able to provide. Wright et al. inform that only 3% of companies using RPA are able to scale their digital workforce and only 14% of the 424 executives interviewed expressed familiarity with RPA [23]. In addition, choosing suitable tasks to be robotized may be difficult as each company has a huge amount of data that should be analyzed before adoption [24]. A wrong decision may result in work disorganization and chaos. If RPA fails, it may cause several risks for a company, including (1) rapid mistakes without sufficient control, (2) using robots to cover symptoms, rather than root cause, of a problem, and (3) significant manual rework, overcompensating the automation benefits [25]. In addition, LLamberton et al. suggest that when RPA fails it is due to the internal environment of an organization, and point out the following reasons: (1) wrong processes targeted for robotization, (2) wrong methodologies used, (3) robot prototypes moved to full production without sufficient consideration, (4) too much of a process is being automated, (5) the IT infrastructure of the company is not taken into account, (6) the thinking that RPA is perceived as the only way to achieve a great ROI, (7) RPA being IT-owned, whereas it's best being owned by the business, (8) scaling past proof of concepts or pilots is not considered, (9) robots are left unsupervised after processes have been automated, and (10) RPA is not treated as a change program, with a focus on realizing benefits [26]. While industry reports suggest why RPA fails on a macro scale, they neither show what the consequence of robot's withdrawal is nor what logical strings lead to this.

## 3   Research Design

Two primary factors triggered this work: observations of RPA's reports [22], and feedback received during interviews with participants for a separate project, which pointed to the re-manualization phenomenon. Hence, the foundation for this research became a so called, 'window of opportunity', described by Czarniawska [27] as a situation when researchers observe the field and start posing questions about a reality that he/she does not understand. In line with the methodological approach proposed by Czarniawska [27], the authors aimed to understand and describe the phenomenon observed in business. After reading the literature the authors found, surprisingly, that none of the papers focused on process re-manualized yet.

To overcome this gap, the research procedure depicted in Fig. 1 was conducted. It consisted of six stages, with the use of mixed methods: individual interviews, group discussions, and secondary data analysis.

In the first stage of the research, data related to the companies' digital transformation and robotics was explored by analysing industry reports to find which of the biggest companies adopted RPA, and when they did it. We identified nine international companies in Poland that implemented RPA, three of which agreed to take part in this research on the condition their names would not be disclosed.

We, authors, signed a confidentiality agreement which regulated the conditions of the research works, paying attention to the code of research ethics

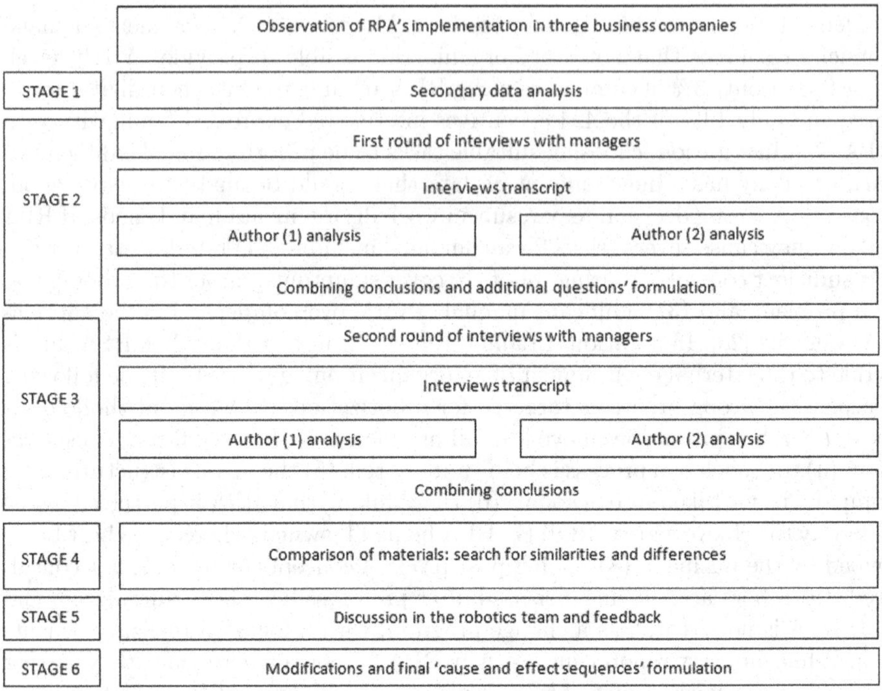

| STAGE 1 | Observation of RPA's implementation in three business companies |
|---|---|
| | Secondary data analysis |

| STAGE 2 | First round of interviews with managers |
|---|---|
| | Interviews transcript |
| | Author (1) analysis · Author (2) analysis |
| | Combining conclusions and additional questions' formulation |

| STAGE 3 | Second round of interviews with managers |
|---|---|
| | Interviews transcript |
| | Author (1) analysis · Author (2) analysis |
| | Combining conclusions |

| STAGE 4 | Comparison of materials: search for similarities and differences |
|---|---|
| STAGE 5 | Discussion in the robotics team and feedback |
| STAGE 6 | Modifications and final 'cause and effect sequences' formulation |

**Fig. 1.** Research process.

suggested by Taylor [28]. The common features of the companies which agreed to take part in the research were that they are (1) international, (2) business-oriented corporations, (3) employ over 250 people, and (4) RPA was used for at least 5 years. The primary difference was in their individual field of operations, including banking, IT-services, and production (cf. Table 1).

The second stage of the research started with the first round of interviews. It was conducted by one author of the paper with managers of the companies. The interviews were conducted in Poland; one took place in the headquarters

**Table 1.** Participants of the research

| Participant | Gender | Age | Company | Company profile | Tenure (years) | Interview duration (minutes) |
|---|---|---|---|---|---|---|
| A | Female | 36 | 1 | Banking | 10–15 | 47 |
| B | Male | 39 | 1 | Banking | 10–15 | 31 |
| C | Female | 34 | 1 | Banking | 5–7 | 25 |
| D | Male | 44 | 2 | IT-Service | 5–7 | 42 |
| E | Male | 42 | 3 | Production | 5–7 | 21 |
| F | Female | 31 | 3 | Production | 5–7 | 29 |

of the company and the other five outside of office space. The interviews were recorded with consent and lasted between 21 and 47 min. The format for the interviews was constructed according to the guidelines for unstructured questionnaires, which consisted of a set of research questions determined by the problem and situation in the field (cf. Table 2). This approach allowed for in-depth questions to be asked when new or unknown information appeared.

**Table 2.** Interview guide

| |
| --- |
| **RQ** What are the reasons for RPA withdrawal from the company? |
| - How did you organize the robot's withdrawal process? |
| - Why did you withdraw the robot? |
| - Who decided that the robot should be withdrawn and why? |
| - How did you organize your work after the robot was withdrawn? |

The transcripts of the interviews were made, the material was analyzed independently by two researchers to decrease the subjectivity of qualitative research [29], and conclusions were subsequently combined. That resulted in finding three types of narrations linked to 'cause and effect' sequences leading to re-manualization: a) overenthusiasm for RPA, b) low awareness and fear, and c) legal and supply changes. However, after the first round of interviews, other questions surfaced as some facets were not explained fully. Therefore, the topics needing more explanation were listed and a next round of interviews planned.

During the third stage, the second round of interviews was conducted, all of which took place outside the office, via Skype, Teams, or mobile phone. Again, transcripts of the interviews were made and analyzed independently. As a result, it was identified the fourth narration linked to 'cause and effect' sequences leading to re-manualization: d) robot failure.

During the fourth stage, a comparative analysis was conducted of all the empirically collected material. This allowed for the similarities and differences in the perception of re-manualization to be explored from the perspective of each interviewee. During the fifth stage, the conclusions were shared with the interviewees to reflect and discuss their accuracy. Ultimately, in the sixth stage, the feedback from interviewees was analyzed and included in the analysis. Eventually, all interviewees received the final results of the paper to assess if the anonymity conditions were kept properly.

## 4   Results

The outcome of the research process is discussed in this section to answer our Research Question: "What are the reasons for RPA withdrawal from a company?". The current study identified four cause-related narrations concerning why robots are withdrawn (also referred to as "retired" during the interviews).

More precisely, overenthusiasm for RPA in the company (cf. Sect. 4.1), low awareness and fear linked with robotics (cf. Sect. 4.2), legal and supply changes that
have an impact on particular processes or tasks in the company (cf. Sect. 4.3),
and robot's failure (cf. Sect. 4.4). Consequently, the cause and effect sequences
representing employees' experiences have been constructed.

Nonetheless, besides answering the RQ, the conducted study gave rise to
general findings regarding the complexity and the implications of process re-
manualization (cf. Sect. 4.5)

### 4.1   Cause 1: Overenthusiasm for RPA

Adopting innovation may help a company gain a competitive advantage over
the market, but it may expose business to trouble as well, especially when the
downsides and upsides of innovation are not fully considered:

> *Some people perceived RPA as such an exciting process that they did not think
> logically about its consequences. We have seen just the positive side of it [robot]. (B)*

The overenthusiasm of innovators and early adopters may blunt business
reality. Some managers do not carefully consider the value a robot is going to
bring, as well as the consequences of its introduction:

> *We did not think if it [task] is a good option for automation. I'm not sure
> if I was aware that something is a wrong option for automation at all. It [task
> description purposefully hidden] was something we did not like to do, so we
> decided to use robots. (F)*

Our interlocutor suggested that his team faced two consequences related
to overenthusiasm for robotics: (a) underestimation of robot costs and/or (b)
misunderstanding of robot capacities. For some employees, robotics is associated
with a total reduction of costs. Hence, managers may perceive the investment in
robots just as a one-time expense. However, its maintenance often generates an
unexpected expenditure:

> *We did not check how much it costs - at first, there was some money for robots
> because companies have money for innovations. They invest in a robot's creation,
> but they did not take into account how much money they will need to run a robot in
> the company. It was an extremely visible tendency [to not consider the total cost of
> robot] at the beginning - when the company started investing in robotics. (A)*

Some managers with low awareness about robotics may be so fascinated with
the promises of RPA implementation that they do not consider the specifics of
robotics and propose to automate tasks which robots should not perform:

> *We did not know what are the barriers for robots. I experienced that managers
> think that bots can be used for any task and process. But it is not really true. (F)*

Both underestimation of RPA costs and misunderstanding of RPA limitations
may lead to disappointment:

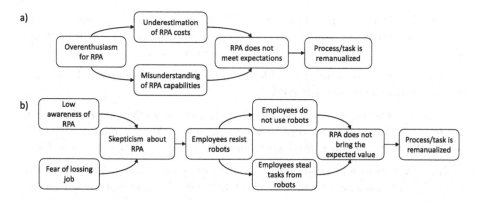

**Fig. 2.** Cause and effect sequence leading to re-manualization: a) Overenthusiasm for RPA. b) Low awareness and fear (source: own elaboration).

*Robot did not bring the value that we expected before its implementation and it caused frustration in the team. (D)*

The question about the value that people expect from RPA is of inordinate importance. We believe this value may be both objective as well as subjective. In the literature, no clear proposals on how managers should measure RPA value and classify robot success or failure was found. Even though it was not an intended subject of this research, we believe it could be a starting point for further investigation. As a result of the disappointment from unmet expectations for robot performance, the tasks were moved back to human processing:

*After implementation, however, they [managers] discovered that they couldn't spend so much money [on robots] and [tasks] had to be manualized. Robots could not bring the values everybody expected. (A)*

To conclude, overenthusiasm for robotics can lead to failure during implementation, especially if it is not supported by thorough knowledge of the robot's cost and capabilities. Ultimately, the company bears double costs, that of creating and retiring a bot. This cause-related narration, summarized in Fig. 2a), was mentioned by three companies.

## 4.2   Cause 2: Low Awareness and Fear

In collective imaginations, narrations related to threats posed by robots to humans are widely encountered [30,31]. One of our interlocutors claims that, among employees with low awareness of robotics, bots are perceived as their direct competitors:

*At the beginning [of RPA implementation] some people were terrified that robots would take their place. They knew nothing about robots but were really terrified and they were those ones the most skeptic about this idea [RPA]. (F)*

For some employees, the fear of RPA may derive from their belief that RPA is a complicated technology, reserved only for technically advanced users:

*Robots are really easy and even a child is able to learn working with them. But before they came to our company, some people associated robotics with math and physics that a person with humanistic background would not understand. (E)*

Both low awareness and fear of losing one's job may lead to employee skepticism and resistance towards RPA:

*Sometimes people's awareness about robots was low so they did not use them. We found that they did not want to use them [robots] because they expected that the more a robot produces, the bigger the chance they will lose their jobs. (A)*

High awareness and trust towards robots does not have to lead directly to lower skepticism towards them. There are 'pure skeptics' who oppose any change because of their values and/or previous experiences [32]. This approach was noticeable in the narrations provided by the interviewees:

*It happened that people just did not like robots and nothing could change their views. (B)*

The interviewees claimed that people who are reluctant to robots do not use them even if they are already implemented. Surprisingly, we found that they even take back the task from robots purposely:

*We experience that sometimes people who performed some task that was taken [by a robot], they still try to perform this task on their own, stealing the job from the robot. (A)*

Consequently, robots do not fulfill their function or are not used at their whole capacity. In such cases, double costs may be generated by employees who stay in the company and are assigned to perform a different task, but still take the work from robots. Ultimately, humans cannot focus on their new assignments and work subpar, while the robots generate maintenance costs and work subpar as well. As a result, the task returns to human operators and the robot is withdrawn:

*We decided to retire our robot after some time because people did not use them as we planned, and the robot generated costs we had to shoulder (C)*

This cause-related narration, represented in Fig. 2b), was mentioned by three companies.

### 4.3   Cause 3: Legal and Offering Changes

We found that the external and internal environment of an organization can have an impact on a company's decision to withdraw a robot. Our interlocutors experienced situations where a process or task had to be modified due to (1) new legal regulations on the market and (2) new products being introduced by the company. Both became triggers to finally withdraw a robot:

*Both systems and processes change. In our company a product was modified, and robot got outdated. We intended to rebuild it. However, it was not an easy story. (D)*

*It was the law introduced by the government that started the whole story with retirement. (A)*

Each human organization is an open system that adapts to its environment to survive [33]. Governments which act as an organization's stakeholder impose legal frames and borders which regulate how an organization fulfills its functions [34]. Companies adapting to their external environment need to modify their internal environment as well. As a result, tasks or processes must be adjusted to new realities:

*The task was changed so the robot was not valid anymore. We started to think how to modify it and we found some challenges in that process. (B)*

Adapting an existing robot to a new or modified task demands both money and knowledge. It was found that rebuilding a robot in accordance with new regulations may be too expensive for a company:

*The cost to rebuild the robot was too high and we decided to not cover it. So, we gave it [a task] to a human, which was cheaper because we did not have to pay to make another robot. (B)*

The other factor which prevents a company from adjusting existing robots to a modified task is the lack of capacity of people (or technicians) who prepared the prototype:

*They [people who constructed the robot] were not working anymore with us and the new ones did not know how robot was made. We had codes and maps but it's visible that they [the new robotic team] were not the robot's creator and preferred to do the new one that rebuild the existing one. But it [making a new robot] costs money and time. So, it was better to do it manually. (C)*

From this statement we concluded that people who created robots are perceived to have more capabilities or readiness to rebuild them. This assumption may be linked to the fact that people who design a product are more willing to modify it than abandon it [35]. This suggests that maintaining trustworthy members in robotic team could provide some benefits for the company as they may be more willing to modify an existing robot than new machine. To conclude, we found that if a company does not have people who are able to rebuild a robot or managers perceive the cost of a new robot as too high, it may lead to re-manualization. This cause-related narration, depicted in Fig. 3a), was mentioned by two companies.

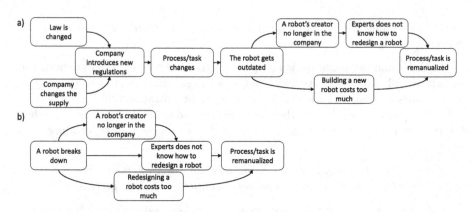

**Fig. 3.** Cause and effect sequence leading to re-manualization: a) New law and new supply. b) Robot failure (source: own elaboration).

### 4.4    Cause 4: Robot Failure

The last of the reasons for process re-manualization identified in this research relates to situations when a robot stops working as expected or breaks down:

*With time, our robot lost its functionality and we decided that it is cheaper to retire it. (A)*

*All started [manualization] when one day robot reported an error and we were not able to revive it. (F)*

A robot may work for a company for a long time and suddenly loses its functionality. It may be related to human-related or coding errors, as well as system hacking. In the review of literature, no report was found which summarized the most frequent reasons for robot accidents. Our interlocutors experienced a situation when a robot failed and there was no one in the company cable of repairing it, and/or the cost of such repair was perceived as too high:

*It was impossible to repair the robots by a person who was not building it. And X [name anonymized] who was building it, was no longer working in the company anymore. (E)*

The other situation experienced by the interviewees was that experts capable of rebuilding the robot were ready to do it, but the cost of repair was perceived as too high:

*The scale of error was enormous, and it cost more money than to build a new one or manualize the task/process (F).*

As a consequence, the robot was withdrawn, and the task returned to manual processing. This cause-related narration was mentioned by two companies and can be seen in Fig. 3b).

## 4.5    Implications of Re-Manualization

It is important to note that is is complicated to precisely fix the scale of robot withdrawal from companies or/and managers who were not willing to disclose such information, perceiving it as internal taboo:

*It's hard to say how many robots [are withdrawn], we do not calculate it. It was our defeat, but we learned this lesson. (C)*

Just one of our interviewees (A) estimated that about 10% of all robots are withdrawn. However, it was not our intention to investigate in which tasks RPA fails or how much time passes from robot implementation to withdrawal. We found, however, that this phenomenon is widespread enough to have its own name. The interviewees used collocations 'process re-manualization' (person D) and 'robot's retirement'(person A, B, C, E) to name a situation when a robot gets replaced by human. The narrations identified during the research suggest that RPA's withdrawal may be linked to both loss of control and financial risks, especially if such actions were not foreseen by managers during the robot's implementation phase:

*In consequence, we were completely lost and did not known what to do in that case. (D)*

When a robot gets withdrawn, a company may lose the money that was invested to build, test, and deploy the robot, in addition to the resources needed to train employees developing or configuring the code. The companies of the interviewees used two approaches for a robot's withdrawal, either replacing the retiring robot with (1) a new robot, or (2) with a human operator. In the second scenario, a process or task is re-manualized and comes back to human operators who start processing it manually again. We found that re-manualization process may bring about concern:

*It's not easy to manualize the task, as it's not easy to understand a robot. You should know how the task was made by humans before the robot was implemented. However, the people who did it may not work with us anymore. So, we need to map the process and design it from the beginning. (A)*

It would also be interesting to study good practices when implementing RPA that might later be useful to guide companies in situations of re-manualization.

## 5    Discussion and Limitations

The intention of this paper was to explore the logics of unsuccessful RPA implementation, resulting in 're-manualization'. It was not an intention of this research, however, to investigate the quantitative data linked to these phenomena, nor to make general conclusions about RPA or robotic process re-manualization, but rather to signal that such a phenomenon as robot withdrawal exists and needs further investigation.

This paper contributes to both research and practice in three areas. First, it addresses phenomena which had not been described in the literature before. It was demonstrated that people may not be aware that robots are not a universal technology for any task and team. This lack of awareness may pose risks and even be harmful for organizations. To avoid this, training for employees presenting both downsides and upsides of RPA should be provided. Robotics may offer benefits to the company only if it is applied according to proved and objective methodologies. Secondly, the perceptions depicted in our study suggest that robot withdrawal is assessed by employees in terms of failure. It was visible that our interviewees felt uncomfortable talking about task re-manualization and referred to it as something embarrassing. We believe this was mainly due to the fact that robot withdrawal was an unexpected event that a team had to face unprepared. We discovered that robot withdrawal may be caused by errors as well as change coming from the environment. Even if an organization cannot influence the environmental changes, it may prepare itself for them by fixing procedures and rules concerning eventual re-manualization process. Thirdly, we found that robot redesign may be impeded due to the human fluctuations in the robotic team. People who had not constructed the robot's prototype may not know how to repair, redesign or simmply maintain it. To avoid robot withdrawal, managers should elaborate strategies that will ensure 'knowledge continuance' in their robotic teams. The maps of processes and tasks should be made before they are robotized. It may help to re-manualize the task based on precise data even if people who mapped the processes are no longer working for the company. We believe that there are many potential strategies and further research is needed. Undoubtedly, the RPA's scope within the company, internal and external environmental factors, as well as a task's specifications will play an important role.

All cause-related sequences provided in our paper are in line with findings provided by Lamberton et al., presented in the research background [26]. We found, however, that RPA withdrawal may be caused not only by people who adopt this technology, but by environmental factors as well. Therefore, the locus of risk should be considered when crisis plans are constructed by teams.

The results obtained in this study are subject to certain limitation and threats to validity. First, the results reflect the experiences of three companies. According to social science methodology, there are no premises to generalize our conclusions toward other companies, where such phenomena may not occur or be perceived differently. Nonetheless, these initial results will help to create some hypothesis to be validated in broader studies. In addition, the direct observation of this phenomenon faces several constraints. First, it is a challenge for employees to foresee which robot will be withdrawn. Hence, it is also a challenge for researchers to capture and describe the exact moment when decisions are made by the team when a robot is being withdrawn. Retrospective, post-factum studies are always linked to the recall limitation as there is always a chance that research participants may not recall all the details of an incident they experienced. Future studies are encouraged to try to depict ethnographically the process re-manualization just in time it occurs. Secondly, the access to data linked to robot implementations is limited due to internal policies and legal restrictions throughout many organizations. Signing confidentiality agreements regulating which data may be disclosed was a substantial limitation of this work. Thirdly, robot withdrawal is perceived by employees as an emergency situation. According to Coombs [36], research to be conducted on phenomena perceived by employees to be an emergency is particularly challenging, as interviewees are less willing to share data and spend time on consultations with researchers when they feel they must work under pressure. What is more, companies may wish not to disclose situations in which they did not succeed. We are aware that the narrations, which became the base for the cause and effect sequences leading to re-manualization, are not the objective constructs. They were made according to subjective experiences of employees who faced such problems in the past. This subjectivism juxtaposing of the narrations of independent managers from various companies, as well as analyzing interviews independently from each other, was a challenge in this research. Nonetheless, to mitigate this threat, the material presented in the paper is based on secondary observations and reflects the experiences of certain groups of employees.

## 6    Conclusions and Future Work

This paper presents an exploratory study to shred light to the end-of-life phase on RPA. Three Polish companies agreed to participate in this study. As a result four main cause-related sequences of events to explain the process re-manualization were identified. Furthermore, an extensive discussion of the factual contribution of this study to industry and academia is included.

Nonetheless, in order to address the limitations identified in Sect. 5, some future works are planned. First, to work on a common model that includes the different cause and effect sequences to help managers identify the main roots of 'failure'. Secondly, to generalize the results, we plan to replicate this study including a broader set of companies from different countries so that the previous model can be validated or updated. Lastly, an additional study is planned to complete the current results with guidelines for preventing these situations or address them when they happen.

## References

1. Syed, R., et al.: Robotic process automation: contemporary themes and challenges. Comput. Ind. **115**, 1–15 (2020)
2. Aguirre, S., Rodriguez, A.: Automation of a business process using robotic process automation (RPA): a case study. In: Figueroa-García, J.C., López-Santana, E.R., Villa-Ramírez, J.L., Ferro-Escobar, R. (eds.) WEA 2017. CCIS, vol. 742, pp. 65–71. Springer, Cham (2017). https://doi.org/10.1007/978-3-319-66963-2_7
3. Lacity, M., Khan, S., Carmel, E.: Employing U.S. military families to provide business process outsourcing services: a case study of impact sourcing and reshoring. Commun. Assoc. Inf. Syst. **39**, 150–175 (2016)

4. Richards, G., Yeoh, W., Chong, A.Y.L., Popovič, A.: Business intelligence effectiveness and corporate performance management: an empirical analysis. J. Comput. Inf. Syst. **59**, 188–196 (2019)
5. Sun, H., Ha, W., Teh, P.L., Huang, J.: A case study on implementing modularity in software development. J. Comput. Inf. Syst. **57**, 130–138 (2017)
6. Choi, J., Nazareth, D.L., Ngo-Ye, T.L.: The effect of innovation characteristics on cloud computing diffusion. J. Comput. Inf. Syst. **58**, 325–333 (2018)
7. Stople, A., Steinsund, H., Iden, J.: Lightweight it and the it function: experiences from robotic process automation in a norwegian bank. Bibsys Open J. Syst. **25**, 1–11 (2017)
8. Kedziora, D., Kiviranta, H.M.: Digital business value creation with robotic process automation (RPA) in northern and central Europe. Management **13**, 161–174 (2018)
9. Hallikainen, P., Bekkhus, R., Pan, S.L.: How opuscapita used internal RPA capabilities to offer services to clients. MIS Q. Exec. **17**, 41–52 (2018)
10. Ratia, M., Myllärniemi, J., Helander, N.: Robotic process automation - creating value by digitalizing work in the private healthcare?, vol. 18, pp. 222–227 (2018)
11. Anagnoste, S.: Robotic automation process - the next major revolution in terms of back office operations improvement. In: Proceedings of the International Conference on Business Excellence, vol. 11, pp. 676–686 (2017)
12. Fung, H.P.: Criteria, use cases and effects of information technology process automation (ITPA). Adv. Robot. Autom. **3** (2013)
13. Mendling, J., Decker, G., Reijers, H.A., Hull, R., Weber, I.: How do machine learning, robotic process automation, and blockchains affect the human factor in business process management? Commun. Assoc. Inf. Syst. **43**, 297–320 (2018)
14. Clair, C.L.: Building a center of expertise to support robotic automation preparing for the life cycle of business change. Technical report, Forrester (2014)
15. Tsaih, R.H., Hsu, C.C.: Artificial intelligence in smart tourism: a conceptual framework, vol. 89, pp. 2–6 (2018)
16. Czarnecki, C., Auth, G.: Prozessdigitalisierung durch robotic process automation (2018)
17. Holder, C., Khurana, V., Hook, J., Bacon, G., Day, R.: Robotics and law: key legal and regulatory implications of the robotics age (Part II of II). Comput. Law Secur. Rev. **32**, 557–576 (2016)
18. Schatsky, D., Muraskin, C.: Robotic process automation. a path to the cognitive enterprise. Technical report, Deloitte University Press (2016)
19. Kanellou, A., Spathis, C.: Accounting benefits and satisfaction in an ERP environment. Int. J. Account. Inf. Syst. **14**, 209–234 (2013)
20. EverestResearchGroup: Robotic process automation (RPA) annual report 2018 - creating business value in a digital-first world. Technical report, ERG (2016)
21. Report: artificial intelligence is winning more hearts & minds in the workplace. Technical report, Oracle and Future Workplace (2019)
22. IT-CentralStation: Why do RPA project fail? And how to avoid it? Paper peer report. Technical report, ITCS (2019)
23. Wright, D., Witherick, D., Gordeeva, M.: The robots are ready. Are you? Untapped advantage in your digital workforce. Technical report, Deloitte (2017)
24. Leopold, H., van der Aa, H., Reijers, H.A.: Identifying candidate tasks for robotic process automation in textual process descriptions, vol. 318, pp. 67–81 (2018)
25. Kirchmer, M.: Robotic process automation - pragmatic solution or dangerous illusion. Technical report, University of Pennsylvania (2017)

26. Lamberton, C., Brigo, D., Hoy, D.: Impact of robotics, RPA and AI on the insurance industry: challenges and opportunities. J. Financ. Perspect. **4**, 8–20 (2017)
27. Czarniawska, B.: Social Science Research, From Field to Desk (2014)
28. Taylor, S., Land, C.: Organizational anonymity and the negotiation of research access. Qual. Res. Organ. Manag. **9**, 98–109 (2014)
29. Burnard, P.: A method of analysing interview transcripts in qualitative research. Nurse Educ. Today **11**, 461–466 (1991)
30. Jones, S.E., Park, R.L.: Against technology: from the luddites to neo-luddism. Phys. Today **60**, 59 (2007)
31. Baum, S.D.: Superintelligence skepticism as a political tool. Information **9**, 1–16 (2018)
32. Lennerfors, T.T., Fors, P., van Rooijen, J.: ICT and environmental sustainability in a changing society: the view of ecological world systems theory. Inf. Technol. People **28**, 758–774 (2015)
33. Scott, W.R., Davis, G.F.: Organizations and organizing: rational, natural and open systems perspectives (2015)
34. Amaeshi, K.M., Crane, A.: Stakeholder engagement: a mechanism for sustainable aviation. Corp. Soc. Responsib. Environ. Manag. **13**, 245–260 (2006)
35. Norton, M.I., Mochon, D., Ariely, D.: The IKEA effect: when labor leads to love. J. Consum. Psychol. **22**, 453–460 (2012)
36. Coombs, W.T., Holladay, S.J.: The Handbook of Crisis Communication (2010)

# Steering the Robots: An Investigation of IT Governance Models for Lightweight IT and Robotic Process Automation

Vincent Borghoff[✉] ⓘ and Ralf Plattfaut ⓘ

South Westphalia University of Applied Sciences, Soest, Germany
{borghoff.vincent,plattfaut.ralf}@fh-swf.de

**Abstract.** Robotic Process Automation opens up new possibilities for organizations to automate processes that were previously not worth automating for technical or financial reasons. Furthermore, the fact that RPA does not require deep IT skills allows for the development within business departments. However, RPA, and lightweight IT in general, not only creates new opportunities, but also poses new challenges for organizations, especially for IT Governance. Based on existing IT Governance frameworks, we investigate the design of RPA-specific Governance models in practice and elaborate respective advantages and disadvantages. Based on this, we identify organizational internal context factors that influence the implementation mechanisms of a lightweight-specific IT Governance. We build our research on a qualitative approach based on in-depth interviews with practitioners, consultants, and a vendor representative.

**Keywords:** Lightweight IT · Robotic Process Automation · RPA · IT Governance

## 1 Introduction

Robotic Process Automation (RPA) is a relatively new information technology (IT). However, the interest in RPA still rises [1]. This rise is, among others, reflected by the initial public offering of "UIPath", one of the leading providers of RPA technology, closing with a 23% increase on market closure [2]. RPA is adopted by a broad range of organizations, ranging from large corporations to small and medium-sized enterprises [3].

RPA can be discussed using the academic stream of lightweight IT [4], as no or at least no intense coding and IT skills are needed to develop and implement the respective bots, concentrating on front-end solutions [5]. As such, RPA allows the business departments to take automation, at least partly, into their own hands to increase efficiency [4, 6].

With the adoption of RPA solutions, organizations require steering the application and scope of the running bots to maintain control and avoid the rise of a growing body of shadow IT [7]. While RPA, as part of IT, might fall under the traditional regime, Willcocks et al. [8] state that RPA should not be governed with approaches similar to those applied to legacy Enterprise Resource Planning Systems (ERP). The authors base their claim on

© Springer Nature Switzerland AG 2022
A. Marrella et al. (Eds.): BPM 2022, LNBIP 459, pp. 170–184, 2022.
https://doi.org/10.1007/978-3-031-16168-1_11

the RPA-inherent dynamics, implying a specific governance mindset and methodology. Other than heavyweight systems, lightweight systems focus more on innovation than stability [4]. With this finding, the traditional IT Governance approaches become obsolete regarding their applicability for RPA, and the work on factors influencing the design of IT Governance needs to be questioned.

While the Governance of heavyweight systems, like ERP solutions, was the subject of various publications and studies, the contributions concerning the governance of lightweight systems are scarce [9]. The framework proposed by Bygstad and Iden [10] depicts the only holistic approach. While this framework reflects appropriate variants of governance models for lightweight IT, it only relies on two dimensions, determining the application of one specific governance model. A clear analysis of contingencies determining the design of an appropriate IT Governance is lacking.

With our paper, we set out to close this gap by substantiating the understanding of the fit of IT Governance models for lightweight IT – using the example of RPA. Therefore, we will answer the following research questions:

RQ1: What archetypes of RPA-specific IT Governance can be found in practice?
RQ2: Which factors determine the design of an RPA-specific IT Governance?

To answer these research questions, we conducted a qualitative multi-perspective study, relying on in-depth interviews with 11 knowledgeable experts stemming from three different domains (i.e., RPA customers, consultants, and vendors). We tested the fit of the found RPA IT Governance approaches against established governance theories in consideration of the peculiarities of RPA and its locus between business and IT.

The remainder of this paper is structured as follows. In the following section, we present the theoretical background we build our research on. Subsequently, we explain the chosen method and present details on our data sample, while presenting the research findings. We then discuss the findings in light of the introduced theoretical background to conclude our paper with a synopsis, highlighting our contribution both to theory and practice, as well as the limitations of our approach.

## 2 Background

### 2.1 IT Governance

While governance, in general, is concerned with the control over certain parts of an organization to align with a given strategy, IT Governance includes mechanisms and structures specifically for IT-related decisions and processes [11, 12].

Multiple models have been proposed in literature describing IT Governance approaches which are, to a high degree, built around the design of structural governance mechanisms [e.g., 13, 14]. Structural mechanisms reflect organizational units as well as roles and responsibilities for decision-making [14]. These models primarily address the issue of centralization of IT decisions and authority (e.g., central, federal, decentral) [15]. While in a central IT Governance, all decision-making authority is centralized in one organizational unit, in a decentral IT Governance, the authority is distributed among individual operating units [16, 17]. Federal IT Governance approaches divide the authority between multiple instances. The division can be based on responsibilities for specific systems, for example, or on organizational grounds. In practice, both concepts are often applied simultaneously. While a central unit sets certain framework conditions that regulate interaction with core systems, for example, freedom is granted for internal departmental systems [11].

Research also identified contingency factors that determine the adoption of one specific IT Governance approach, following the assumption that the governance differs according to internal and external factors [17–20].

However, the described IT Governance approaches were initially developed against a heavyweight IT background. Therefore, they might not incorporate the specific peculiarities of RPA and lightweight IT in general (see below). For this reason, academia calls for a revised IT Governance, internalizing specific concepts for lightweight IT [21].

## 2.2  RPA as an Example of Lightweight IT

The concept of lightweight IT is based on the growing trend of IT consumerization [4]. The trend of using consumer-oriented IT artifacts within professional contexts affects the relationship between the IT departments and the business as users of the respective systems [6]. Bygstad [4], therefore, sees lightweight IT more as a new knowledge regime, shifting the focus to business-driven exploration and innovation [4]. In contrast, he describes traditional systems (e.g., ERP systems) as heavyweight IT. Heavyweight IT has a clear focus on security and stability and is driven by IT professionals, mostly centralized within IT departments. On the contrary, lightweight IT is mostly driven and owned by the business side, with a clear frontend focus and a rather non-invasive nature [22].

RPA is a form of software, in which bots mimic user action to automate business processes [21]. Thereby the focus lies on repetitive, rules-based and frequent processes [5]. The developed bots are able to work on the user interface and act as human workers by mimicking their actions but with significantly higher efficiency [23].

RPA is easy to implement (i.e., it requires low software development know-how) and considered minimal-invasive (i.e., it requires no change on underlying core systems). Authors agree that non-programmers can conduct the majority of RPA development, e.g., business specialists [21, 24, 25]. The main reasons for this include the configuration of RPA bots via drag-and-drop interfaces [21] and the integration with other software over the graphical user interface [23, 24, 26].

RPA is, as lightweight IT is in general, often business-led. Authors have observed that RPA is introduced in single business units or departments, often with the help of external consultants [21, 25]. As such, RPA can be conceptualized as being business-driven or business-led, focusing on local innovations [4, 21].

The application of RPA in organizations typically follows a distinct sequence, starting with a pilot phase, followed by a scaling phase and concluding in an RPA maturity. Likewise, specific RPA projects can be described with a typical phase approach, starting with the idea phase in which candidate processes are identified, the definition phase in which the processes are analyzed, and requirements for the bot are defined, and the implementation phase, in which the actual development takes place and the bot is put into production [9, 21, 25].

Building on these observations, RPA is located between business and IT departments and needs skill and support from both sides. This split leads to multiple governance questions, e.g., "how to control RPA [bots] and avoid security, compliance, and economic risks" [23]. As Lacity and Willcocks argue, "RPA must still be consistent with IT Governance" [21]. Plattfaut strengthened this argument and calls for "joint IT and business commitment" [25]. In a literature review on RPA, Syed et al. [9] argue that the systematic design, development, and evolution of bots, together with seamless handling of exceptions and proactive monitoring and control, are challenges contemporary RPA research needs to cover. The governance of RPA can be one part of the answer to these challenges.

### 2.3  Implications for IT Governance

Both academia and practice claim that the development of RPA-specific IT Governance approaches is one of the most pressing future issues [27]. Willcocks et al. [8] describe specific challenges for the IT department induced through RPA and the peculiarities RPA has as a lightweight process automation technology. Three of these challenges directly affect the design of an IT Governance [28].

(1) As RPA is non-invasive, it is possible to implement RPA-based automation solutions without the involvement of the IT department, making the technology appealing to multiple business units [28]. The potentially multiple arising implementations then may lead to a rising body of shadow IT, confronting the IT with an increasing effort related to maintaining the stability and compliance of the core systems [7]. With this often business-driven nature of RPA, the decision about who is in the lead of the running and future developed bots becomes urgent. (2) As RPA is often triggered by business problems grounded in perceived deficiencies within an organization's IT infrastructure, the lead may also be set within the responsible IT department [8]. RPA is closely related to occurring business needs and a business inherent process view. (3) This makes respective business-related skills, potentially unavailable in traditional IT departments, necessary for successfully developing RPA-based automation solutions [8].

As RPA can be subsumed under the umbrella of lightweight IT, governance approaches from this field might be appropriate to be applied to respective solutions. Bygstad and Iden [10] proposed a framework for the governance of lightweight IT, consisting of 4 distinct models structured along two variables reflecting innovation and systems stability. The framework orients on three structural IT governance models, ranging from a completely unsupervised implementation and operation to harsh control (laissez-faire, platform, bimodal and central control). These models can be handled as an expansion of the traditional central, federal, and decentralized IT Governance forms [28]. The degree of centrality is thereby defined as the extent of decision-making and control authority the IT has regarding business-driven lightweight IT. Other than traditional IT Governance frameworks, the model of Bystad and Iden determines the application of one of the four models on a combination of just two variables wanted security and innovation. Other factors or contingencies are not in the scope of the approach. Practically oriented literature in particular, tries to compensate but usually concentrates on partial aspects. While Schmitz emphasizes the importance of including governance early in the planning of RPA solutions [29], Noppen et al. focus primarily on the design of the subsequent maintenance of the implemented bots [30]. Asatiani et al. highlight the difficulties associated with federated governance approaches in RPA [28].

Our research aims to close this gap by identifying lightweight IT-specific contingencies, influencing the design of an appropriate lightweight IT Governance.

## 3    Method

To answer our research questions, we chose an explorative qualitative research approach based on grounded theory [31]. Our approach can be structured into the phases of data collection and analysis. For data collection, we conducted semi-structured in-depth interviews with 11 knowledgeable experts in RPA, which offer distinct perspectives on RPA and lightweight IT in general (see Table 1). We selected interviewees with explicit professional backgrounds in RPA-based and heavyweight automation. We focused on identifying promising interview partners for achieving information-rich cases for later analysis [32]. Here, we aimed at achieving maximum sample variation [33] along the dimensions of domains, sector, and size of the organization, and the role of the interviewee as follows: Since we have tried to reflect the current practice of RPA governance as precisely as possible, we analyzed the used governance structures from three distinct perspectives. These are achieved by selecting interview partners from different domains. Firstly, we directly chose practitioners who are in actual touch with RPA technology by operating, developing, or managing RPA-based automation solutions in their respective organizations. Secondly, we included a consulting perspective in our sample. This decision is mostly grounded in the finding that organizations' RPA solutions are often either introduced to or later scaled up by consultancies [3]. Thirdly, we also interviewed a representative of an RPA technology vendor. Such vendors guide their customers with a multitude of training resources as well as with direct support and advice for implementation and operation. This often includes specific bot operation models, which, among other topics, shed light on structural governance characteristics. To enrich variation within our data sample, we selected experts from organizations differing in size, represented by the number of employees of the specific organization and organizational sector. We additionally aimed at also achieving a broad representation in terms of the role of the individual interviewees within the selected organizations.

For analyzing the collected data, we applied a grounded theory approach with open coding [34]. We coded for the perceived role of governance for the success of RPA usage within both the specific cases and from a consulting and vendor-centric perspective. Within the practitioner interviews, we additionally coded for the general maturity in automation, not only with an RPA focus but also including heavyweight automation solutions to depict the total experience with automation technologies in order to be able to better classify and contrast particularities of lightweight automation. In a second layer, we coded specifically for the advantages and disadvantages of the specific IT Governance approaches found within the cases. All coding was conducted by both researchers individually. In case of conflicting codes, the respective interviews were reviewed and taken into a discussion until consensus was reached. Moreover, we applied axial coding to identify and aggregate categories within our coding structure [30].

**Table 1.** List of interviewed RPA experts

| # | Role of interviewee | Organizational sector | Domain | Employees |
|---|---|---|---|---|
| I1 | Division Manager IT | Statutory Health Insurance | Practice | 3.900 |
| I2 | Head of Automation | Manufacturing Industry | Practice | 36.000 |
| I3 | Automation Lead | Building Material Producer | Practice | 55.000 |
| I4 | Project Manager | Industrial Service Provider | Practice | 34.000 |
| I5 | IT Engineer | Chemical Industry | Practice | 14.000 |
| I6 | CEO | RPA Training and Consulting | Practice/Consulting | 14 |
| I7 | Consultant | Strategic Consulting | Consulting | 30.000 |
| I8 | Senior Consultant | Strategic Consulting | Consulting | 30.000 |
| I9 | Technical Consultant | Strategic IT Consulting | Consulting | 150 |
| I10 | Consultant | Strategic IT Consulting | Consulting | 150 |
| I11 | Business Analyst | RPA Vendor | Vendor | 3.000 |

## 4 Findings

This section presents findings for the identified RPA IT Governance models applied in practice by describing the actual cases, reasons for choosing one specific model, and the internal design and function. In only one of the cases, the chosen RPA IT Governance approach stayed the same over time. The IT Governance approach develops in design and degree of centricity with increasing RPA and automation maturity. Figure 1 depicts the observed developments.

Within our data sample, we found evidence for distinct designs of an RPA governance. **Central governance models** are often applied within organizations that aim to keep systems stable. This can be grounded in a latent fear of rising shadow IT, which implies a harsher control over all artifacts implemented within the overall systems landscape. One interviewee stated that they prevent shadow IT by simply not allowing business-driven system implementation but forcing a completely centric IT approach (I2). Therefore, one digitization center of excellence (CoE) is installed, which develops and operates all lightweight and heavyweight systems and includes a distinct team concerned with identifying potential automation cases. In this case, the governance is designed very top-down to embed the RPA-related control guidelines in the respective higher level of authority. While this allows tightly meshed and comprehensive supervision of all automation activities, it hinders the business from utilizing the main advantages of lightweight solutions, namely velocity and innovativeness: *"Of course, there are also some drawbacks because there might be some bottlenecks in terms of prioritization of the processes, and also identifying the suitable processes for RPA because you will have multiple departments and you do not know exactly where to start, how to prioritize them and another issue might be related to slower RPA deployments. It is just about the focus you set"* (I11). While central control can sustain security, decentral approaches can therefore foster innovation. However, the organization values long-term control of their systems landscape higher than implementation speed, acknowledging that the missing automation experience within the business units also influenced the decision for a central mode.

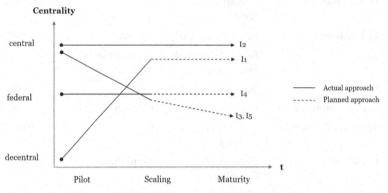

**Fig. 1.** Development of the RPA governance within the cases over time and increasing maturity

One of the interviewees stated that his organization started RPA-based automation with a **decentral governance** approach, keeping all authority within the business (I1). According to the interviewees with a consulting background, this procedure is typical for cases where RPA is newly introduced through external consulting. Intending to increase efficiency, the organization was striving to digitize. For this reason, they commissioned a consulting firm to identify potentials within their processes and structures. In this term, RPA was part of the overall digitization toolkit. While the organization is still relatively new to RPA, they are planning to broaden the use of the technology. Within this process, compliance was getting more important. As statutory health insurance, the organization handles sensitive data, which implies stricter internal and legal regulations. A growing shadow IT would increase the risk of fraud and malfunction, leading the organization to centralize its governance model: *"I advocate a completely centralized unit. That makes it easier because, of course, as a health insurance company, we have to take the issue of co-determination into account very intensively. From an economical perspective, it is easier to have it [governance ed.] in one central place and then also to advance the issue with certain rules than if each department does it. Because then you lose the overview"* (I1). In the same vein, the growing complexity, reflected in multiple RPA instantiations running, influences the choice to refrain from decentral models. This also reflects a typical process for externally induced RPA automation, as one consultant stated: *"If they develop it further, if they touch all the processes in the business function* *that you can automate, that actually becomes centralized again, so to speak, that's no longer decentralized"* (I7). This organization started with a very decentralized, business-driven approach when piloting RPA and is now shifting to a complete centralization of governance structures.

Three interviewed practitioners describe their organizations applying a federal governance for RPA (I3, I4, I5) but with different internal arrangements. These arrangements differ on the one hand in the lead responsibility areas of the business and the IT, and on the other in the processual IT Governance setup. All approaches have in common that parts of the responsibilities are distributed to individual business units or functions besides a central control unit. Two of the three interviewee organizations (I3, I4) apply a model in which, besides a central CoE, responsible employees are enabled to act as RPA champions, either for business units or in specific global business functions, enabling the business to take the initializing lead: *"If we now say we have a purchasing RPA champion who then wants to take care of every purchase process globally, then it's fine for us, but we also accept if someone now says okay I come from company XY I only want to do this for the company, then it's also fine"* (I4). In one case, automation projects are initiated both by the CoE and the business: *"We also have the so-called champions concept with us, because we also give countries the opportunity to automate themselves, to become active themselves and some also use that in the meantime"* (I3). The other case (I5) involves direct but non-structural enablement within the departments. This more decentralized approach should leave RPA structures' specifications to the business. Interviewees from the consulting domain also emphasized that when choosing a federal governance model, the business units must have the respective capabilities (I6, I8, I10). The choice for a federal model was initially influenced by the structural diversity the respective organizations show. This diversity is grounded in international divisions and the resulting cultural differences: *"Above all, we see the point of covering different cultures through the ambassadors. When I talk to a Chinese colleague about a process, he understands it differently than I do"* (I5). The regional factor can also manifest in an organizational structure with a magnitude of separate subsidiaries, all with specific processual and application needs: *"I think in one country alone we have three hundred plants. You can calculate how many plants we have globally if we are active in 60 countries around the world. Then we have something like shared service centers or similar aggregations in each region - everything is very decentralized"* (I3). Moreover, in all cases, the choice was influenced by the focus on innovation and velocity to solve occurring business problems: *"The problem with the core systems is that they are expensive, and resources must first be created in the company's own IT. Often there are no interfaces between systems and yes, then RPA can quickly provide relief"* (I4).

For the case of the building material producer, the interviewee reported that for piloting of RPA, they stated by choosing a very centralized governance approach, concentrating all responsibilities and decision making in the leading CoE, located in one shared service center, globally initiating all RPA developments (I3). At the same time, they started with the training of RPA champions across several decentral business units to develop towards a more federal governance model, in which RPA initiatives are initiated and defined decentral but developed and implemented centrally by specific RPA experts. In this case, the business side would take the lead in the running RPA initiative.

With this shift, they adapted their initial governance model in the direction of a federal model. This development is the same for the case of an organization originating from the chemical industry (I5). In this case, the central CoE enables the business units to initiate and implement RPA themselves: *"We have an ambassador there. These are the extended arm to the regions. They are authorized to develop processes"* (I5). We could observe a similar approach for an industrial service provider, with the difference of not specifically enabling business units but also local and global business functions (I4). The interviewee reported that keeping the definition phase centrally helps in ensuring scalability: *"On the topic of scalability, we will then also make sure that the bots are programmed in such a way that they can be implemented not only in the US but also in India and that we can all scale accordingly"* (I4). Another facet of scaling is the CoE becoming a bottleneck when global decentral RPA initiatives increase in number and scope: *"Exactly, so currently that is purely centralized, purely implemented from the CoE. But we want to move away from this centralized approach and move towards a decentralized approach because we want to avoid making the CoE the bottleneck"* (I4). The decision at this point would be either to enlarge the CoE capacities, which would be cost intense or to switch to a more decentral approach. This means that central governance structures might be able to foster the scaling of RPA-based automation when combined with a decentral project lead.

All three cases (I3, I4, I5) have in common that RPA projects can be initiated decentrally. All three organizations have designed a specific process for initiating an automation project, which includes defining clear criteria for the applicability of RPA and setting technical and processual standards in case the bots are also developed decentral. To push through these standards, the designed process includes a reporting option. This includes that business employees can propose candidate processes that are analyzed together with the automation CoE. Both sides can directly assess the profitability of the potential automation as well as the fit of respective automation technologies, guided by a support system. This procedural approach is also backed by the consulting perspective, which additionally emphasizes the coordination function of the CoE between decentral RPA initiatives and the IT: *"I can well imagine going back to decentralization in the implementation and saying okay, then we have discussed it centrally and decided strategically that it will first be an RPA project in the implementation mode and then it will be implemented across the board. What I believe we still need to keep together centrally is which IT systems are affected"* (I10).

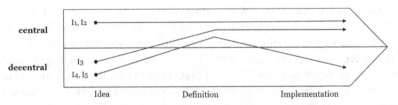

**Fig. 2.** Identified processual IT Governance mechanisms

Five cases show specific process IT Governance mechanisms. These reflect the responsibility within a certain RPA initiative in the course of the different phases of an RPA project (see Fig. 2). We can observe three distinct operational manifestations. The first, applied in cases I1 and I2, centralizes all phases of the RPA project. Automation is initiated, developed, and implemented centrally. The second approach lets the RPA project be initiated decentrally within individual business units. The definition and implementation of the project, including process analysis and profitability check, are done by or by standards from a central unit (I3). The third approach, applied by I4 and I5, works similarly but gives the responsibility for RPA implementation back to the respective business unit or the respective decentral CoE spoke. According to one of the interviewees stemming from a strategical IT consulting, this approach would best fit the initial idea of RPA: *"So the goal of RPA is to relieve the IT or to save scarce IT resources, in the sense of giving responsibilities to the business departments. That's where it originates, or where it wants to originate"* (I9).

## 5  Discussion

RPA is most often subject to specific governance approaches within practice, differing from the core IT Governance. These approaches use different mechanisms and models to steer the use of RPA. Our study identified different organization internal context factors that determine the design of an RPA governance by affecting the applicable governance mechanisms. While some of the factors are covered within established governance frameworks, we identified three factors specific to RPA. The identified factors show distinct dimensions in which the internal context of organizations can vary. Table 2 shows a brief summary.

**Table 2.** Summary of the proposed factors and their dimensions

| Factor | Dimension | Source |
|---|---|---|
| Lead | Business lead – IT lead | Lightweight specific |
| Automation Maturity | Mature - Immature | Lightweight specific |
| System Complexity | Complex - Simple | Lightweight specific |
| Structural Diversity | Heterogeneous – Homogenous | [16, 20, 35] |
| Focus | Security - Innovation | [10, 20, 36, 37] |
| Scale | Local - Global | [20] |

**Lead:** With RPA being low-code, the business is getting able to develop bots without or with little involvement of the IT [21, 24, 25]. This enables the business side to completely take over the lead, while IT needs to conduct the development and maintenance with heavyweight automation. Therefore, the possibility of business-led automation structures and a shift of responsibility arises. Fürstenau et al. [38] describe different approaches for handling user or business-introduced systems, broadly distinguishing between leaving the responsibility within the business or assigning the responsibility to the IT department but reflecting more about shadow IT in general.

**Automation Maturity:** This factor reflects the experience and knowledge of automation technology and RPA within the business units [39]. A decentral RPA approach requires technical knowledge within the business. The maturity can be fostered by enabling business units by change and knowledge management initiatives [40]. Depending on the sector and organizational structure, the business can, to a specific degree, anyways be mature, enabling decentral governance from the start. As business internal development is lightweight and RPA-specific, the factor is not covered within the field of IT Governance.

**System Complexity:** RPA solutions can reach high levels of complexity, foremost in the sense of touching a multitude of other (core) systems, and therefore have an equally high number of permissions and access rights [30]. In our study, we also found evidence for this case, with one interviewee stating: *"The more powerful a tool becomes, the more you have to worry about security and the more alert administrators, for example, become. The more rights a robot gets, the more applications it is allowed to operate, the more IDs and passwords it manages. And of course, this is where people who have a clue about this, i.e., myself in case of doubt, can do the most mischief"* (I9). When setting up governance, the complexity of the to-be-governed systems must be considered [41]. The complexity can either be grounded in the age of the running solution or specific business needs [42]. Leaving the business more freedom increases the risk of fraud and malfunction. Therefore, the factor is strongly interconnected with the factor focus described below. The more complex systems get, the more centralized the governance should be. The factor does not appear in IT Governance literature, as heavyweight IT is complex by definition. Therefore, it seems specific to lightweight IT and RPA, where the complexity is subject to greater variation.

**Structural Diversity:** Structural diversity refers to the degree of heterogeneity within the organization. This heterogeneity can be based on regional and cultural differences (e.g., between different country organizations) but also on the prevailing organizational structure, reflected in the needs of specific industries [35]. Organizations with a large number of subsidiaries or are active in several industries with specific characteristics (i.e., having a higher structural diversity) will tend to choose more decentralized approaches to take advantage of the speed and innovation benefits of RPA. Weill and Ross [20] mention the comparable factor size and diversity.

**Focus:** Focus describes the balance between stability and innovation. While stability lowers the risk of malfunctions and fraud, it hinders innovation, and therefore potential efficiency increases [36, 43]. Organizations have to decide where to set the focus. The factor is influenced by organizational form, sector, and legal requirements. When focusing more on stability, organizations should choose a more centralized governance approach, hindering shadow IT and malicious systems from occurring. To foster innovation, organizations should tend to more decentral or federal approaches. The factor is part of all three other frameworks presented in the background section. Bygstad and Iden (2017) build their framework along the variables securing and resourcing, directly reflecting facets of stability and innovation. Györy et al. [36] also make the choice of a governance model dependent, among other things, on the accepted risk. Weill and Ross [20] partly cover the factor as Strategic and Performance Goals.

**Scale:** The factor reflects the intention to apply once developed RPA solutions on multiple parts of the organization, e.g., transferring a bot-based accounting process between multiple regional subsidiaries. In this scenario, a decentral approach would ease the process analysis due to direct process knowledge. Conversely, centralizing the governance would increase standardization between all regions. Another facet of this factor is the steering of the actual development. In a centralized model, there might be no capacity for controlled scaling. One interviewee stated that in his experience, a federated model would be the best fitting for scaling RPA solutions: *"In the first phase, rather control centrally to create structures. The load becomes too great when scaling up, so decentralize and steer more in the direction of a federated model"* (I9). These factors overlap with the factor size and diversity proposed by Weill and Ross [20].

Our findings show that a holistic governance framework for RPA and lightweight IT is yet not present in the literature or practice. With our research, we conducted a first attempt to close this gap. We could show that a specific governance approach for RPA and lightweight IT is necessary. We identified six specific factors influencing the design of an RPA governance by affecting the applicability of governance mechanisms. Practitioners can use our findings to ground their decision on what form of governance is appropriate for elaborating on the individual manifestation of these factors to improve the later fit of the chosen model. Moreover, we contributed to specific lightweight IT theory by providing insights into the role and design of a center of excellence in the field of lightweight IT. At the same time, these insights are directly relevant for practitioners facing the question of how a CoE for automation should be designed and where its focus should lie.

## 6    Conclusion

This paper contributes to the understanding of RPA and lightweight IT Governance. We used a research approach based on grounded theory to conduct in-depth interviews with 11 RPA experts from the practical, consulting, and domain of an RPA vendor. We condensed six organization-internal factors determining the choice and design of a specific lightweight IT Governance. We found that existing, valuable frameworks do not cover all relevant aspects of lightweight IT. Additionally, we found evidence that a pure RPA CoE is not common in practice, as most practitioners and consultants suggest a holistic automation and digitization CoE.

Future research can build upon our results in multiple directions. First, we identified factors influencing the design of an RPA governance by affecting the mechanisms and institutional structures. However, the actual effect size is yet unclear. Future research can therefore assess the actual effect of the respective factors. Secondly, we use RPA as an example for lightweight IT, but the generalizability to other instantiations of lightweight IT is not clear yet. Future research can close this gap by analyzing the identified factors related to multiple lightweight solutions.

Our research is limited for multiple reasons. Firstly, we set a regional focus on Germany. Despite all case organizations except one being multinational organizations, this could influence the results in terms of a cultural and regional bias. Secondly, although we tried to sample our respondents purposefully, the generalizability of our results might be influenced by the composition of that sample. Thirdly, we use RPA as an example of lightweight IT in general. Although the literature suggests that RPA is part of this knowledge regime, the findings might not be completely generalizable for all types of lightweight solutions.

# References

1. Kregel, I., Koch, J., Plattfaut, R.: Beyond the hype: robotic process automation's public perception over time. J. Organ. Comput. Electr. Commer. **31**, 1–21 (2021). https://doi.org/10.1080/10919392.2021.1911586
2. Roof, K., Tse, C.: UiPath Rises in Trading Debut After $1.3 Billion Software IPO. Bloomberg (2021)
3. Madakam, S., Holmukhe, R.M., Kumar Jaiswal, D.: The future digital work force: robotic process automation (RPA). J. Inf. Syst. Technol. Educ. Mange. **16**, 1–17 (2019). https://doi.org/10.4301/s1807-1775201916001
4. Bygstad, B.: The coming of lightweight IT. In: Becker, J., vom Brocke, J., Marco, M. de (eds.) Proceedings of the 23rd European Conference on Information Systems (2015). https://doi.org/10.18151/7217282
5. Penttinen, E., Kasslin, H., Asatiani, A.: How to choose between robotic process automation and back-end system automation? Research Papers (2018)
6. Niehaves, B., Köffer, S., Ortbach, K.: IT consumerization – a theory and practice review. In: AMCIS 2012 Proceedings (2012)
7. Osmundsen, K., Iden, J., Bygstad, B.: Organizing robotic process automation: balancing loose and tight coupling. In: Bui, T. (ed.) Proceedings of the 52nd Hawaii International Conference on System Sciences. Proceedings of the Annual Hawaii International Conference on System Sciences. Hawaii International Conference on System Sciences (2019). https://doi.org/10.24251/HICSS.2019.829
8. Willcocks, L., Lacity, M., Craig, A.: The IT function and robotic process automation. The Outsourcing Unit Working Research Paper Series, pp. 1–38 (2015)
9. Syed, R., et al.: Robotic process automation: contemporary themes and challenges. Comput. Ind. **115**, 103162 (2020). https://doi.org/10.1016/j.compind.2019.103162
10. Bygstad, B., Iden, J.: A governance model for managing lightweight IT. In: Rocha, Á., Correia, A.M., Adeli, H., Reis, L.P., Costanzo, S. (eds.) WorldCIST 2017. AISC, vol. 569, pp. 384–393. Springer, Cham (2017). https://doi.org/10.1007/978-3-319-56535-4_39
11. Peterson, R.: Crafting information technology governance. Inf. Syst. Manag. **21**, 7–22 (2004). https://doi.org/10.1201/1078/44705.21.4.20040901/84183.2
12. Webb, P., Pollard, C., Ridley, G.: Attempting to define IT governance: wisdom or folly? In: Proceedings of the 39th Annual Hawaii International Conference on System Sciences (HICSS 2006), pp. 194a–194a. IEEE (2006). https://doi.org/10.1109/HICSS.2006.68
13. Haes, S. de, van Grembergen, W.: IT Governance and its mechanisms. Inf. Syst. Control J. **1** (2004)
14. Weill, P., Ross, J.: A matrixed approach to designing IT governance. MIT Sloan Manag. Rev. **46** (2005)
15. Schmidt, N.-H., Kolbe, L.: Towards a contingency model for green IT Governance. In: ECIS 2011 Proceedings (2011)
16. Sambamurthy, V., Zmud, R.W.: Arrangements for information technology governance: a theory of multiple contingencies. MIS Q. **23**, 261 (1999). https://doi.org/10.2307/249754

17. Xue, L.: Boulton: information technology governance in information technology investment decision processes: the impact of investment characteristics, external environment, and internal context. MIS Q. **32**, 67 (2008). https://doi.org/10.2307/25148829
18. Weber, K., Otto, B., Österle, H.: One Size Does not fit all—a contingency approach to data governance. J. Data Inf. Qual. **1**, 1–27 (2009). https://doi.org/10.1145/1515693.1515696
19. Brown, A.E., Grant, G.G.: Framing the frameworks: a review of IT governance research. Commun. Assoc. Inf. Syst. **15** (2005). https://doi.org/10.17705/1CAIS.01538
20. Weill, P., Ross, J.W.: IT Governance. How top Performers Manage IT Decision Rights for Superior Results. Harvard Business School Press, Boston (2010)
21. Lacity, M., Willcocks, L.: Robotic process automation at Telefonica O2. MIS Q. Exec. **15** (2016)
22. Bygstad, B.: Generative innovation: a comparison of lightweight and heavyweight IT. J. Inf. Technol. **32**, 180–193 (2017). https://doi.org/10.1057/jit.2016.15
23. van der Aalst, W.M.P., Bichler, M., Heinzl, A.: Robotic process automation. Bus. Inf. Syst. Eng. **60**(4), 269–272 (2018). https://doi.org/10.1007/s12599-018-0542-4
24. Hallikainen, P., Bekkhus, R., Shan, L.: Pan: how opuscapita used internal RPA capabilities to offer services to clients. MIS Q. Exec. **17**, 41–52 (2018)
25. Plattfaut, R.: Robotic process automation - process optimization on steroids? In: Krcmar, H., Fedorowicz, J., Boh, W.F., Leimeister, J.M., Wattal, S. (eds.) Proceedings of the 40th International Conference on Information Systems, ICIS 2019. Association for Information Systems (2019)
26. Mendling, J., Decker, G., Hull, R., Reijers, H.A., Weber, I.: How do machine learning, robotic process automation, and Blockchains affect the human factor in business process management? Commun. Assoc. Inf. Syst. **43**, 297–320 (2018). https://doi.org/10.17705/1CAIS.04319
27. Alberth, M., Mattern, M.: Understanding robotic process automation (RPA). Capco Inst. J. Fin. Transform. 54–61 (2017)
28. Asatiani, A., Kämäräinen, T., Penttinen, E.: Unexpected problems associated with the federated IT Governance structure in robotic process automation (RPA) Deployment. Aalto University publication series BUSINESS + ECONOMY (2019)
29. Schmitz, M., Dietze, C., Czarnecki, C.: Enabling digital transformation through robotic process automation at deutsche Telekom. In: Urbach, N., Röglinger, M. (eds.) Digitalization Cases. MP, pp. 15–33. Springer, Cham (2019). https://doi.org/10.1007/978-3-319-95273-4_2
30. Noppen, P., Beerepoot, I., van de Weerd, I., Jonker, M., Reijers, H.A.: How to keep RPA maintainable? In: Fahland, D., Ghidini, C., Becker, J., Dumas, M. (eds.) BPM 2020. LNCS, vol. 12168, pp. 453–470. Springer, Cham (2020). https://doi.org/10.1007/978-3-030-58666-9_26
31. Strauss, A.L., Corbin, J.M.: Grounded theory in practice. Sage Publ, Thousand Oaks, Calif (1997)
32. Paré, G.: Investigating information systems with positivist case research. CAIS **13** (2004). https://doi.org/10.17705/1CAIS.01318
33. Palinkas, L.A., Horwitz, S.M., Green, C.A., Wisdom, J.P., Duan, N., Hoagwood, K.: Purposeful sampling for qualitative data collection and analysis in mixed method implementation research. Admin. Policy Mental Health Mental Health Serv. Res. **42**(5), 533–544 (2013). https://doi.org/10.1007/s10488-013-0528-y
34. Bryant, A., Charmaz, K.: The SAGE Handbook of Grounded Theory. SAGE (2007)
35. Choe, J.: The consideration of cultural differences in the design of information systems. Inf. Manag. **41**, 669–684 (2004). https://doi.org/10.1016/j.im.2003.08.003
36. Györy, A., Cleven, A., Uebernickel, F., Brenner, W.: Exploring the shadows: IT governance approaches to user-driven innovation. In: ECIS 2012 Proceedings (2012)

37. Jacobson, D.: Revisiting IT governance in the light of institutional theory. In: 2009 42nd Hawaii International Conference on System Sciences, pp. 1–9. IEEE (2009). https://doi.org/10.1109/HICSS.2009.374

38. Fürstenau, D., Rothe, H., Sandner, M.: Leaving the shadow: a configurational approach to explain post-identification outcomes of shadow it systems. Bus. Inf. Syst. Eng. **63**(2), 97–111 (2020). https://doi.org/10.1007/s12599-020-00635-2

39. Cochran, M.: Proposal of an operations department model to provide IT Governance in organizations that don't have IT C-level executives. In: 2010 43rd Hawaii International Conference on System Sciences, pp. 1–10. IEEE (2010). https://doi.org/10.1109/HICSS.2010.309

40. Asatiani, A., Penttinen, E., Rinta-Kahila, T., Salovaara, A.: Organizational implementation of intelligent automation as distributed cognition: six recommendations for managers. In: ICIS 2019 Proceedings (2019)

41. Beetz, K.R., Kolbe, L.: Towards managing IT complexity: an IT governance framework to measure business-IT responsibility sharing and structural IT organization. In: AMCIS 2011 Proceedings - All Submissions (2011)

42. Mocker, M.: What is complex about 273 applications? Untangling application architecture complexity in a case of European investment banking. In: 2009 42nd Hawaii International Conference on System Sciences, pp. 1–14. IEEE (2009). https://doi.org/10.1109/HICSS.2009.506

43. Lyytinen, K., Newman, M.: Explaining information systems change: a punctuated socio-technical change model. Eur. J. Inf. Syst. **17**, 589–613 (2008). https://doi.org/10.1057/ejis.2008.50

# Identifying the Socio-Human Inputs and Implications in Robotic Process Automation (RPA): A Systematic Mapping Study

Harmoko Harmoko[1]([⊠]), Andrés Jiménez Ramírez[2],
José González Enríquez[2], and Bernhard Axmann[1]

[1] Technische Hochschule Ingolstadt, Ingolstadt, Germany
{harmoko.harmoko,bernhard.axmann}@thi.de
[2] Universidad de Sevilla, Seville, Spain
{ajramirez,jgenriquez}@us.es

**Abstract.** Recent studies show that the success rate of Robotic Process Automation (RPA) projects is between 50 and 70%. In cases where they are not successful, most failures are caused by organizational factors, i.e., people and their social environment. People in the organization have a significant influence on the success of a RPA project. Likewise, the implementation of RPA will bring some implications to them as well. This research is a preliminary study on visualization socio-human implications (VoSHI) in the RPA projects. As one of the office automation technologies, the application of RPA provides economic benefits and social implications. Unfortunately, most studies highlight the benefits and lack discussions on the impact of RPA on humans. This paper presents a systematic mapping study to analyze the current state-of-the-art on this topic, recognizing the socio-human implications of RPA implementations described in the literature. The research analyzed 56 primary studies selected from both academic digital libraries and grey literature. The results showed 16 positive and 6 negative implications of RPA implementations for humans and their social environment. Furthermore, this research also found 6 positive and 13 negative inputs contributed by humans, which can influence RPA implementations.

**Keywords:** Robotic Process Automation (RPA) · Human · Social · Input · Implication

## 1 Introduction

Automation technology and robots have played a significant role in the Industrial Revolution 4.0. They have spanned many sectors, including manufacturing, trade, and services. As a part of automation technology, Robotic Process

This research is part of the project PID2019-105455GB-C31 funded by MCIN/AEI/ 10.13039/501100011033.

A. Marrella et al. (Eds.): BPM 2022, LNBIP 459, pp. 185–199, 2022.
https://doi.org/10.1007/978-3-031-16168-1_12

Automation (RPA) allows the software robot to emulate human behavior in carrying out repetitive tasks in the office [37]. Instead of using manual workers, organizations have deployed RPA bots to increase efficiency and shift the workers from mundane tasks to higher-value tasks. Nevertheless, Ernst & Young study shows that 30–50% of RPA projects have failed. Most failures are caused by the organization, not by technology [7]. The people in the organization have a significant influence on the success of the RPA project. Conversely, RPA implementation will bring some implications to people's lives.

Economically, RPA's investment is promising, but its relationship with people in the organization must be explored more deeply to create harmony between people, organizations, and technology. When implementing RPA, organizations sometimes ignore social and human impacts. As a result, a project can be financially viable, but have dire implications for people and their social environment. Therefore, it is crucial to identify all the effects and consequences of implementing RPA for humans. With this knowledge, organizations can comprehensively assess RPA projects and place humans' interests at the center of technology assessment.

In academia, the trend for RPA studies has increased in the last five years [11]. The discussion topics that were presented are also diverse. It ranges from the challenges of RPA implementation [8,36], the benefits of RPA [1,21,36], process selection [4,40], human & RPA collaboration [9,22,30,33], RPA assessment [2, 41] and others. However, there are further areas that have not been exposed properly. For instance, in the area of RPA assessment, most studies still focus on technology & economic perspectives [3,40]. However, the human perspective is scarcely researched and mainly reflects the implications on employees [33], while broader social & human perspectives are still ruled out. In the future (Industrial Revolution 5.0), the success of new technology is seen not only by financial benefit but also by human acceptance and the way technology supports human life [10].

Inspired by the Visualization of Financial Implication (VoFI) framework [13], that has been successfully applied to study the viability of projects from an economic perspective [12,26], we will conduct research series to develop a framework called VoSHI (Visualization of Socio-Human Implication). VoSHI will assist organizations in evaluating projects from the human perspective. The following research steps are considered to implement the whole VoSHI framework (cf. Fig. 1): (1) Identifying VoSHI parameters revealed by the collection of literature, (2) validating & quantifying VoSHI parameters, and (3) developing a tool to measure VoSHI parameters. The current paper focuses on the first step. For this, a systematic mapping study is conducted to analyze the current state of the art, to identify both (1) the human influence on RPA projects and (2) the implications of the project for humans. These will be the parameters of the VOSHI framework. Meanwhile, the validation and quantification and the tool development will be carried out in further research.

The current paper contributes to the RPA field by laying the foundations of an innovative assessment approach that places the human at the center, as introduced in Sect. 2. In addition, Sect. 3 reports the literature review conducted

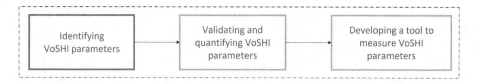

**Fig. 1.** The series of VoSHI research. The current paper addresses the *identification* phase.

whose implications are discussed in Sect. 4. Finally, Sect. 5 concludes the paper and sets the future research agenda.

## 2   Background

The automation project failure is characterized by the inability to achieve goals and mitigate risks. It is caused by technology failure [23], human failure [30], and economic failure [3]. Economic failure can be detected by visualizing the financial implications of a project using the VoFI method. The Visualization of Financial Implication (VoFI) is a comprehensive financial plan that considers the economic consequences of an investment project. It includes all cash flows from internal funds, debt & redemption capital, various forms of loans (with various redemption & interest rates), and tax implications investment [12]. VoFI offers accountability, transparency, and a long-term perspective on the quantitative aspects of investment decisions [38].

VoFI was first introduced by Heister in 1962 and explained in more depth by Professor Heinz Lothar Grob in 1993 and 2006 [13]. Over time, the VoFI method has been used in various ways, such as:

- LIVO (Liquiditätplannung VoFI), where the VoFI method is used to measure company liquidity to minimize the risk of bankruptcy [32].
- BSC-VoFI (The Balanced scorecard) measures company performance by modifying the balanced scorecard with VoFi [14].
- TCO-VoFI (Total Cost Ownership-VoFI), which utilizes the VoFI method to correct the weaknesses of traditional TCO (Total Cost Ownership) analysis [12].
- CBA-VoFI (Cost-Benefit Assessment based on VoFI), which was developed by Oesterreich & Teuteberg (2017), to evaluate the application of augmented reality in the construction area [26].

Although VoFI has evolved into various forms, the basic principle has never changed. VoFI always has two sides, adding and deducting factors, that affect the profitability of the investment. In the model of CBA-VoFI, a project is considered financially profitable if the amount of benefits achieved by the organization or company is greater than the incurred costs. On the other hand, a project is financially unprofitable if the benefits outweigh the cost [26]. However, the financial implication is not the only dimension to measure the success of an RPA project.

Harmoko & Axmann determined the assessment of an RPA project into five dimensions, two financial dimensions (i.e., cost and benefit), two technological dimensions (i.e., technology readiness and usability), and one organizational dimension that involves humans and social environment (i.e., company readiness) [2]. This idea is in line with the principle of Industrial Revolution 5.0, where human implications should be a primary consideration in the implementation of technology [10].

Unfortunately, there is no method to measure the socio-human implications of an RPA project. Whereas RPA and most other automation technologies are met with strong challenges from humans, especially workers [18, 23, 31]. The issues of losing a job and collaboration failure between humans and robots are still frightening specters for some people [30, 33, 43]. Therefore, we will develop a method to measure and visualize the socio-human implications of an RPA project through the series of research. The Visualization of Socio-Human Implication (VoSHI) will be designed with the same basic principles as CBA-VoFI. As illustrated in Fig. 2, the balance of VoSHI is influenced by adding factors (i.e., positive inputs and implications) and deducting factors (i.e., negative inputs and implications).

- **Positive Input:** The supporting factors from humans and their social environment to the RPA project, starting from the initiation stage until the project is implemented.
- **Negative Input:** The inhibiting factors from humans and their social environment to the RPA project, starting from the initiation stage until the project is implemented.
- **Positive Implication:** all benefits gained by humans and the environment as a result of RPA implementation.
- **Negative Implication:** all negative consequences suffered by humans and their environment as a result of the RPA implementation.

In the VoSHI method, the project is considered a success if the benefits for people (e.g., workers, management, and other stakeholders) are greater than their struggles and sacrifices for the project or, in other words, if the adding factor (i.e., positive inputs and implications) is greater than deducting factor (i.e., negative inputs and implications).

## 3   Systematic Mapping Study Method

In this paper, we propose to review the studies related to the socio-human implications of RPA implementation using the Systematic Mapping Study (SMS) method, proposed by Peterson et al. [27]. This method is used to build a structured classification scheme in the software engineering area. Following the guidelines described by Petersen and the procedures suggested by Khanra et al. and Wewerka & Reichert [17, 27, 41], we design a protocol (cf. Fig. 3) that describes the formulation of research questions and the definition of rules for conducting

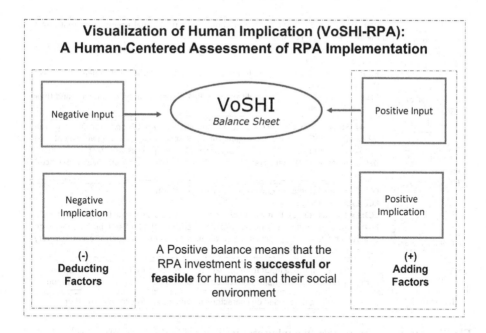

**Fig. 2.** Pre-illustration of VoSHI model.

the searches (cf. Sect. 3.1), the selection of primary studies (cf. Sect. 3.2), and the data extraction and analysis methods (cf. Sect. 3.3).

### 3.1 Planning

Before starting the SMS, it is necessary to formulate the research questions (RQ), which guide the search for relevant studies. RQ keeps research focused on topics and predetermined goals. As the initial stage of VoSHI studies, this research will answer the following questions:

- RQ1: What are positive inputs from humans and their social environment that affect the implementation of RPA?
- RQ2: What are the positive implications of RPA implementation for humans and their social environment?
- RQ3: What are negative inputs from humans and their social environment that affect the implementation of RPA?
- RQ4: What are the negative implications of RPA implementation for humans and their social environment?

In addition, two steps have been defined to search the relevant studies: (1) Search relevant literature in both scientific digital libraries and general search engines (i.e., Google). In our study, six digital libraries have been selected, i.e., Scopus, IEEExplore, Web of Science, ACM, ScienceDirect, and AIS eLibrary. (2) Select keywords that can accelerate the searching process. The keywords will

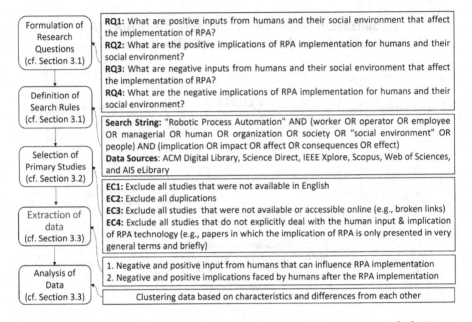

**Fig. 3.** Protocol for systematic mapping study (inspired by Petersen et al. [27], Khanra et al. [17], and Wewerka and Reichert [41]).

**Table 1.** Keywords.

| Keywords | Similar term |
|---|---|
| Robotic process automation | RPA |
| Human | People, worker, operator, employee, managerial, organization, society, social environment |
| Implication | impact, affect, consequence, effect |

help to build queries for digital libraries. Besides that, similar terms of a keyword (e.g., employees, operators, managerial, and social environment), which refer to the "humans", are also used to ensure the relevant studies in the digital library are not missed (cf. Table 1).

After determining the keywords, the authors search for relevant studies with different queries. It is possible to have different queries since each digital library has a different search input form. At this step, authors targeted titles and abstracts directly. In addition, the exclusion criteria are determined to eliminate studies that use non-English language and briefly discuss RPA or beyond the socio-human aspects of RPA (cf. Fig. 3).

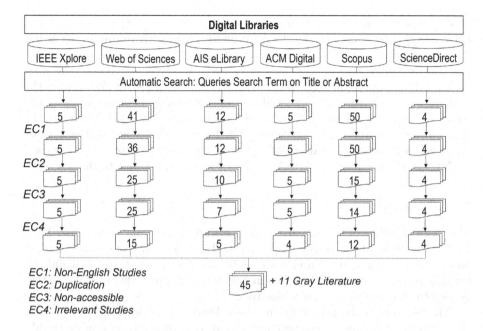

EC1: Non-English Studies
EC2: Duplication
EC3: Non-accessible
EC4: Irrelevant Studies

45  + 11 Gray Literature

**Fig. 4.** The primary studies selection process.

## 3.2  Conducting

The Primary Studies Selection Process is divided into automated searching and detailed review. The automated searching uses queries and targets metadata: title, keywords, and abstract. From this step, 117 studies were selected and screened using the first exclusion criteria (EC1), leaving 112 studies. Of the 112 studies, there are 48 duplications to be excluded (EC2). Consequently, there are 64 studies remaining.

Furthermore, the 64 studies were reviewed manually by the authors, but before that, the authors had to ensure that the studies were accessible based on EC3. In this process, 4 studies cannot be accessed. Thereafter, the authors conducted a "detailed review" of the 60 remaining studies. In a detailed review, the authors read the entire section of the reviewed paper to find information that can answer research questions. In this research, the authors focused on (1) negative and positive input from humans that can influence RPA implementation and (2) negative and positive implications faced by humans after the RPA implementation. From this step, the authors found that 15 studies were irrelevant to the topic of humans and RPA. Therefore, the authors excluded them according to EC4. In the end, there are 45 primary studies (cf. Fig. 4).

In addition, following the recommendations of Wieringa et al. [42], this research also searches and reviews the studies in grey literature. In this paper, the authors reinforce the fact of including this kind of literature. By definition, grey literature consists of publications produced at all levels by government, academia, business, and industry, whether in print or electronic format, but not controlled by commercial publishing interests, and where publishing is not the organization's primary business activity [24]. The searching process uses the Google search engine with keywords "Robotic Process Automation" and "Human Implication" in PDF format. In the first 10 pages of the results, the authors found 12 titles that pointed to the relationship between humans and RPA. The authors did not find such a title from page 11 onward. Therefore the grey literature search was discontinued. Of the 12 studies selected, the authors reviewed them in detail and excluded one study based on exclusion criteria 4 (EC4). Finally, the authors considered only 11 studies relevant to the research.

## 3.3    Reporting

The main objective of this research is to find four types of parameters as a basis for further development of the VoSHI method (i.e., positive and negative input and positive and negative implication). The detailed review found that only 56 primary studies, including 11 studies in grey literature, provided the information needed to achieve the research objectives.

The sources for the grey literature have been marked with an * every time they are cited to increase transparency. The authors plan further studies to validate and search the empirical evidence of revealed parameters and statements in the grey literature. In the research, the authors clustered the inputs and implications into four clusters (i.e., worker skill, worker health, company health, and society health). This is because not all identified inputs and implications relate directly to the workers as individuals. Some are more related to the social environment, such as company (company health) and even society (society health). In the context of the individual worker, the authors found the input and implication are not only on worker's pleasure (worker health), but also on the worker's skills (worker skills).

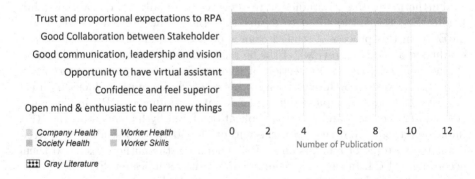

**Fig. 5.** Positive inputs for RPA projects.

**RQ1:** *What are positive inputs from human and their social environment that affect the implementation of RPA?* Companies and workers may believe that robots can perform repetitive tasks more effectively [25]. Nevertheless, the leading indicator of RPA performance is not the number of fired workers but the number of repetitive tasks that can be delegated to robots [8]. Sometimes not all repetitive tasks can be done by robots. Therefore, workers should not feel inferior [19]*. Companies and workers have to be more realistic and manage their expectations proportionally. The open-mindedness [6], trust, and proportional expectations of workers will positively impact the implementation of RPA [9,30].

The other positive inputs (influences) come not from individual workers, but purely from the social environment (company health): good leadership from top management [44], and good collaboration between stakeholders [28]. Good leadership is reflected in a strong vision, shared knowledge, and good communication from the top to the lowest levels in the organization. The absence of leadership causes uncertainty and anxiety among people within the organization [39] (cf. Fig. 5).

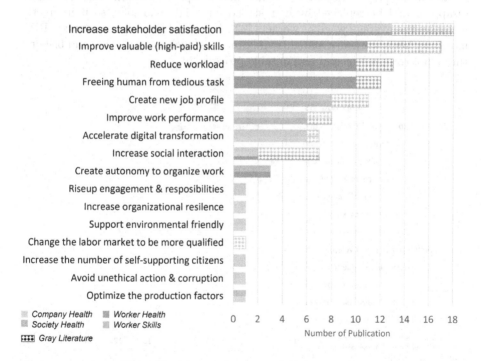

**Fig. 6.** Positive implications of RPA projects.

**RQ2:** *What are the positive implications of RPA implementation for humans and their social environment?* From the detailed review, the authors found the majority of publications reveal that improvement skills [6], reduction of workload [9], and freeing workers from tedious tasks [19,22], are the positive implication of RPA projects for the worker. By not performing tedious tasks, workers' workload can be reduced [9], so they can focus on value-added tasks that improve working performance [8] and social interactions, especially with customers [16]. The good and intense interactions will make customers feel valued and prioritized [22], and may potentially increase their satisfaction and loyalty. The presence of RPA as a reliable virtual assistant will increase worker satisfaction and motivate them to learn new skills, especially in the field of automation technology and digitization. These skills are essential in a digital working ecosystem, where humans and robots work side by side [30].

Several studies also reveal that the social environment (company & society health) is also affected in the form of organizational resilience [35], changing labor market [34]*, and digital society [35] (cf. Fig. 6). For society, RPA will change the labor market by creating new job profiles that support digital transformation in organizations [34]*. The old low-skilled and low-paid job profile will gradually disappear and be replaced by high-skilled and high-paid jobs, so it indirectly increases the average wage of workers [22]. In a good governance context, RPA indirectly helps workers avoid unlawful practices, such as corruption and bribery, through a transparent and accountable process [29].

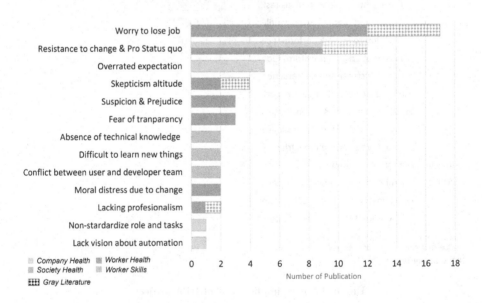

**Fig. 7.** Negative inputs for the RPA project.

**RQ3:** *What are negative inputs from humans and their social environment that affect the implementation of RPA?* From the detailed review, the authors found that the most negative input comes from the worker. Publications capture job loss concerns, rejection, bad attitude, lack of skill & professionalism, prejudice, and skepticism as obstacles to the RPA project (cf. Fig. 7).

When a RPA project begins, human responses can vary. It is based on different perspectives, skills, and past experiences. Humans tend to feel negative and feel rejected by automation [30]. Negative feelings or prejudices are usually triggered by fear of being unemployed [43] and the inability to meet new job requirements [29]. Another prejudice comes from IT workers or people who will develop RPA. They do not believe that RPA is better than traditional automation [9], and they also doubt whether RPA can adapt easily to any changes in the workflow [43]. The prejudices are generated resistance among people in the organization. Most of them do not want to change their old behavior and working culture [18]*. They are satisfied with the status quo and think they are too old to learn new technology [9]. The reluctance to change will lead people to an unprofessional attitude, while reluctance to learn will make people stuck as low-skilled or low-paid workers.

Several studies explain that negative inputs come not only from humans but also from the social environment (companies). Management that does not design standard workflows and human roles in the organization also slows down the process of implementing RPA [5]. Another negative input is the lack of vision and leadership from top management. Without clear direction from the leader, implementing RPA will trigger confusion and conflict within the organization [9].

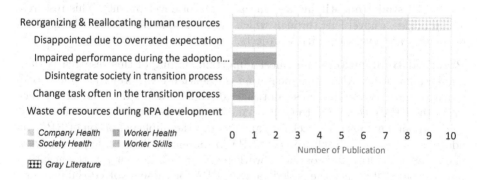

**Fig. 8.** Negative implications of RPA projects

**RQ4:** *What are the negative implications of RPA implementation for humans and their social environment?* Most studies see that the negative implication of RPA implementation is reorganizing and reallocating human resources (cf. Fig. 8). Implementing RPA will change the structure and role of humans in the organization. Unfortunately, it is difficult for individuals to adapt to new roles, systems, and work cultures [23]. The change requires workers to adjust to new working norms and key performance indicators (KPI). Instead of speeding up the process and helping workers, the adjustments to automation hinder the process and frustrate workers [23]. There will be a disaster if human labor is replaced prematurely without ensuring that humans and robots can work together in a digital environment [20,30]. The shift of human role in the organization can generate internal disputes between workers and management [15]. While for society, implementing RPA will liquidate some familiar jobs such as data entry, administration staff, and others. These are the negative effects of implementing RPA on individuals in the organization and society.

## 4    Discussion

The VoSHI method is not only limited to RPA implementation but is also open to various technology close to humans and their social environment. To develop VoSHI, the appraiser must be able to define humans' influence (input) on technology implementation and the impact of technology on humans. Therefore, this research result comes with theoretical and practical implications, and limitations.

**Theoretical Implications:** This research contributes to the visualization of the added value of RPA to humans and their social environment. Although the amount of academic literature is limited, the author's survey on the socio-human implications of the RPA project is the first study on this emerging topic. In addition, the practice reports in the grey literature have enriched and confirmed survey findings. With this combination, the identified socio-human input & implications in RPA projects can be readily used in the subsequent series of the VoSHI study (quantifying and measuring tool development). This research also provides a novel lens through which organizations can consider humans in estimating and evaluating RPA projects.

**Practical Implications:** The most important finding of this research is that human influence in RPA implementation is essential. Fear, prejudice, and excessive expectations from humans can trigger disruption and reduce humans' interest in the RPA project. On the other hand, enthusiasm, trust, and collaborative spirit among humans can accelerate and expand the implementation of RPA. In addition, the positive implications of RPA for humans will increase productivity and create a healthy and conducive work environment. In contrast, the negative implications will reduce the added value of RPA for those involved (humans) in the organization. In practice, organizations can use the findings of this study to better understand the socio-human perspective of the RPA project.

**Limitations:** This research is limited to literature research, so that the findings may be just the tip of the iceberg. The grey literature used as a reference also lacks solid empirical evidence. In the future, direct observation is needed to collect new information and to validate and confirm the VoSHI parameters revealed in this study.

# 5   Conclusion

It is undeniable that humans have an essential role in the success of RPA projects. They are not only users, developers, or decision-makers in the company but also RPA's partners who collaborate in the digital working environment. As the current state of the art scarcely covers how humans influence and have implications from RPA [33], the current paper lays the foundation of the VOSHI framework, which will assist organizations in comprehensively evaluating RPA projects from the human perspective. For this, a systematic mapping study is conducted to review the literature and identify negative and positive influences and implications.

The literature review shows that the positive implications of RPA projects for humans outweigh the negative ones. In contrast, human influence in RPA projects is dominated by negative inputs rather than positive inputs. It means two different things: (1) It is a fact that the RPA project was refused by humans in the initial step and accepted in the end; or (2), most authors are only interested in the negative influence and positive implications on humans. Therefore, this research should not be stopped in the literature review, but must be followed by field observations or case studies. In the future, research should not only continue the VoSHI study (quantifying and developing tools) but also verify the result of today's research.

# References

1. Axmann, B., Harmoko, H.: Robotic process automation: an overview and comparison to other technology in industry 4.0. In: 2020 10th International Conference on Advanced Computer Information Technologies (ACIT), pp. 559–562. IEEE (2020)
2. Axmann, B., Harmoko, H.: The five dimensions of digital technology assessment with the focus on robotic process automation (RPA). Tehnički Glasnik **15**(2), 267–274 (2021)
3. Axmann, B., Harmoko, H., Herm, L.-V., Janiesch, C.: A framework of cost drivers for robotic process automation projects. In: González Enríquez, J., Debois, S., Fettke, P., Plebani, P., van de Weerd, I., Weber, I. (eds.) BPM 2021. LNBIP, vol. 428, pp. 7–22. Springer, Cham (2021). https://doi.org/10.1007/978-3-030-85867-4_2
4. Beetz, R., Riedl, Y.: Robotic process automation: developing a multi-criteria evaluation model for the selection of automatable business processes (2019)
5. Chuong, L.V., Hung, P.D., Diep, V.T.: Robotic process automation and opportunities for vietnamese market, pp. 86–90. Association for Computing Machinery (2019). https://doi.org/10.1145/3348445.3348458
6. Dias, M., Pan, S., Tim, Y.: Knowledge embodiment of human and machine interactions: robotic-process-automation at the Finland government (2019)
7. Ernst & Young Limited: Get ready for robots: Why planning makes the difference between success and disappointment (2016). https://assets.ey.com/content/dam/ey-sites/ey-com/en_gl/topics/emeia-financial-services/ey-get-ready-for-robots.pdf

8. Figueiredo, A.S., Pinto, L.H.: Robotizing shared service centres: key challenges and outcomes. J. Serv. Theory Pract. **31**, 157–178 (2021). https://doi.org/10.1108/JSTP-06-2020-0126

9. Flechsig, C., Anslinger, F., Lasch, R.: Robotic process automation in purchasing and supply management: a multiple case study on potentials, barriers, and implementation. J. Purch. Supply Manag. **28**(1), 100718 (2022)

10. George, A.S., George, A.: Industrial revolution 5.0: the transformation of the modern manufacturing process to enable man and machine to work hand in hand. Seybold Rep. **15**, 214–234 (2020)

11. González Enríquez, J., Jiménez Ramírez, A., Domínguez Mayo, F.J., García García, J.A.: Robotic process automation: a scientific and industrial systematic mapping study. IEEE Access **8**, 39113–39129 (2020)

12. Götze, U., Northcott, D., Schuster, P.: Compounded cash flow methods. In: Investment Appraisal: Methods and Models, pp. 93–110 (2008)

13. Grob, H.L., Langenkämper, C., Wieding, A.: Unternehmensbewertung mit vofi. Schmalenbachs Zeitschrift für betriebswirtschaftliche Forschung **51**(5), 454–479 (1999)

14. Gust, E.M.: Balanced Scorecard und VOFI-Kennzahlen. [electronic ed.] edn. (2002). https://nbn-resolving.de/urn:nbn:de:hbz:6-85659524744

15. Hofmann, P., Samp, C., Urbach, N.: Robotic process automation. Electron. Mark. **30**(1), 99–106 (2019). https://doi.org/10.1007/s12525-019-00365-8

16. Hollebeek, L.D., Sprott, D.E., Brady, M.K.: Rise of the machines? Customer engagement in automated service interactions. J. Serv. Res. **24**, 3–8 (2021). https://doi.org/10.1177/1094670520975110

17. Khanra, S., Dhir, A., Islam, A.N., Mäntymäki, M.: Big data analytics in healthcare: a systematic literature review. Enterp. Inf. Syst. **14**(7), 878–912 (2020)

18. Kirkwood, G.: The impact of RPA on employee engagement, a forrester consulting thought leadership paper. https://www.uipath.com/blog/rpa/impact-of-rpa-on-employee-engagement-forrester

19. KONICAMINOLTA: Workers' little helpers: how robotic process automation (RPA) supports every company and its employees. https://www.konicaminolta.eu/getmedia/9b5a5113-83c3-43ec-901b-ec4ccfb66fc8/Robotic-Process-Automation-RPA-Infographic.pdf.aspx

20. Korhonen, T., Selos, E., Laine, T., Suomala, P.: Exploring the programmability of management accounting work for increasing automation: an interventionist case study. Account. Audit. Account. J. **34**, 253–280 (2021). https://doi.org/10.1108/AAAJ-12-2016-2809

21. Meironke, A., Kuehnel, S.: How to measure RPA's benefits? A review on metrics, indicators, and evaluation methods of RPA benefit assessment (2022)

22. Mendling, J., Decker, G., Hull, R., Reijers, H.A., Weber, I.: How do machine learning, robotic process automation, and blockchains affect the human factor in business process management? Commun. Assoc. Inf. Syst. **43**(1), 19 (2018)

23. Mishra, S., Devi, K.K.S., Narayanan, M.K.B.: People & process dimensions of automation in business process management industry. Int. J. Eng. Adv. Technol. **8**, 2465–2472 (2019). https://doi.org/10.35940/ijeat.F8555.088619

24. Nahotko, M.: Some types of grey literature: a polish context. In: Grey Foundations in Information Landscape, vol. 67 (2008)

25. Nauwerck, G., Cajander, A.: Automatic for the people: implementing robotic process automation in social work (2020). https://doi.org/10.18420/ecscw2019_p04

26. Oesterreich, T., Teuteberg, F.: Evaluating augmented reality applications in construction-a cost-benefit assessment framework based on vofi (2017)

27. Petersen, K., Feldt, R., Mujtaba, S., Mattsson, M.: Systematic mapping studies in software engineering. In: 12th International Conference on Evaluation and Assessment in Software Engineering (EASE) 12, pp. 1–10 (2008)
28. Plattfaut, R., Borghoff, V., Godefroid, M., Koch, J., Trampler, M., Coners, A.: The critical success factors for robotic process automation. Comput. Ind. **138**, 103646 (2022). https://doi.org/10.1016/j.compind.2022.103646
29. Ranerup, A., Henriksen, H.Z.: Digital discretion: unpacking human and technological agency in automated decision making in Sweden's social services. Soc. Sci. Comput. Rev. **40**, 445–461 (2022). https://doi.org/10.1177/0894439320980434
30. Ruiz, R.C., Ramírez, A.J., Cuaresma, M.J.E., Enríquez, J.G.: Hybridizing humans and robots: an RPA horizon envisaged from the trenches. Comput. Ind. **138**, 103615 (2022)
31. Saukkonen, J., Kreus, P., Obermayer, N., Ruiz, Ó.R., Haaranen, M.: AI, RPA, ML and other emerging technologies: anticipating adoption in the HRM field. In: Proceedings of the European Conference on the Impact of Artificial Intelligence and Robotics ECIAIR 2019 Oct-1 Nov (2019)
32. Schulenburg, K.: Kurzfristige Liquiditätsplanung mit VOFI. [electronic ed.] edn. (1997). https://nbn-resolving.de/urn:nbn:de:hbz:6-85659524791
33. Seiffer, A., Gnewuch, U., Maedche, A.: Understanding employee responses to software robots: a systematic literature review. In: ICIS (2021)
34. Sharma, M.: How RPA will impact the future workplace. https://www.fahr.gov. ae/Portal/assets/442b2f56
35. Sobczak, A.: Robotic process automation as a digital transformation tool for increasing organizational resilience in polish enterprises. Sustainability **14**, 1333 (2022). https://doi.org/10.3390/su14031333
36. Syed, R., et al.: Robotic process automation: contemporary themes and challenges. Comput. Ind. **115**, 103162 (2020)
37. Taulli, T.: The Robotic Process Automation Handbook (2020). https://doi.org/ 10.1007/978-1-4842-5729-6
38. Trost, R., Fox, A.: Investitionsplanung bei unvollkommenen kapitalmärkten: die vofi-methode (2017). https://doi.org/10.1515/9783110517163-020
39. Wallace, E., Waizenegger, L., Doolin, B.: Opening the black box: exploring the socio-technical dynamics and key principles of RPA implementation projects (2021)
40. Wanner, J., Hofmann, A., Fischer, M., Imgrund, F., Janiesch, C., Geyer-Klingeberg, J.: Process selection in RPA projects-towards a quantifiable method of decision making (2019)
41. Wewerka, J., Reichert, M.: Robotic process automation-a systematic literature review and assessment framework. arXiv preprint arXiv:2012.11951 (2020)
42. Wieringa, R., Maiden, N., Mead, N., Rolland, C.: Requirements engineering paper classification and evaluation criteria: a proposal and a discussion. Requir. Eng. **11**(1), 102–107 (2006)
43. Willcocks, L., Lacity, M., Craig, A.: Robotic process automation: strategic transformation lever for global business services? J. Inf. Technol. Teach. Cases **7**, 17–28 (2017). https://doi.org/10.1057/s41266-016-0016-9
44. Zhang, N., Liu, B.: Alignment of business in robotic process automation. Int. J. Crowd Sci. **3**, 26–35 (2019). https://doi.org/10.1108/IJCS-09-2018-0018

# A Human-in-the-Loop Approach to Support the Segments Compliance Analysis

Simone Agostinelli[✉], Giacomo Acitelli, Michela Capece,
and Massimo Mecella

Sapienza Universitá di Roma, Rome, Italy
{agostinelli,mecella}@diag.uniroma1.it,
{acitelli.1643776,capece.1694700}@studenti.uniroma1.it

**Abstract.** Robotic Process Automation (RPA) is an emerging automation technology in the field of Business Process Management (BPM) that creates software (SW) robots to partially or fully automate rule-based and repetitive tasks (a.k.a. routines) previously performed by human users in their applications' user interfaces (UIs). Nowadays, successful usage of RPA requires strong support by skilled human experts, from the discovery of the routines to be automated (i.e., the so-called segmentation issue of UI logs) to the development of the executable scripts required to enact SW robots. In this paper, we present a human-in-the-loop approach to filter out the routine behaviors (a.k.a. routine segments) not allowed (i.e., wrongly discovered from the UI log) by any real-world routine under analysis, thus supporting human experts in the identification of valid routine segments. We have also measured to which extent the human-in-the-loop strategy satisfies three relevant non-functional requirements, namely effectiveness, robustness and usability.

**Keywords:** Robotic Process Automation · Segmentation of UI logs · Declarative constraints

## 1 Introduction

Robotic Process Automation (RPA) is an emerging automation technology in the Business Process Management (BPM) domain [17] that creates software (SW) robots to partially or fully automate rule-based and repetitive tasks (or simply *routines*) performed by human users in their applications' user interfaces (UIs) [1] of their computer systems.

To date, the identification of the routine steps to robotize from a UI log require the support of skilled human experts, which need to [16]: *(i)* preliminary observe how routines are executed on the UI of the involved SW applications (by means of walkthroughs, etc.), *(ii)* convert such observations in explicit flowchart diagrams, which are specified to show all the potential behaviours of the routines of interest, and *(iii)* finally implement the SW robots that automate the routines

© Springer Nature Switzerland AG 2022
A. Marrella et al. (Eds.): BPM 2022, LNBIP 459, pp. 200–214, 2022.
https://doi.org/10.1007/978-3-031-16168-1_13

enactment on a target computer system. However, the current practice is time-consuming and error-prone, as it strongly relies on the ability of human experts to correctly interpret the routines to automate.

For tackling this challenge, in their Robotic Process Mining framework [20], Leno et al. propose to exploit the User Interface (UI) logs recorded by RPA tools to automatically discover the candidate routines that can be later automated with SW robots. UI logs are sequential data of user actions performed on the UI of a computer system during many routines' executions. Typical user actions are: opening a file, selecting/copying a field in a form or a cell in a spreadsheet, read and write from/to databases, open emails and attachments, etc.

Nowadays, when considering state-of-the-art RPA technology, it is evident that the RPA tools available in the market are not able to learn how to automate routines by only interpreting the user actions stored into UI logs [4]. The main trouble is that in a UI log there is not an exact 1:1 mapping among a recorded user action and the specific routine segment it belongs to. *Routine segments* describe the different behaviours of the routine(s) under analysis, in terms of repeated patterns of performed user actions. In fact, the UI log usually records information about several routines whose actions are mixed in some order that reflects the particular order of their execution by the user. The issue to automatically understand which user actions contribute to a particular routine segment inside a UI log and cluster them into well-bounded *routine traces* (i.e., complete execution instances of a routine) is known as *segmentation* [4,20].

The majority of state-of-the-art segmentation approaches are able to properly extract routine segments from unsegmented UI logs when the routine executions are not interleaved from each others. Only few works are able to partially untangle unsegmented UI logs consisting of many interleaved routines executions, but with the assumption that any routine provides its own, separate universe of user actions. This is a relevant limitation, since it is quite common that real-world routines may share the same user actions (e.g., copy and paste data across cells of a spreadsheet) to achieve their objectives. The limitations mentioned above are addressed in [3], where we proposed a new approach to the discovery of routine traces from unsegmented UI logs, that is able to segment a UI log that records in an interleaved fashion many different routines with shared user actions but not the routine executions, thus losing in accuracy when there is the presence of interleaving executions of the same routine.

Specifically, the technique presented in [3] may discover routine segments that represent not allowed routine behaviours. This happens because a UI log combines the execution of several routines that are usually interleaved from each others. In addition, in case of routines that make use of the same kinds of user actions to achieve their goals, it may happen that new patterns of repeated user actions, which represent potential not allowed routine segments, are rather detected as valid ones within the UI log. In this paper, starting from [3], we present: (1) a human-in-the-loop approach together with its implemented tool called SCAN[1] (Segments Compliance ANalysis), that allows users to filter out

---

[1] The tool can be downloaded at: https://github.com/bpm-diag/SCAN.

those routines' segments not compliant with any real-world routine behaviours (thus supporting human experts in performing the segmentation task), and (2) the results of the multi-step evaluation conducted on SCAN.

The rest of the paper is organized as follows. Section 2 introduces a running example that will be used to explain our approach, then it discusses literature works on the segmentation of UI logs. Section 3 presents the required steps to enact the human-in-the-loop strategy through SCAN, instantiating it over the RPA use case of Sect. 2. Section 4 measures the impact of the human-in-the-loop strategy to filter out the wrongly discovered routine segments. Specifically, we present the results of SCAN to investigate to which extent it satisfies three relevant non-functional requirements, namely *effectiveness, robustness* and *usability*. The target is to understand if SCAN can potentially complement the traditional solutions provided by open-source Process Mining tools for helping users to perform the segmentation task in RPA. Finally, Sect. 5 draws conclusions and traces future works.

## 2    Background

### 2.1    Running Example

In this section, we describe an RPA use case inspired by a real-life scenario at the Department of Computer, Control and Management Engineering (DIAG) of Sapienza Università di Roma, which has already proved to be effective in ours previous works [3,5]. The scenario concerns the filling of the travel authorization request form made by professors, researchers and PhD students of DIAG for travel requiring prior approval. The request applicant must fill a well-structured Excel spreadsheet (cf. Fig. 1(a)) providing some personal information, such as her/his bio-data and the email address, together with further information related to the travel, including the destination, the starting/ending date/time, the means of transport to be used, the travel purpose, and the envisioned amount of travel expenses, associated with the possibility to request an anticipation of the expenses already incurred. When ready, the spreadsheet is sent via email to an employee of the Administration Office of DIAG, which is in charge of approving and elaborating the request. Concretely, for each row in the spreadsheet, the employee manually copies every cell in that row and pastes that into the corresponding text field in a dedicated Google form (cf. Fig. 1(b)), accessible just by the Administration staff. Once the data transfer for a given travel authorization request has been completed, the employee presses the "Submit" button to submit the data into an internal database.

We denote this routine procedure as $R_{example}$. In particular, the path of user actions performed by the Administration employee in the UI to properly enact $R_{example}$ is as follows:

- loginMail, to access the client email;
- accessMail, to access the specific email with the travel request;
- downloadAttachment, to download the Excel file including the travel request;

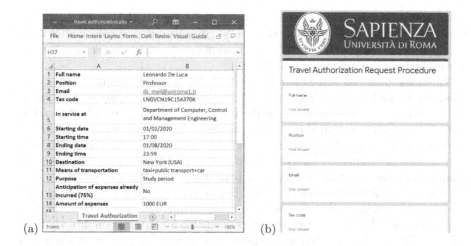

**Fig. 1.** UIs involved in the running example

- openWorkbook, to open the Excel spreadsheet;
- openGoogleForm, to access the Google Form to be filled;
- getCell, to select the cell in the i-th row of the Excel spreadsheet;
- copy, to copy the content of the selected cell;
- clickTextField, to select the specific text field of the Google form where the content of the cell should be pasted;
- paste, to paste the content of the cell into the corresponding text field of the Google form;
- formSubmit, to press the button to finally submit the Google form to the internal database.

Note that, the user actions openWorkbook and openGoogleForm can be performed in any order. Moreover, the sequence of actions ⟨getCell, copy, clickTextField, paste⟩ will be repeated for any travel information to be moved from the Excel spreadsheet to the Google form.

For example, a valid routine segment of $R_{example}$ is ⟨loginMail, accessMail, downloadAttachment, openWorkbook, openGoogleForm, getCell, copy, clickText Field, paste, formSubmit⟩. Valid routine segments are also those ones where: *(i)* loginMail is skipped (if the user is already logged in the client email); *(ii)* the pair of actions ⟨openWorkbook, openGoogleForm⟩ is performed in reverse order; *(iii)* the sequence of actions ⟨getCell, copy, clickTextField, paste⟩ is executed several time before submitting the Google form.

## 2.2   Segmentation in RPA

Given a UI log that consists of events including user actions with the same granularity[2] and potentially belonging to different routines, in the RPA domain

---

[2] The UI logs created by generic action loggers usually consist of low-level events associated one-by-one to a recorded user action on the UI (e.g., mouse clicks, etc.).

*segmentation* is the task of clustering parts of the log together which belong to the same routine. In a nutshell, the challenge is to automatically understand which user actions contribute to which routines and organize such user actions in well-bounded routine traces [4,20].

In [6] we identified three main forms of UI logs and their segmentation variants, which can be categorized according to the fact that: *(i)* any user action in the log exclusively belongs to a specific routine; *(ii)* the log records the execution of many routines that do not have any user action in common; *(iii)* the log records the execution of many routines, with the possibility that some performed user actions are shared by many routines at the same time. In the following, we analyze literature works in terms of supported segmentation variants.

Concerning RPA-related techniques, Bosco et al. [10] provide a method that exploits rule mining and data transformation techniques, able to discover routines that are fully deterministic and thus amenable for automation directly from UI logs. The method combines a technique for compressing a set of sequences of user actions into an acyclic automaton using rule mining techniques and data transformations. This approach is effective in the case of UI logs that keep track of well-bounded routine executions and becomes inadequate when the UI log records information about several routines whose actions are potentially interleaved. In this direction, Leno et al. [19] propose a technique to identify execution traces of a specific routine relying on the automated synthesis of a control-flow graph, describing the observed directly-follow relations between the user actions. The technique in [19] loses in accuracy in the presence of recurrent noise and interleaved routine executions while it is not able to handle UI logs that record in an interleaved fashion shared user actions of many different routines.

Even if more focused on traditional business processes in BPM rather than on RPA routines, Bayomie et al. [9] address the problem of correlating uncorrelated event logs in process mining in which they assume the model of the routine is known. Since event logs allow to store traces of one process model only, this technique is able to handle logs recording user actions belonging to a specific routine. In the field of process discovery, Mărușter et al. [24] propose an empirical method for inducing rule sets from event logs containing the execution of one process only. Therefore, as in [9], this method is able to partially achieve the first case, thus making the technique ineffective in the presence of interleaved and shared user actions. A more robust approach, developed by Fazzinga et al. [12], employs predefined behavioural models to establish which process activities belong to which process model. The technique works well when there are no interleaved user actions belonging to one or more routines since it cannot discriminate which event instance (but just the event type) belongs to which process model. This makes [12] effective to partially tackle all three cases. Closely related to [12], there is the work of Liu [21]. The author proposes a probabilistic approach to learn workflow models from interleaved event logs, dealing with noises in the log data. Since each workflow is assigned with a disjoint set of operations, it means the proposed approach is able to partially achieve first two cases (the approach can lose accuracy in assigning operations to workflows).

Differently from the previous works, Time-Aware Partitioning (TAP) techniques cut event logs based on the temporal distance between two events [18,28]. The main limitation of TAP approaches is that they rely only on the time gap between events without considering any process/routine context. For this reason, such techniques cannot handle neither interleaved user actions of different routine executions nor interleaved user actions of different routines.

There exist other approaches whose target is not to exactly resolve the segmentation issue. Many research works exist that analyze UI logs at different abstraction levels, which can be potentially valuable for realizing segmentation techniques. With the term *"abstraction"* we mean that groups of user actions to be interpreted as executions of high-level activities. Baier et al. [8] propose a method to find a global one-to-one mapping between the user actions that appear in the UI log and the high-level activities of a given interaction model. This method leverages constraint-satisfaction techniques to reduce the set of candidate mappings. Similarly, Ferreira et al. [13], starting from a state-machine model describing the routine of interest in terms of high-level activities, employ heuristic techniques to find a mapping from a "micro-sequence" of user actions to the "macro-sequence" of activities in the state-machine model. Finally, Mannhardt et al. [23] present a technique that maps low-level event types to multiple high-level activities (while the event instances, i.e., with a specific timestamp in the log, can be coupled with a single high-level activity). However, segmentation techniques in RPA must enable to associate low-level event instances (corresponding to user actions) to multiple routines, making abstractions techniques ineffective to tackle all those cases where is the presence of interleaving user actions of the same (or different) routine(s).

The analysis of the related work has pointed out that the majority of literature approaches are able to properly extract routine traces from unsegmented UI logs when the routine executions are not interleaved from each others, which is far from being a realistic assumption. Only a few works [5,12,19,21] have demonstrated the full or partial ability to untangle unsegmented UI logs consisting of many interleaved routine executions, but with any routine providing its own, separate universe of user actions. However, we did not find any literature work able to properly deal with user actions potentially shared by many routine executions in the UI log. This is a relevant limitation since it is quite common that a user interaction with the UI corresponds to the executions of many routine steps at once. Moreover, it is worth noticing the majority of the literature works rely on the so-called *supervised* assumption, which consists of some a priori knowledge of the structure of routines. Of course, this knowledge may ease the task of segmenting a UI log. But, as a side effect, it may strongly constrain the discovery of routine traces only to the "paths" allowed by the routines' structure, thus neglecting that some valid infrequent routine variants may exist in the UI log.

Finally, we want to underline that process discovery techniques [7] can also play a relevant role in tackling the segmentation issue, as demonstrated by some literature works [9,12,21]. However, the problem is that most discovery techniques work with event logs containing behaviours related to the execution of a single process model only. And, more importantly, event logs are already segmented into traces, i.e., with clear starting and ending points that delimitate any recorded process execution. Conversely, a UI log consists of a long sequence of user actions belonging to different routines without any clear starting/ending point. Thus, a UI log is more similar to a unique trace consisting of thousands of fine-grained user actions. With a UI log as input, the application of traditional discovery algorithms seems unsuited to discover routine traces and associate them to some routine models, even if more research is needed in this area.

## 3   Segments Compliance Analysis

The main limitations of state-of-the-art segmentation techniques are tackled in [3]. Here, we have presented a new approach to the automated segmentation of UI logs which is able to extract routine traces from unsegmented UI logs that record in an interleaved fashion many different routines. Specifically, in [3] when routine segments have been discovered from a UI log, there exists the possibility that many of them represent not allowed routine behaviours. This happens because a UI log combines the execution of several routines that are usually interleaved from each others. In addition, in case of routines that make use of the same kinds of user actions to achieve their goals, it may happen that new patterns of repeated user actions, which represent potential not allowed routine segments, are rather detected as valid ones within the UI log.

Towards this direction, we realized a stand-alone web application called SCAN[3] (*Segments Compliance ANalysis*), which allows to support human experts in performing the segmentation task. The tool enables to visualize the declarative constraints (i.e., the temporally extended relations between user actions) that must be satisfied throughout the discovered routine segments from the UI log. The constraints are represented using Declare, a well-known declarative process modeling language introduced in [25]. This knowledge allows human experts to identify and remove those constraints that should not be compliant with any real-world routine behaviour. Detecting and removing these constraints means to filter out all the not allowed (i.e., wrongly discovered) routine segments from the UI log, as shown in Fig. 2. Declare constraints can be divided into four main groups: existence, relation, mutual and negative constraints. We notice that the use of declarative notations has been already demonstrated as an effective tool to visually support expert users in the analysis of event logs [26].

For instance, if we consider the following valid routine segment of $R_{example}$ (cf. Sect. 2.1): ⟨loginMail, accessMail, downloadAttachment, openWorkbook, open GoogleForm, getCell, copy, clickTextField, paste, formSubmit⟩, then these Declare constraints must hold:

- *Start*(loginMail)
- *Precedence*(getCell,copy), *Precedence*(clickTextField,paste)
- *End*(formSubmit)

---

[3] SCAN can be downloaded at: https://github.com/bpm-diag/SCAN.

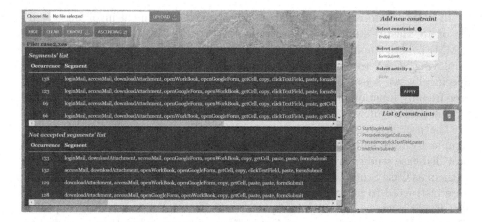

**Fig. 2.** GUI of SCAN

An expert user that is aware of the behaviour of the real-world routines under analysis can immediately understand that those segments not satisfying the above Declare constraints should be filtered out. For this reason, the above Declare constraints can be considered representative of $R_{example}$. As a consequence, all the discovered segments for which one of the above Declare constraints does not hold can be immediately discarded.

We point out that the iterative analysis of the Declare constraints associated to the discovered segments will support the human experts to easily detect and filter out those segments that must not be later emulated by SW robots.

## 4 Evaluation

In this section, we present the results of a multi-step evaluation performed on SCAN to investigate the extent to which our approach satisfies three relevant non-functional requirements, namely *effectiveness*, *robustness*, and *usability*. The target is to understand if SCAN can potentially complement the traditional solutions provided by open-source Process Mining tools for performing the segmentation task in RPA.

### 4.1 Evaluating the Effectiveness of SCAN

An approach that simplifies the segmentation task in RPA, and in particular the inspection of routine segments required to filter out the not allowed ones, can be considered as a relevant artefact to investigate. Consequently, the research question (**RQ**) we aim to investigate is the following one: *"What is the effectiveness of employing an approach that semi-automatically filters out the not allowed routine segments, thus neglecting the (manual) identification stage of the not allowed real-word routine behaviour, through declarative constraints?"*.

In order to address **RQ** we enacted a controlled experiment involving a sample of 18 Master students of the course of Process Management and Mining (PMM) held at Sapienza University of Rome, to investigate the effectiveness of employing SCAN to perform the segmentation task when compared to Disco[4]. Specifically, we selected Disco as target Process Mining tool since it provides user-friendly functionalities, integrated with filtering facilities that allows to filter out the not allowed routine segments as stored into event logs.

The user study was conducted as follows. Two case studies of increasing complexity were submitted to two different user groups of PMM students. The provided case studies are inspired by the one presented in Sect. 2.1 and we refer to them as Case Study #1 and Case Study #2. A first group of 9 PMM students were instructed to perform the case studies #1 and #2 exclusively with Disco. We denote with $p_1$ this first group of users. In parallel, a second group of 9 PMM students received the same instructions of group $p_1$ but they are asked to use SCAN rather than Disco. We denote with $p_2$ this second group of users. It is worth noticing that all the PMM students involved in the user study can be considered expert users in business process modelling and automation.

To assess the effectiveness of SCAN in filtering out the not allowed segments, we investigated the following experimental hypothesis $H_1$: *Employing SCAN, thus neglecting the manual identification stage of the not allowed real-word routine behaviour through declarative constraints, is more effective than employing traditional approaches (e.g. Disco) that require to manually identify and filter out the not allowed routine segments.* To validate $H_1$, a *between-subject approach* was used, i.e., each user in $p_1$ ($p_2$, respectively) was assigned to a different experimental condition, related to the exclusive use of SCAN ($c_1$) or Disco ($c_2$) to perform the required steps for accomplishing both case studies. Any user in $p_1$ was preliminarily instructed about the functionalities of SCAN through a short training session, while the users in $p_2$ already know how to use Disco.

We evaluated the validity of $H_1$ by asking any user expert that completed the user study the following three questions:

- $Q_1$: *The segment's filtering process required to filter out the not allowed routine segments is a complex task. Do you agree?*
- $Q_2$: *The inspection of the routine segments is a complex task. Do you agree?*
- $Q_3$: *SCAN (Disco, respectively) makes the segmentation task feasible. Do you agree?*

Questions are rated with a 5-point Likert scale ranging from 1 ("strongly disagree") to 5 ("strongly agree"). To validate $Q_1$, $Q_2$ and $Q_3$ we performed a comparison of the rates obtained from the questionnaire, respectively in the cases of $c_1$ and $c_2$. Specifically, for each question, we employed a *2-Sample t-test* with a 95% confidence level to determine whether the means between the two distinct populations (i.e., $p_1$ and $p_2$) involved in $c_1$ and $c_2$ differ. We measured the level of statistical significance by analyzing the resulting p-value. We remind that a $p-value \leq 0.05$ is considered to be statistically significant, while a $p-value \leq 0.01$ indicates

---

[4] https://fluxicon.com/disco/.

that there is substantial evidence in favour of the experimental hypothesis. The results of the analysis are summarized in Table 1 that shows the values sorted in descending order, assigned to the responses of each user.

**Table 1.** Effectiveness of SCAN: p-values associated to each question

| $Q_1$ | | $Q_2$ | | $Q_3$ | |
|---|---|---|---|---|---|
| DISCO | SCAN | DISCO | SCAN | DISCO | SCAN |
| 5 | 4 | 5 | 4 | 4 | 5 |
| 4 | 4 | 5 | 3 | 4 | 5 |
| 4 | 3 | 5 | 3 | 3 | 5 |
| 4 | 2 | 5 | 3 | 3 | 5 |
| 4 | 2 | 4 | 2 | 2 | 5 |
| 3 | 2 | 4 | 2 | 2 | 4 |
| 3 | 2 | 4 | 1 | 1 | 4 |
| 2 | 2 | 3 | 1 | 1 | 4 |
| 1 | 2 | 2 | 1 | 1 | 4 |
| p-value: 0.1443957 | | p-value: 0.0018155 | | p-value: 0.0005373 | |

It appears evident that the experimental hypothesis $H_1$ is statistically supported by the results obtained for $Q_2$ and $Q_3$, while it is rejected for $Q_1$. Concerning $Q_1$, *it seems that the segment's filtering process was relatively easier in SCAN with respect to Disco.* Still, there is no statistical difference among the two distinct populations since for $Q_1$, the p-value obtained is 0.1443957, which is greater than 0.05, and this means that hypothesis $H_1$ is rejected on $Q_1$. *On the other hand, the inspection of routine segments in Disco seems to be more complex than SCAN* since, for $Q_2$, the p-value obtained is 0.0018155, which is less than 0.05, and this means that the hypothesis $H_1$ is accepted on $Q_2$. Finally, for $Q_3$, we got a p-value equal to 0.0005373, which is less than 0.05, and this means the hypothesis $H_1$ is accepted on $Q_3$. In particular, this value is less than 0.01, meaning that there is a substantial difference between the means of the two distinct populations. This is reflected in higher values associated with SCAN and lower values associated with Disco, *thus making the segmentation task more feasible in SCAN with respect to Disco.* Therefore, $H_1$ can be considered partially accepted since it is validated for both $Q_2$ and $Q_3$ but rejected for $Q_1$, where there is no statistical evidence that the use of SCAN is more effective than traditional process mining solutions (e.g., Disco) in the process of segment's filtering.

### 4.2 Assessing the Robustness of SCAN

To investigate the robustness of SCAN to the achievement of user tasks specified in both Case Study #1 and Case Study #2, we collected the event logs resulting as an output of the user study, and then we compared them with the ground truth event logs (i.e., we computed a priori the event logs as results of the case studies). Precisely, the *robustness* is measured as the ratio between the number of logs compliant with the ground truth logs and the total number of logs, both for $p_1$ (i.e., SCAN) and $p_2$ (i.e., Disco) grouped by Case (i.e., Case Study #1 and Case Study #2).

In the following, we will show the results obtained both for Case Study #1 and for Case Study #2. Note that both the populations $p_1$ and $p_2$ first executed Case Study #1 in a limited time of 10 min and then Case Study #2, considered more complex, in 20 min.

- **Case Study #1.** Both $p_1$ and $p_2$ had 10 min to read the assigned track and run the task either on Disco (i.e., $p_2$) or SCAN (i.e., $p_1$) respectively. For $p_2$, it is important to remember that users already know how to use the tool. The results obtained in this case is that 8 people out of 9 have executed the task arriving at the right event log, while 1 has obtained a wrong result. Thus, the robustness in case of $p_2$ is as follows $Robustness_{p_2} = \frac{8}{9} = 0.88$.

  On the other hand, for $p_1$, we remind the reader that the users experienced SCAN for the first time during this experiments session. In this case, the number of users who achieved the right result is 6 out of 9, while 3 have computed a wrong event log. Therefore, the robustness in case of $p_1$ is $Robustness_{p_1} = \frac{6}{9} = 0.66$.

- **Case Study #2.** This case was executed immediately after the first one. The time allowed for achieving the task was 20 min due to the major complexity with respect to the previous one. For the class of users belonging to $p_2$, the result obtained was that 4 out of 9 people have computed the right result while 5 the wrong one. It follows that the robustness in case of $p_2$ is $Robustness_{p_2} = \frac{4}{9} = 0.44$.

  On the contrary, users assigned to $p_1$ performs much better. Indeed, 7 users among 9 computed the right result, while 2 the wrong one. As a consequence, the correctness for the users that used SCAN is $Robustness_{p_1} = \frac{7}{9} = 0.77$.

If we make a comparison between the degree of robustness for both SCAN and Disco in each case study, it can be stated that:

- *For Case Study #1, better results are achieved with Disco.* This is because the original log contains solely 8 routine segments, and among these only 4 were correct. For this reason, they were easily identifiable and therefore easy to be manually filtered. Regarding SCAN, we can say that since this was the first time the users experienced the tool, it is possible that the limited time of 10 min was not enough for completing the task. In addition, it is also possible that users had not yet settled into using SCAN even if they had been instructed during the short training session.
- *On the other hand, for Case Study #2, better results are achieved with SCAN.* Since the original log presents more than 80 routine segments, the manual identification stage of the wrong routine segments makes the filtering steps even more challenging with Disco (that required the users to filter the wrong routine segments one by one) rather than with SCAN. Indeed, through SCAN, it is possible to apply a limited number of declarative constraints to filter out

a large number of wrong routine segments, thus neglecting the manual identification stage of Disco. In addition, the learning effective plays an essential role in the achievement of good results since users trained their-self while completing the task outlined in Case Study #1. This learning experience is thus reflected in the accomplishment of Case Study #2.

## 4.3  Quantifying the Usability of the UI of SCAN

We investigated the degree of *usability* of the UI developed for SCAN through the administration of the SUS (Software Usability Scale) questionnaire (which is one of the most widely used methodologies to measure the users' perception of the usability of a tool [11]) to the 9 PMM students that were involved in the experimental condition $c_1$, i.e., that used SCAN. The questionnaire consists of 10 statements, adapted to SCAN and, evaluated with a Likert scale that ranges from 1 ("strongly disagree") to 5 ("strongly agree"):

q1) I think that I would like to use SCAN frequently.
q2) I found SCAN unnecessary complex.
q3) I thought SCAN was easy to use.
q4) I think that I would need the support of a technical person to be able to use SCAN.
q5) I found the various functions in SCAN well integrated.
q6) I thought there was too much inconsistency in SCAN.
q7) I would imagine that most people would learn to use SCAN very quickly.
q8) I found SCAN very awkward to use.
q9) I felt very confident using SCAN.
q10) I needed to learn a lot of things before I could get going with SCAN.

**Table 2.** Computation of the SUS overall score

| Participant | q1 | q2 | q3 | q4 | q5 | q6 | q7 | q8 | q9 | q10 | SUS score | Average |
|---|---|---|---|---|---|---|---|---|---|---|---|---|
| p1 | 5 | 1 | 5 | 1 | 5 | 1 | 5 | 1 | 5 | 1 | 100 | 82,5 |
| p2 | 5 | 2 | 4 | 1 | 4 | 2 | 5 | 2 | 3 | 3 | 77,5 | |
| p3 | 5 | 1 | 4 | 1 | 4 | 2 | 2 | 1 | 4 | 2 | 80 | |
| p4 | 4 | 3 | 4 | 3 | 3 | 2 | 4 | 2 | 4 | 2 | 67,5 | |
| p5 | 4 | 1 | 4 | 3 | 4 | 2 | 5 | 1 | 4 | 3 | 77,5 | |
| p6 | 4 | 2 | 5 | 2 | 4 | 2 | 5 | 1 | 4 | 2 | 82,5 | |
| p7 | 5 | 4 | 5 | 2 | 5 | 1 | 5 | 4 | 5 | 1 | 82,5 | |
| p8 | 4 | 2 | 5 | 1 | 4 | 2 | 5 | 2 | 5 | 2 | 85 | |
| p9 | 5 | 1 | 5 | 1 | 4 | 2 | 4 | 2 | 5 | 1 | 90 | |

At the end of the questionnaire, an overall score is assigned to the questionnaire. The score can be compared with several benchmarks presented in the research literature to determine the degree of usability of the tool being evaluated. In our test, we made use of the benchmark given in [27], which associates to each range of the SUS score a percentile ranking varying from 0 to 100, indicating how well it compares to other 5,000 SUS observations performed in the literature. The collection of the ranks associated with any statement of the SUS is reported in Table 2, calculated following the steps discussed in [27].

Since the average SUS score obtained by SCAN was 82.5, according to the selected benchmark [27], the usability of the tool corresponds to a rank of A, which indicates a degree of usability almost excellent.

*The result shows that the UI implemented has been comprehensive and straightforward since the first use of the tool. And also that the use of the tool has been found effective and performing in achieving the required tasks.*

## 5   Conclusion

RPA recently gained a lot of attention in the BPM domain [1]. Since RPA operates at the UI level, rather than at the system level, it allows applying automation without any changes in the underlying information system. However, the current generation of RPA tools is driven by predefined rules and manual configurations made by expert users rather than by automated techniques [22], preventing widespread adoption of these tools in the BPM domain.

Still, to date, a great deal of time is required to identify the routines for automation and manually program the SW robots. Even if RPA tools are able to automate a wide range of routines, they cannot determine which routines should be automated in the first place. Indeed, in the early stages of the RPA life-cycle it is required to: (1) identify the candidate routines to automate through interviews and detailed observation of workers conducting their daily work, (2) record the interactions that take place during the routines' enactment on the UI of software applications into dedicated UI logs, and (3) manually specify their conceptual and technical structure (often in form of flowchart diagrams) for identifying the behaviour of SW robots. Towards this direction, the presented work tries to improve the process of segments identification (cf. step 1) performed by skilled human experts, throughout the development of a human-in-the-loop approach that support human experts in visualizing and filtering out all those segments discovered from a UI log not satisfying specific declarative constraints. We implemented our approach as a stand-alone web application called SCAN which has been evaluated through the measurement of three non-functional requirements, namely, *effectiveness, robustness* and *usability*.

The presented approach can be leveraged by segmentation techniques that are able to discover from scratch the structure of the routines under analysis that were previously captured in a UI log, thus increasing the quality of discovered routine segments. However, the main limitation of our approach relies on the involvement of human experts when the automated discovery of the routine segments is completed as required by the approach itself, given the possibility of complementing the unsupervised assumption with the experts' knowledge.

For this reason, we think that an important step towards the development of a more complete and unsupervised technique to the segmentation of UI logs is to shift from current semi-supervised learning approaches to completely unsupervised ones [2,14,15].

**Acknowledgements.** This work has been supported by the H2020 project Data-Cloud (grant ID 101016835), the Sapienza grant BPbots, by the Italian projects Social Museum and Smart Tourism (CTN01_00034_23154) and RoMA - Resilience of Metropolitan Areas (SCN_00064).

# References

1. van der Aalst, W.M.P., Bichler, M., Heinzl, A.: Robotic process automation. Bus. Inf. Syst. Eng. **60**(4), 269–272 (2018). https://doi.org/10.1007/s12599-018-0542-4
2. van der Aalst, W.M.P., Bose, R.P.J.C.: Abstractions in process mining: a taxonomy of patterns. In: Dayal, U., Eder, J., Koehler, J., Reijers, H.A. (eds.) BPM 2009. LNCS, vol. 5701, pp. 159–175. Springer, Heidelberg (2009). https://doi.org/10.1007/978-3-642-03848-8_12
3. Agostinelli, S., Leotta, F., Marrella, A.: Interactive segmentation of user interface logs. In: 19th International Conference on Service-Oriented Computing, ICSOC 2021, vol. 13121, pp. 65–80 (2021). https://doi.org/10.1007/978-3-030-91431-8_5
4. Agostinelli, S., Marrella, A., Mecella, M.: Research challenges for intelligent robotic process automation. In: Di Francescomarino, C., Dijkman, R., Zdun, U. (eds.) BPM 2019. LNBIP, vol. 362, pp. 12–18. Springer, Cham (2019). https://doi.org/10.1007/978-3-030-37453-2_2
5. Agostinelli, S., Marrella, A., Mecella, M.: Automated Segmentation of User Interface Logs. In: RPA. Management, Technology, Applications. De Gruyter (2021). https://doi.org/10.1109/EDOC.2019.00026
6. Agostinelli, S., Marrella, A., Mecella, M.: Exploring the challenge of automated segmentation in robotic process automation. In: Cherfi, S., Perini, A., Nurcan, S. (eds.) RCIS 2021. LNBIP, vol. 415, pp. 38–54. Springer, Cham (2021). https://doi.org/10.1007/978-3-030-75018-3_3
7. Augusto, A., et al.: Automated discovery of process models from event logs: review and benchmark. IEEE Trans. Knowl. Data Eng. **31**(4), 686–705 (2019). https://doi.org/10.1109/TKDE.2018.2841877
8. Baier, T., Rogge-Solti, A., Mendling, J., Weske, M.: Matching of events and activities: an approach based on behavioral constraint satisfaction. In: ACM Symposium on Applied Computing, pp. 1225–1230 (2015). https://doi.org/10.1145/2695664.2699491
9. Bayomie, D., Di Ciccio, C., La Rosa, M., Mendling, J.: A probabilistic approach to event-case correlation for process mining. In: Laender, A.H.F., Pernici, B., Lim, E.-P., de Oliveira, J.P.M. (eds.) ER 2019. LNCS, vol. 11788, pp. 136–152. Springer, Cham (2019). https://doi.org/10.1007/978-3-030-33223-5_12
10. Bosco, A., Augusto, A., Dumas, M., La Rosa, M., Fortino, G.: Discovering automatable routines from user interaction logs. In: Hildebrandt, T., van Dongen, B.F., Röglinger, M., Mendling, J. (eds.) BPM 2019. LNBIP, vol. 360, pp. 144–162. Springer, Cham (2019). https://doi.org/10.1007/978-3-030-26643-1_9
11. Brooke, J.: SUS: a retrospective. J. Usability Stud. **8**(2), 29–40 (2013)
12. Fazzinga, B., Flesca, S., Furfaro, F., Masciari, E., Pontieri, L.: Efficiently interpreting traces of low level events in business process logs. Inf. Syst. **73**, 1–24 (2018). https://doi.org/10.1016/j.is.2017.11.001

13. Ferreira, D.R., Szimanski, F., Ralha, C.G.: Improving process models by mining mappings of low-level events to high-level activities. J. Intell. Inf. Syst. **43**(2), 379–407 (2014). https://doi.org/10.1007/s10844-014-0327-2
14. Folino, F., Guarascio, M., Pontieri, L.: Mining multi-variant process models from low-level logs. In: Abramowicz, W. (ed.) BIS 2015. LNBIP, vol. 208, pp. 165–177. Springer, Cham (2015). https://doi.org/10.1007/978-3-319-19027-3_14
15. Günther, C.W., Rozinat, A., van der Aalst, W.M.P.: Activity mining by global trace segmentation. In: Rinderle-Ma, S., Sadiq, S., Leymann, F. (eds.) BPM 2009. LNBIP, vol. 43, pp. 128–139. Springer, Heidelberg (2010). https://doi.org/10.1007/978-3-642-12186-9_13
16. Jimenez-Ramirez, A., Reijers, H.A., Barba, I., Del Valle, C.: A method to improve the early stages of the robotic process automation lifecycle. In: Giorgini, P., Weber, B. (eds.) CAiSE 2019. LNCS, vol. 11483, pp. 446–461. Springer, Cham (2019). https://doi.org/10.1007/978-3-030-21290-2_28
17. Kirchmer, M.: Robotic Process Automation-Pragmatic Solution or Dangerous Illusion. BTOES Insights, June'17 (2017)
18. Kumar, A., Salo, J., Li, H.: Stages of user engagement on social commerce platforms: analysis with the navigational clickstream data. Int. J. Electron. Commer. **23**(2), 179–211 (2019). https://doi.org/10.1080/10864415.2018.1564550
19. Leno, V., Augusto, A., Dumas, M., La Rosa, M., Maggi, F.M., Polyvyanyy, A.: Identifying candidate routines for robotic process automation from unsegmented UI logs. In: 2nd International Conference on Process Mining, pp. 153–160 (2020). https://doi.org/10.1109/ICPM49681.2020.00031
20. Leno, V., Polyvyanyy, A., Dumas, M., La Rosa, M., Maggi, F.M.: Robotic process mining: vision and challenges. Bus. Inf. Syst. Eng. **63**(3), 301–314 (2020). https://doi.org/10.1007/s12599-020-00641-4
21. Liu, X.: Unraveling and learning workflow models from interleaved event logs. In: 2014 IEEE International Conference on Web Services, pp. 193–200 (2014). https://doi.org/10.1109/ICWS.2014.38
22. Lohr, S.: The Beginning of a Wave: A.I. Tiptoes Into the Workplace (2018). https://www.nytimes.com/2018/08/05/technology/workplace-ai.html/
23. Mannhardt, F., de Leoni, M., Reijers, H.A., van der Aalst, W.M., Toussaint, P.J.: Guided process discovery - a pattern-based approach. Inf. Syst. **76**, 1–18 (2018). https://doi.org/10.1016/j.is.2018.01.009
24. Mărușter, L., Weijters, A.T., Van Der Aalst, W.M., Van Den Bosch, A.: A rule-based approach for process discovery: dealing with noise and imbalance in process logs. Data Mining Knowl. Discov. **13**(1), 67–87 (2006). https://doi.org/10.1007/s10618-005-0029-z
25. Pesic, M., Schonenberg, H., van Der Aalst, W.M.: Declarative workflows: balancing between flexibility and support. Comput. Sci.-Res. Dev. **23**(2), 99–113 (2009). https://doi.org/10.1007/s00450-009-0057-9
26. Rovani, M., Maggi, F.M., de Leoni, M., van der Aalst, W.M.: Declarative process mining in healthcare. Expert Syst. Appl. **42**(23), 9236–9251 (2015)
27. Sauro, J., Lewis, J.R.: Quantifying the User Experience: Practical Statistics for User Research. Morgan Kaufmann, Burlington (2016). https://doi.org/10.1145/2413038.2413056
28. Srivastava, J., Cooley, R., Deshpande, M., Tan, P.: Web usage mining: discovery and applications of usage patterns from web data. SIGKDD Exp. **1**(2), 12–23 (2000). https://doi.org/10.1145/846183.846188

# Recommending Next Best Skill in Conversational Robotic Process Automation

Avi Yaeli[(✉)], Segev Shlomov, Alon Oved, Sergey Zeltyn, and Nir Mashkif

IBM Research - Haifa, Haifa, Israel
{aviy,sergeyz,nirm}@il.ibm.com, {segev.shlomov1,alon.oved}@ibm.com

**Abstract.** In recent years, Robotic Process Automation (RPA) has been widely adopted across the industry as an important enabler for business process automation and digital transformation. Recent advancements suggest that next generation RPA will require advanced human-robot collaboration capabilities for providing a more natural conversational interface and supporting more complex automation orchestration needs. Our work focuses on the nascent field of conversational RPA bots that are able to dynamically orchestrate automation tasks through natural language. In this context, recommending possible utterances and next steps to the user is an important capability to enhance human-bot collaboration. We take an exploratory approach to the problem of next-best-skill recommendation in human-robot collaboration. We highlight key characteristics of this problem, examine existing approaches, and call out specific challenges in implementing a solution. We suggest that this problem calls for an integrated strategy for recommendation, and illustrate a possible implementation architecture that can integrate multiple recommendation strategies.

**Keywords:** Intelligent process automation · Robotic process automation · Prescriptive process monitoring · Multi-agent orchestration · Collaborative RPA · Conversational interfaces for RPA

## 1   Introduction

Over the past decade, Robotic Process Automation (RPA) has become widely adopted across the industry as an important and affordable enabler for business process automation and digital transformation [1]. Traditionally, RPA provided a low-code development environment for citizen-developers to program front-end automations of repetitive tasks. In recent years, RPA includes support for collaboration and reuse, centralized governance, connectors and integrations, and a community marketplace for automation reuse and customization [11].

More recently, the term Intelligent Process Automation (IPA) is used to describe the next generation of RPA and digital workers. IPA combines trends and technologies such as document processing, natural language processing, process

© Springer Nature Switzerland AG 2022
A. Marrella et al. (Eds.): BPM 2022, LNBIP 459, pp. 215–230, 2022.
https://doi.org/10.1007/978-3-031-16168-1_14

mining, human-to-bot collaboration, predictive and prescriptive analytics, conversational Artificial Intelligence (AI), and dynamic process orchestration [22]. IPA promises to support a wider and more intelligent form of automation, and enable developers to build, deploy, and engage naturally with digital workers.

Despite many advancements, RPA is still mostly rule-based and capable only of handling repetitive tasks. Programming RPA includes capturing user interface (UI) interactions, generalizing those into a program, or customizing a prebuilt template. To support the long tail of automation use cases, RPA will also need to support more flexible and adhoc workflows, understand work at a higher level of semantics, and support additional patterns of human-robot collaboration. Chakraborti et al. [6,7] describe a method for dynamic composition and orchestration of automation through natural language interface and the use of an AI planner. Similarly Watson Orchestrate [15] provides a large catalog of automation skills and a conversational interface through which the dynamic orchestration of these skills is possible.

As employees spend more time in messaging platforms, it is essential to design an effective conversational interface for them to engage with digital workers. In the modality of conversation, a typical interaction pattern includes the user clicking a button or sending text message to a bot, and the bot providing a response to the user. Depending on the automation type and dialog design, responses may include answers to user questions; requests for additional inputs; requests for clarification of the user's intent (disambiguation); menu options for follow up actions; error message from performing an action, or request for feedback.

As RPA bots become more intelligent and include a larger set of automation skills, it may be very difficult for a user to understand what the bot is capable of doing in general, what new useful orchestrations are worth trying out, what are the next best steps for moving the process forward, and what not to ask, as it might trigger an error or result in frustration.

In this work, we identify the need for recommending the next-best-skill (NBS) in RPA automation. We distinguish between two types of recommendations that play different roles: (1) the NBS to invoke from the loaded skills of a digital employee; and (2) the NBS to load to a digital employee from the available skills in the catalog. In this paper, we present both types but mainly focus on the former, which is our current research focus.

The contribution of this paper is three-fold. First, we describe a new problem of next-best-skill recommendations in the context of conversational RPA. Second, we analyze the key characteristics of the problem, relate to existing approaches, and call out specific challenges. Third, we illustrate a possible architecture that can implement an integrated recommendation approach.

**Outline.** The paper is organized as an opinion paper. Section 2 introduces key concepts and representative use case in conversational RPA. Section 3 presents and discusses key dimensions of the problem, related art, and challenges. Section 4 illustrates an architecture for next-best-skill in RPA. Finally, Sect. 5 summarizes the paper.

## 2    Definitions and Use Cases

To present and discuss the problem of next-best-skill recommendations, we first introduce some common concepts and terms. A summary of these definitions is provided in Table 1. We also present three use cases that exemplify these concepts.

A *digital employee* (hereinafter referred to as a *digi*) is a software bot that collaborates with humans, systems, and other bots to automate work. A digi has an identity that is used for authorization and access to enterprise systems. Digis are built from a collection of *skills* and an *orchestration* logic. A skill is the most granular automation building block that a digi can use, and the orchestration is a configuration of how skills can be orchestrated together and how to evaluate user utterances vis-à-vis multiple skills. A digi can communicate with humans over multiple *channels* such as web and mobile. When a digi is deployed on a conversational channel, a user can send it an utterance representing a question, task, parameter, or any dialog act type[1].

The configuration of the orchestration determines the routing of the utterance to the right skill or sequence of skills for execution (aka actuation) in a single turn or multiple turns of conversation. The *short-term memory* of a digi is a runtime context object that is used to store interim results from skills across the user's conversation *session* with a digi. This enables automatic passing of parameters between skills without the need of the user to explicitly mention them in the subsequent utterance.

Digis can be created by instantiating a pre-configured template, or through manual configuration of an orchestration and picking specific skills to add to it from a *catalog*. The task of adding skills from the catalog is referred to as *upskill*. The task of making a skill available to the catalog is referred to as *bootstrapping*.

The actual digis may vary quite a lot in their goals, the type of processes they support, their skills, how users engage with them, and how they are built, deployed, and updated. To illustrate this variance and to set the stage for discussing the requirements of NBS recommendations, we present three use cases of digis that are inspired by real-world implementations.

***HR Recruitment Sidekick***[2] ***("HR Jerry").*** HR Jerry is a digi focused on automating tasks in employee recruitment. HR Jerry was built after interviewing HR recruiters in several customer organizations and discovering that they use a combination of many disparate tools such as file sharing, mail, task management, messaging, social network and internal HR systems. An example flow is creating a job requisition[3] from a document template and creating tasks for internal review with managers before posting the job to social networks. Another flow concerns receiving resumes from candidates via email, saving the attachment in file sharing, and performing the review and selection process. HR Jerry is built from an initial set of curated skills that perform relevant tasks in the

---

[1] An utterance that serves a function in the dialog (e.g., questions and request).

[2] Sidekick - common RPA slang for a close companion or personal assistant.

[3] A job requisition is document describing the required skills for a job.

**Table 1.** Terms and Definitions.

| Term | Description |
|---|---|
| Application | An application exposing backend APIs, or frontend UI actions that can be accessed by an RPA bot, e.g., Gmail, Box |
| Attended bot | A form of RPA bot that requires human interaction, such as a command to trigger the bot, or the bot collecting inputs from the user |
| Automation templates | Pre-built automation flows that are customized by a bot developer |
| Bootstrapping skills | Loading application skills to a catalog from which they can be loaded to a digital employee with proper credentials and natural language interface |
| Catalog | A repository of automation skills and templates |
| Channel | A platform such as web, mobile, or messaging through which a user can engage with a bot |
| Digital employee (digi) | A bot that has an identity and can collaborate with humans to automate work |
| Entity | Pieces of information from a user utterance that can be used as parameters for the task, e.g., a job-id for submitting a job task |
| Intent | A classification of a user utterance to needs that can be actuated by the bot |
| Orchestration | A configuration of how a set of skills should be composed to perform a task |
| Session | A group of conversation turns with a digi that take place within a given scope of time frame or between login and logout |
| Short-term memory | A context object associated with a digi and a session that can be used to connect and pass parameters between skills |
| Skill | The most granular automation block. Can be used for composition and orchestration |
| Skill set (skill group) | A group of skills that are logically associated |
| Upskill | The act of adding a skill from the catalog to a digital employee |

recruitment process. Some tasks might represent primitive actions, such as saving attachments from Gmail to Box, while others represent more complex actions, such as analyzing resumes and matching them with job requisition keywords. While HR Jerry was built as a collection of related skills that are designed to automate the recruitment process, it is expected that organizations and teams will further tailor its behavior to the de-facto processes being used and may even add or remove specific skills from its collection. HR Jerry is deployed on Slack so teams can easily engage with it through natural language conversations.

***Long Tail Sidekick ("Luis").*** Luis is a personal digi that includes a large set of primitive skills for common enterprise tools such as mail, file sharing, Microsoft Office, and Github. Luis is published in the catalog as a skill group, so every employee can quickly upskill a personal digi from it. The idea of Luis is to support the long tail of automation use cases in organizations, so it is expected that users will make substantial changes to it as they learn which skills provide them more value and how to customize them. It is also expected that Luis will include skills that were developed by citizen developers through low-code/no-code RPA tools and incrementally upskilled by specific employees to their digis.

***HR Promotion ("HR Pro").*** The HR Pro use case is related to a bi-yearly promotion process in a large software company. Three user types are engaged in the process. First-level managers (coaches) select candidates for promotion and salary increase among their team members. They have access to a number of reports and charts that keep track of the performance of team members and their eligibility for promotion. Upline managers review nominations and either accept or reject them. Finally, the HR talent team launches and closes the promotion cycle. HR users also have access to reports and charts that provide overall data on the promotion process. They track numbers of promotions by different categories according to company goals and diversity criteria. In contrast to the other two use cases, HR Pro is governed by a well-defined business process with a focus on the process deadline. Developed by IT, some process elements are strongly governed and rigid, but there is some variance in how teams perform the process.

## 3    Next Best Skill Recommendation

### 3.1    Problem Formulation and Requirements

While RPA enables users to create rule-based task automations, the conversational channels enable them to trigger, orchestrate, and interact with RPA bots in natural language interface. A common way to integrate RPA and chatbot technologies is to integrate the RPA automation after performing the NLU classification of the action to be performed [8]. The experience of human-to-bot conversation often mimics the chat experience of humans in a messaging channel. Interaction takes place across conversation turns, and the UI is somewhat limited in terms of real-estate[4], navigation, and interaction patterns.

Typically, chatbot UI includes a text input control for user utterances, a messaging-like area to present the chatbot response and the history of the conversation. In a typical interaction pattern, the user provides an initial utterance that once classified into an intent triggers the right RPA automation which can perform a task, including collecting additional parameters from the user via the same conversational channel. To help the user navigate around options, an initial welcome message is presented, which may include some navigational menu steps and suggestions of what to ask. This pattern represents a common design practice in chatbots that can be designed by conversation analysts to lead a user down a specific path of automation.

In conversational RPA bots, which include a larger set of automation skills, it may be very difficult for a user to understand what the bot is capable of doing in general, what new useful orchestrations might be worth trying out, what are the best next steps to engage the bot to move the process forward, as well as what not to ask, which may yield an error and result in frustration. Ideally, we would like to be able to always present a few options to the user, e.g., in the form of UI buttons or text completion, so that the user is never lost in these types of conversational RPA bots.

---

[4] The amount of space available on a display for an application to provide output.

**Requirements.** In the following, we present the requirements for a new class of recommendations that we believe are important for creating an effective user experience in conversational RPA. Collectively we refer to these requirements as *next-best-skill recommendation*. We highlight the motivation for each requirement and suggest the possible basis for recommendation. Next, we present several user needs for next-best-skill recommendations.

**Welcome and what's new.** Providing suggestions for utterances (and skills) that can be used to start a conversation with an RPA bot during a new session. These suggestions can be useful for new users who do not necessarily know what to ask the bot. A variation of this requirement is for existing users, after the skillset of a digi has been recently updated.

**Next task in an activity flow.** Given the state of conversation session or process, suggest the possible next step to perform. There are several cases where this might be useful: (1) a user performed a task and is not sure what can follow next; (2) a user performed a task in one conversation session and may want to follow up in the next session; (3) the result of the previous conversation step is an object in the execution context of the bot, which a user may want to use in a follow up action, e.g., "Send it to John to review", right after "create a job requisition for a devops engineer". Note that the pronoun "it" plays a similar role of a contextual entity in dialog systems.

**Required task in process.** Tasks that must be done, given the state of a process and its goals, e.g., "submit candidate review now to meet deadline".

**Recovery from user struggle.** Recommend options to recover the conversation, when the bot does not understand the user intent.

**Potential new skills to upskill.** Recommend a potential skill that can be added (upskilled) to the digi to be more productive for the user tasks. Note this recommendation is substantially different in nature than the previous ones since it doesn't produce a recommended utterance for the next turn, but rather an action to add or change the skills of the digi.

**Basis for Recommendations.** While these user needs may vary in their specific requirements, it is important to consider them all together as they compete for the shared and limited real-estate of chatbot UI, as well as for the user's attention. But what is the basis for these recommendations, and what type of data can be used for learning? We identify the following main sources of learning:

**Crowd wisdom.** Leverage the experience of other digis and teams to recommend common and successful next skills given a similar context. This approach requires establishing criteria for similarity of digi, teams, and context, as well as criteria for success.

**Personalization.** Given user preferences, habits, and history of interactions with digi across sessions, suggest next skills that are common and successful.

**Discovery.** Given that there is no history for some skills at all, suggest some of them to users and with time gather the knowledge necessary for personal or crowd recommendation.

**Conversation context.** Given recent objects in conversation context and short-term memory of digi, suggest next tasks that can take these objects as parameters for automation tasks.

**Process.** Given a definition of a process and goals in the form of rules, suggest actions that lead the user towards accomplishing the goals.

Besides the user need and the basis for recommendation, there are additional user experience considerations, such as how do we decide which recommendation to present and when? What constitutes a good recommendation? How do we collect feedback on recommendations? And how do we synthesize multiple recommendations into a common UI? Answers to these questions may require empirical studies. However, we strongly believe that an integrated approach is required for an effective user experience - one that can present the right set of recommendations at the right time, be able to synthesize and sort recommendation types, turn recommendations into user-friendly buttons, and provide explainability on the sources of recommendations.

### 3.2 Technical Approaches

In this section, we provide an overview of five technical approaches to next-best-skill recommendations and comment on their relationship with the use cases, user needs, and learning sources described earlier.

*The Rule-Based Approach.* Using this basic Business Process Management (BPM) approach [14], NBS recommendations are provided via deterministic rules. These rules are designed by process stakeholders before the system launch. For example, a rule for HR Pro suggests reviewing an "Eligibility criteria for promotion" report, if not done so before. A rule-based approach is neither scalable nor flexible. It is hardly applicable for large or rapidly changing systems. However, it can constitute a starting point to address the cold start problem. Once enough data is collected, the rule-based approach can be replaced by more advanced methods.

*The Crowd-Wisdom Approach.* These methods help inexperienced users get implicit guidance from more experienced users. For example, HR Pro users could be suggested to view reports that were often viewed by other users. Luis users could receive recommendations concerning popular primitive skills and their sequences. The methods usually start with a prediction of the next action or a sequence of actions from historical data. Typically, deep learning models, such as LSTM, provide the best results for this type of tasks [19] in the classical BPM environment. These methods can also be extended to other domains. For example, Weinzierl et al. [23] use BPM prediction methodology in recommender systems. The sequence of user clicks is modeled as a process and then a natural language processing embedding is applied to recommend the next best click.

After the prediction stage, several top skills with the highest prediction probabilities are recommended to a user. However, these recommendations should not lead to failures. Filtering out such "failure paths" can be addressed via several alternative approaches. In the simplest case, one can discard skills that correspond to undesirable system states, such as fallback or disambiguation question. Alternatively, the data of inexperienced or unsuccessful users can be filtered out from the model training dataset all together. Crowd wisdom methods can be combined with other approaches, such as personalization. Instead of learning from a general historical sample, one can learn from similar users in the spirit of collaborative filtering. In addition, parameters for the recommended skills can be extracted from the context information in short-term memory.

The approach can also be used to recommend the NBS to upskill. We can consider adapting techniques such as semantic-aware content-based filtering [13]. In using such adaptations, we would need to find a good mapping for key concepts in the IPA system. For example, skills could play the role of items; the skill metadata and possibly user utterances invoking these skills could play the role of semantic attributes of the items; and skill catalog rating, usage statistics, and outcome of execution of the skills could play the role of feedback.

*The Personalization Approach.* In the case of one-off personal digis with unique skills, NBS recommendations may need to draw only from the personal experience and preferences of an individual user. In such a case, we can build process-aware personal models at the level of the individual user and process.

*The Goal-Driven Approach.* Often, crowd-wisdom is not sufficient for next-best-skill recommendations and should be combined with the "process wisdom". Business processes that incorporate RPA and conversational bots typically have objectives related to process time, cost, quality, and outcomes [16]. For example, the HR Pro process has a deadline for promotion cycle completion.

The alarm-based approach [21] is a two-step methodology that addresses these process goals. First, prediction of the process performance metrics is performed for the current process instance. Instances with unsatisfactory predictions generate alarms and, then a corrective action is implemented. Similar to a prediction of the next action, deep learning methods provide the best results for process performance metrics [19]. In some papers, prior art research does not elaborate on corrective action modeling once an alarm was generated [18]. Alternatively, corrective interventions are modeled explicitly [5].

*The Reinforcement Learning (RL) Approach.* RL is used in many applications, including gaming, robotics manipulation, self-driving cars, and others. However, it has still not been widely used in BPM. Possible reasons for that include a multi-goal pattern of most business processes and challenges in rigorous definition of all objectives using the corresponding rewards and costs. For example, banking loan applications are popular in predictive and prescriptive process monitoring. Obviously, there are time and cost constraints related to this domain. However, quick rejection or approval of applications without a thorough study of complicated process instances is a suboptimal strategy. One of the first attempt to use RL in BPM has been performed by Agarwal et al. in [2]. They address the problem of multiple goals by combining a time-related reward function with a balancing reward function that tries to mimic the distribution of different process outcomes using the ground truth traces.

### 3.3  Main Challenges

Existing technical approaches can be leveraged to address some aspects of the problem; however, there are additional challenges in implementing the NBS recommendation. In the section below, we point to specific challenges in three main areas: (1) the specification of the learning tasks; (2) handling the cold start problem; and (3) operationalizing an end-to-end pipeline.

**Specification of the Learning Task**

*The Skill Set Drift Problem.* This problem refers to the phenomena that the target space, the skill set, changes over time[5]. In our domain, new skills might be added and old skills might be removed. This makes the predictions progressively less accurate as time passes [12]. To prevent deterioration in prediction accuracy, the system must first track this change and trigger it [3]. Thereafter, the system needs to take proactive action to fix the problem. The solutions for this problem highly depend on the approach taken. For example, in reinforcement learning methods, one must retrain the model, while with the crowd-wisdom approach, the users might stop using the removed skill, so its probability will vanish.

Naturally, the retraining solutions are commonly considered [24]. The literature suggests a frequent retraining on the most recently sample or maintaining several classifiers (with a voting mechanism), where each new classifier that is trained on the new data replaces the oldest one [10]. However, retraining the system every time that the skill set changes might be too demanding and may result in a performance decrease, especially when the skill set changes frequently.

Another solution might come from "digis' wisdom". That is, we observe that different digis sometimes contain similar skill sets. This makes us wonder if and how we can leverage the information "similar" digis have to update the model. For example, when a new skill is added, we can search for other skill sets that contain this skill and increase the a priori probability, according to some portion of its weight, in the different skill sets.

*Boundaries of the Process, and Conversation Session.* Over time, users enter and exit the system frequently. It is common that users want to start a session from the point that they left the previous time. In that case, it can be considered as a natural continuation of a previous conversation. To this end, the data should contain both session id and case id. This suggest a new challenge- should the system makes predictions based on a specific conversation, or should it base it on the overall path of a user, even if a user was engaged in several conversations on different days?

---

[5] In some domains, this problem is also known as concept drift.

Note that this decision mainly affects the features that are used for training the model. Nonetheless, users might behave completely differently over different sessions or forget details of the previous ones. As a result, making such decisions can have a vast impact on the performance. A related feature engineering dilemma is which process features to extract, and "how much" history to use. For example, is information on the previous turn sufficient for prediction, or should it be complemented by process-aware features (e.g., process path statistics such as the order of past activities)?

*Defining and Measuring Success.* This problem is a key challenge for NBS. There are two types of successes that the system should aim for. The first one relates to the recommendation itself. In this type, knowing if users click on the recommendation and if that is considered as a success is a key question for building a good recommendation system. Note that sometimes it is not straightforward to define success. For example, defining success by user clicks is not necessarily the correct way, as users might try different recommendations to explore the system. In addition, even when formally defining success, it is sometimes very hard to measure (e.g., chat-bot containment).

The second type of success relates to the process-path that the users are traversing throughout their sessions. A good system will provide recommendations that help to direct the user to the right end-goal. For example, assume that we are using the crowd-wisdom approach, and that users are frequently using skill A and thereafter skill B. Also, after using the skill pair, we observed that most users abandoned the system. Should we still recommend skill B after skill A?

A few sub challenges appear with respect to that issue. For example, assume that we have a large dataset. Should we train on all of it? Should we train only on "good" paths? Perhaps a weighted combination of "good" and "bad" paths? Can we predict the result of our recommendation to better understand their implications? The latter is also related to explainable AI [9]. Another related challenge is how do we define successful journeys in cases where we do not know the goals of the process.

*The Selection of Learning Dataset.* This issue is crucial for achieving good performance and robust results. The main challenges we observed are as follows: How should we divide the data into test, train, and validation sets so that there is little-to-no dependence between them? Which users should we use for training? How should we correctly divide the data in terms of the temporal aspect (e.g., train on the last year of data or the last month)? Should we use one model per digi or one global model? Should we study multiple similar digis together? Solutions to these challenges may require further studies in real-world settings.

## Cold Start

*The Cold Start Problem.* This problem refers to the inability of a system to draw inferences when the system did not yet gather sufficient information.

The problems has been extensively studied in the literature [4, 17, 20], especially in the domain of recommender systems. In our case, the cold start problem may occur in three main situations: when a new user enrolls in the system, when a new skill enters the catalog, and at the beginning of the system lifecycle. Let's say a new user enters the system. In that case, the system has to provide recommendations without relying on the user's past interactions. The system can rely on other users' interactions (i.e., the wisdom of the crowd) and can use rule-based approaches The system might also try to find users that are relatively "similar" to the new user and suggest recommendations based on their actions.

Similarly, a new skill might be added to the catalog. In that case, the a priori probability that the system will recommend it is small. There are several ways to address that situation. For example, if enough users use the new skill, its probability will increase. The system can also determine a pre-defined rule for recommending new skills (e.g., with a small recommendation on new skills).

The life cycle of a recommender system usually starts when the system is deployed. After the initial startup, the system needs to make predictions but has no information to rely upon. This problem (also known as bootstrapping) is one of the major problems that a recommender system needs to handle, and other strategies should be leveraged until there is sufficient data in the system.

It should be noted that the cold start problem affects most technical approaches. A possible solution to the problem consists of a combination of several technical approaches. First start with the rule-based approach and explore the data, then use prediction-based methods or reinforcement learning methods.

## Operationalizing an End-to-End Pipeline

*Utterance Generation.* This issue is concerned with the challenge of creating a mapping between the recommended skill and an utterance to be displayed to the user corresponding to the next-best-skill. If we have access to the system metadata (e.g., intent examples from conversation engine), then we can use it as the text value for the button. Otherwise, we need to construct these examples from the utterances. In that case, we observe two main challenging tasks. First, make sure that the system preserves its users' privacy (i.e., does not share personal information). Second, construct utterances with minimal disambiguation. That is, we should be careful in choosing utterances that might trigger the wrong skills.

*Integration of Different NBS Approaches.* This issue is essential for achieving a high-quality recommendation system. We hypothesise that using a single NBS approach is not enough due to the complexity of the problem and different properties across the approaches. We need to consider how to synthesize them. Should we merge multiple engines with different confidence? If so, how? Should we present them at different occasions?

One possible option is that models are used in different life cycle stages. Consider the use case of HR Pro. The system can start with a rule-based approach to address the cold-start challenge. Once enough data is collected, crowd-wisdom recommendations can be implemented to help inexperienced users. At the same time, goal-driven model can be trained to predict probabilities of deadline violations and provide alarms to case instances with high probabilities. Finally, the crowd-wisdom and goal-driven recommendations should be supplemented by user feedback on the helpfulness of the recommendations. At that point, the reinforcement learning model can be trained with the feedback incorporated into reward functions.

## 4    Illustration of a System Architecture

We illustrate below a possible architecture for a next-best-skill recommendation (NBSR) system that can integrate recommendations from multiple recommendation approaches. Figure 1 depicts the architecture of an NBSR as a subsystem within a broader conversational RPA system. The system includes the following components.

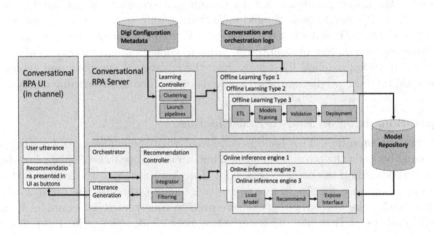

**Fig. 1.** Illustration of the main components of NBSR system and how they integrate within the broader conversational RPA system.

*Learning Controller*: A process that is triggered periodically to manage all the learning pipelines. To support crowd-wisdom recommendations, this controller launches a clustering task that finds and groups similar digis and stores this information for later use. In addition, the controller launches the learning pipelines for each type of recommendation approach discussed in Sect. 3.

*Digi Configuration Metadata*: A persistent data store for the configuration metadata of digis, orchestrations, and the skills they contain.

*Conversation and Orchestration Logs*: A persistent data store containing the complete traces of all the steps taken by the conversational RPA system, from the user utterances and orchestration, including the list of actuated skills, as well as the system's responses.

***Offline Learning Pipeline***: A sequence of steps involved in the learning process. Typically a pipeline would include four steps: (1) fetching the relevant conversation and orchestration logs; (2) training a model; (3) measuring the performance of the model and managing its lifecycle - for example, deciding if a model should be deployed to production and if to replace a previous version of the model; and (4) storing the model in the model repository.

***Online Inference Engine***: Each type of recommendation approach includes an inference engine that leverages a model produced by the corresponding offline learning pipeline. The inference engine typically includes the following steps: (1) loading relevant models from the Model repository; (2) implementing a recommendation logic; and (3) exposing a common recommendation interface.

***Recommendation Controller***: The main recommendation interface through which the conversational RPA server can receive recommendations. The controller is responsible to invoke and integrate recommendations from multiple online inference engines. It may select only a subset of the acquired recommendations to limit their quantity or achieve a desired composition of recommendation types. It should also act as the final guard against negative or invalid recommendations that may yield an error or negative user experience, e.g., when a recommended skill was recently removed from the loaded digi.

***Utterance Generator***: Converts the next-best-skill recommendations into a UI element that a user can click to invoke the next utterance.

***Model Repository***: A persistent data store for the learned models.

Next, we describe the two main execution flows in NBSR: (1) the offline flow of how models are learned; and (2) the online execution that takes place between a user communicating with the system and the final recommendation presentation.

***Offline Flow.*** In this flow, the learning controller periodically reads the digi configuration metadata and triggers the relevant learning pipelines. Each pipeline further reads relevant conversation and orchestration logs and produces a trained model that is deployed to the model repository. Every digi will ultimately be associated with one or more models based on different learning sources and recommendation approaches. Personalized models learn from the specific data of the digi, while shared models utilize data collected from the relevant cluster of digis. Therefore, the learning controller also produce a digis-to-cluster mapping and trigger offline learning pipelines that learn about this scope of data.

***Online Flow.*** This flow starts by a user typing an utterance or clicking on a UI button that sends a text utterance. The utterance is then handled by the orchestrator logic which invokes automation skills. The orchestrator may call the recommendation controller at different points in the dialog, based on different situations and user experience considerations. One such point could be right after a task was completed. Another might be if the user has explicitly asked for help. In these cases, the orchestrator will trigger the recommendation controller, which in turn will consult with the online inference engines to produce integrated recommendations.

The current session's orchestration steps are the required data inputs for producing recommendations by the specific online inference engines (personal, crowd-wisdom, goal-driven, and context) and are passed as parameters to the engines. Each engine returns a sorted list of suggested skills with explainability information concerning the source of the recommendation. The controller synthesizes and filters the recommendations into a single set of results to be presented to the user. Next, the utterance generation component creates a payload for the list of buttons that include a textual utterance and a label to be presented. Finally, this response payload is rendered as user interface buttons in the UI of the channel.

## 5  Summary

In this work, we presented next-best-skill recommendations as a new type of recommendation problem that has emerged in the latest generation of conversational RPA bots capable of dynamically orchestrating automation skills through a natural language interface. We provided definitions and use cases that illustrate different situations in which conversational RPA bots are built and deployed. We highlighted the main characteristics of this problem and explained why it is in fact composed of several types of recommendations, each of which are based on a different user need, source of data for learning, and technical approach. Due to the nature of human-robot collaboration via a single conversational UI, we suggested that this problem calls for an integrated strategy for recommendations.

When considering how to implement a solution to this problem, we considered existing approaches in the literature and called out specific research challenges that emerge in this specific problem domain. We further provided an illustration of a possible implementation architecture that can serve as an integrated recommendation strategy.

We acknowledge that the work presented in this position paper is still exploratory, as this nascent problem domain is still being shaped by technology providers and not yet fully adopted in real-world production settings. This has placed limits on our ability to perform proper validation or evaluation of the problem characteristics and technical approaches. In the near future, we plan to address some of these challenges as part of a solution we are building and planning to deploy in a real-world production setting. We hope that this paper will also help bring awareness to the research community to address these challenges.

**Acknowledgements.** We would like to thank Sebastian Carbajalo, Yara Rizk, Vatche Isahagian, Vinod Muthusamy, Mahmoud Mahmoud, Scott Boag, and Ben Herta for invaluable inputs and feedback.

# References

1. Van der Aalst, W.M., Bichler, M., Heinzl, A.: Robotic process automation. Bus. Inf. Syst. Eng. **60**(4), 269–272 (2018)
2. Agarwal, P., Gupta, A., Sindhgatta, R., Dechu, S.: Goal-oriented next best activity recommendation using reinforcement learning. arXiv preprint arXiv:2205.03219 (2022)
3. Basseville, M., Nikiforov, I.V., et al.: Detection of Abrupt Changes: Theory and Application, vol. 104. Prentice Hall, Englewood Cliffs (1993)
4. Bobadilla, J., Ortega, F., Hernando, A., Bernal, J.: A collaborative filtering approach to mitigate the new user cold start problem. Knowl.-Based Syst. **26**, 225–238 (2012)
5. Bozorgi, Z.D., Teinemaa, I., Dumas, M., La Rosa, M., Polyvyanyy, A.: Process mining meets causal machine learning: discovering causal rules from event logs. In: 2nd International Conference on Process Mining (ICPM), pp. 129–136. IEEE (2020)
6. Chakraborti, T., Agarwal, S., Khazaeni, Y., Rizk, Y., Isahagian, V.: D3BA: a tool for optimizing business processes using non-deterministic planning. In: Del Río Ortega, A., Leopold, H., Santoro, F.M. (eds.) BPM 2020. LNBIP, vol. 397, pp. 181–193. Springer, Cham (2020). https://doi.org/10.1007/978-3-030-66498-5_14
7. Chakraborti, T., et al.: From robotic process automation to intelligent process automation. In: Asatiani, A., et al. (eds.) BPM 2020. LNBIP, vol. 393, pp. 215–228. Springer, Cham (2020). https://doi.org/10.1007/978-3-030-58779-6_15
8. Do, T.T., Tran, K.: The combination of robotic process automation (RPA) and chatbot for business applications (2021)
9. Dumas, M., et al.: Augmented business process management systems: a research manifesto. arXiv preprint arXiv:2201.12855 (2022)
10. Elwell, R., Polikar, R.: Incremental learning of concept drift in nonstationary environments. IEEE Trans. Neural Networks **22**(10), 1517–1531 (2011)
11. Everest: Stepping into the era of digital workers - robotic process automation (RPA) state of the market report 2022. https://www2.everestgrp.com/reportaction/EGR-2021-38-R-4842/Marketing
12. Gama, J., Medas, P., Castillo, G., Rodrigues, P.: Learning with drift detection. In: Bazzan, A.L.C., Labidi, S. (eds.) SBIA 2004. LNCS (LNAI), vol. 3171, pp. 286–295. Springer, Heidelberg (2004). https://doi.org/10.1007/978-3-540-28645-5_29
13. de Gemmis, M., Lops, P., Musto, C., Narducci, F., Semeraro, G.: Semantics-aware content-based recommender systems. In: Ricci, F., Rokach, L., Shapira, B. (eds.) Recommender Systems Handbook, pp. 119–159. Springer, Boston, MA (2015). https://doi.org/10.1007/978-1-4899-7637-6_4
14. Goedertier, S., Haesen, R., Vanthienen, J.: Rule-based business process modelling and enactment. Int. J. Bus. Process. Integr. Manag. **3**(3), 194–207 (2008)
15. IBM: Watson orchestrate. https://www.ibm.com/cloud/automation/watson-orchestrate
16. Kubrak, K., Milani, F., Nolte, A., Dumas, M.: Prescriptive process monitoring: quo vadis? (2021). https://arxiv.org/pdf/2112.01769.pdf
17. Lika, B., Kolomvatsos, K., Hadjiefthymiades, S.: Facing the cold start problem in recommender systems. Expert Syst. Appl. **41**(4), 2065–2073 (2014)
18. Metzger, A., Kley, T., Palm, A.: Triggering proactive business process adaptations via online reinforcement learning. In: Fahland, D., Ghidini, C., Becker, J., Dumas, M. (eds.) BPM 2020. LNCS, vol. 12168, pp. 273–290. Springer, Cham (2020). https://doi.org/10.1007/978-3-030-58666-9_16

19. Neu, D.A., Lahann, J., Fettke, P.: A systematic literature review on state-of-the-art deep learning methods for process prediction. Artif. Intell. Rev. 1–27 (2021). https://doi.org/10.1007/s10462-021-09960-8

20. Rashid, A.M., Karypis, G., Riedl, J.: Learning preferences of new users in recommender systems: an information theoretic approach. ACM SIGKDD Explor. Newsl. **10**(2), 90–100 (2008)

21. Teinemaa, I., Tax, N., de Leoni, M., Dumas, M., Maggi, F.M.: Alarm-based prescriptive process monitoring. In: Weske, M., Montali, M., Weber, I., vom Brocke, J. (eds.) BPM 2018. LNBIP, vol. 329, pp. 91–107. Springer, Cham (2018). https://doi.org/10.1007/978-3-319-98651-7_6

22. UIPath: Uipath. https://www.uipath.com/product

23. Weinzierl, S., Stierle, M., Zilker, S., Matzner, M.: A next click recommender system for web-based service analytics with context-aware LSTMs. In: Proceedings of the 53rd Hawaii International Conference on System Sciences, pp. 1542–1551. IEEE (2020)

24. Widmer, G., Kubat, M.: Learning in the presence of concept drift and hidden contexts. Mach. Learn. **23**(1), 69–101 (1996)

# Process Discovery Analysis
# for Generating RPA Flowcharts

Fabian Rybinski[(✉)] and Selina Schüler[(✉)]

Institute of Applied Informatics and Formal Description Methods (AIFB),
Karlsruhe Institute of Technology (KIT), Kaiserstr. 89, 76133 Karlsruhe, Germany
{fabian.rybinski,selina.schueler}@kit.edu

**Abstract.** Robotic Process Automation (RPA) is a frequently used approach for automation in IT systems. RPA uses existing graphical user interfaces to automate simple, rule-based tasks from a user's perspective. Before automation can be accomplished using RPA, extensive knowledge about individual user interactions must be gathered. This information is represented in a flowchart by a human designer, which is then typically refined in a trial-and-error approach. This approach becomes time-consuming and error-prone as processes become more complex.

In parallel, Process Discovery techniques as part of Process Mining enable the automated generation of process models from event logs. Thus, they are considered appropriate to simplify and automate the described creation of flowcharts in the context of RPA. In this regard, the so-called UI logs are particularly relevant. These can be used to record user interactions with various user interfaces within a process.

This paper examines existing Process Discovery methods for application in the context of RPA. The goal is to clarify what is needed for the automatic creation of RPA flowcharts with Process Discovery on the basis of UI logs.

The results show that existing Process Discovery methods generate process models that are not suitable for immediate translation into RPA flowcharts, as they are mainly control flow oriented. For the creation of RPA flowcharts that are suitable for automation, the integration of linked data is fundamental. Therefore, a prototypical implementation demonstrates a method that enables the automated translation of a process model into an RPA flowchart.

**Keywords:** RPA support · RPA flowchart generation · Process Discovery

## 1 Introduction

Robotic Process Automation (RPA) automates process steps by using software robots that imitate the execution of manual tasks. For this purpose, process steps that take place on graphical user interfaces are recorded and automated. RPA focuses on simple, repetitive tasks that are executed across different IT

© Springer Nature Switzerland AG 2022
A. Marrella et al. (Eds.): BPM 2022, LNBIP 459, pp. 231–245, 2022.
https://doi.org/10.1007/978-3-031-16168-1_15

systems. Current RPA tools offer the ability (a) to drag and drop process paths into flowcharts by using simple rules such as if-then formulations, or (b) to generate such flowcharts by directly recording user interactions within a process path. A flowchart allows the representation of simple, sequential processes in which only one activity is executed at a time [21]. RPA can perform high-volume tasks with high accuracy, consistent quality and high efficiency. Developing RPA-bots requires thorough knowledge of the tasks to be automated, the IT systems involved, and their user interfaces. This knowledge is currently obtained through interviews and recordings and then implemented in trial-and-error approaches [10]. This is time-consuming and error-prone and requires extensive testing, which is complicated by direct integration with live systems [12].

Process Discovery is a Process Mining technique to automatically build process models from process execution data recorded by IT systems [2]. Process Discovery enables the visualization of process data in models to identify improvement potentials with respect to certain attributes such as time. Therefore, the combination of these two research areas is obvious, as Process Discovery could automatically create the relevant process paths of a process for RPA. Process Discovery can provide information regarding important questions for RPA, such as how the process steps are ordered or which process paths are the most frequent or efficient. By creating a complete model with all paths, it becomes possible to assess all variants of the process in regard to the desired output. Although there is existing work on combinations of these topics, there exist several shortcomings: There is no work on which Process Discovery algorithm could be used to generate RPA flowcharts, what the requirements are for these algorithms, and which of these algorithms produce the best results. Furthermore, the existing work on implementing process steps with RPA does not enable the analysis and improvement of processes. As a result, RPA is often seen as a short-term solution. To enable a long-term use of RPA, the process should be analyzed in advance and not only one process variant should be implemented with RPA. Therefore, the decision options of alternatives should be implemented in the RPA solutions, so that several variants of a process are represented. In addition, these variants to be implemented should be freely selectable.

This paper is structured as follows. The second section introduces RPA and related flowcharts, Process Discovery and an overview of related work. Then the limitations and shortcomings of existing Discovery Algorithms for the application in the context of RPA are described to motivate the idea of the prototype. Next, the prototype is presented that generates RPA flowcharts based on UI logs. Finally, an evaluation is given, based on a use case, showing the potential of the prototype, but also some of the existing problems and future research opportunities.

# 2    RPA, Process Mining and Related Work

In this chapter, selected basics and examples of the areas of Robotic Process Automation and Process Mining are presented first. Afterwards, related work is introduced.

## 2.1    Robotic Process Automation (RPA)

Robotic Process Automation (RPA) is a frequently used term in the context of business process automation [34]. According to the Robotic Process Automation Institute's definition, the term RPA covers intelligent software and the application of that software [24]. Thus, RPA is a completely software-based solution of process automation. RPA uses the applications, functionalities and security of a company's existing IT systems, so there is no need for integration with those systems [36]. For this reason, RPA is described in literature as a lightweight automation method and is considered an evolution of macros and scripts [14]. The evolution consists in implementing more complex logics and processes. In doing so, RPA uses the existing graphical user interfaces for working with IT systems and thus only sits on top of these existing systems [6]. This enables RPA to (partially) automate cross-system processes without implementing interfaces.

At the beginning of RPA implementation detailed knowledge about tasks, involved user interfaces and user actions is necessary [9]. Currently, this information is obtained through interviews, workshops, or observation of user interactions. Based on this information, the RPA flowcharts are developed by a human designer [10]. In a trial-and-error approach, a flowchart is created first with all the activities that the software robot is supposed to perform. These activities have to be specified in great detail (e.g. write in Excel cell A3) and possible decisions in case of multiple paths have to be defined in advance [18]. The execution of the actions is then monitored and the flowchart is adjusted in the case of any errors. An exemplary RPA flowchart is shown in Fig. 1. Besides the sequence of the activities, the respective data of the activities must be known in order to create the flowcharts [31]. For example, this is the ID of a specific Excel cell or a browser text field that the robot is supposed to select in the execution.

RPA is suitable for automating extensive, repetitive tasks that are considered both time-consuming and trivial to perform. To be particularly suitable, these tasks need to fulfill the following characteristics: structured data, rule-based process and deterministic results [26]. Additionally, there are other criteria such as frequent execution, infrequent exceptional cases, different systems involved and vulnerability to human error [22]. Tasks that meet these requirements are often found in the so-called back offices, i.e. finance, purchasing or HR departments [13]. For example, the robot is capable of moving the mouse or copying data from one file to another. The capacities freed up by automation with RPA can be used by humans for more unstructured and interesting tasks. This might result in benefits in employee satisfaction as well as scalability and 24-h serviceability [37].

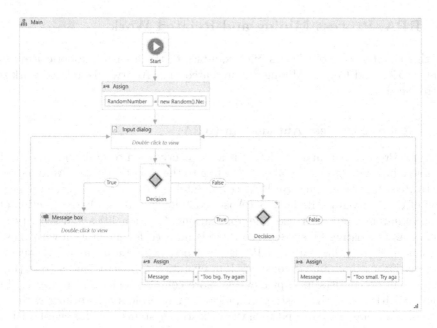

**Fig. 1.** RPA flowchart example [38]

## 2.2  Process Mining

Process Mining is a research area that links the areas of data science and process science [4]. The term Process Mining is used to describe techniques that allow the capture, analysis and improvement of real processes [5]. For example, Process Mining can be deployed to examine deviations from the to-be process or to analyze performance. Companies can use Process Mining to increase the understanding and control of the own processes. The gained insights are based on the real behavior reflected in the event logs and thus enable fact-based business process management [3].

Basically, the area of Process Mining is subdivided into three different categories: Process Discovery refers to the generation of process models, Process Conformance is the comparison of real process executions with modeled processes, and Process Enhancement adds new aspects to the existing model [2]. All three categories have in common that they build on event logs. These event logs record occurring events in IT systems and are often already stored in the company's IT systems (e.g. ERP system, CRM system) [8]. The information associated with an event, such as activity names, timestamps, and the person assigned to it, is recorded in the event logs [7]. For the application of Process Mining techniques they need to be extracted, pre-processed and combined. Fundamental to understanding Process Mining is the assumption that events are recorded sequentially in the event logs [1]. In addition, each event must correspond to a task and a case.

In the following sections, the focus is on Process Discovery as one category of Process Mining and, especially, on the existing algorithms in that category. The goal of Process Discovery is to generate a process model that represents the behavior in the event log without prior knowledge about this model [1]. Thus, the Process Discovery techniques provide the basis for further analysis of the processes within a company. In this context, a Process Discovery algorithm takes an event log with activities as input and generates a process model over the same activities [39].

## 2.3   Related Work

To gain insight in current approaches in the literature that address the research area covered, a *Systematic Literature Review* (SLR) was conducted. The objective was to identify literature that discusses Process Mining and especially Process Discovery in the context of RPA. This also involves identifying the requirements of process models in the context of RPA. Based on this objective, a search string was created containing terms from Process Mining, Process Modeling and RPA: ("RPA" or "Robotic Process Automation") and ("Process Mining" or "Process Discovery" or "Process Learning" or "Workflow Discovery" or "Workflow Learning" or "Model" or "Flowchart" or "Script" or "UI Log"). A forward and backward reference search was then performed with the relevant publications identified. The literature research was conducted in December 2020. Based on the method described, a total of 14 publications could be classified as relevant for further analysis. The following section briefly summarizes current approaches to Process Discovery in the context of RPA.

Nowadays, Process Mining methods are already used in the early stages of the RPA lifecycle to support the human designer [25]. The selection of the process to be automated also takes place in these phases [15,29]. In particular, Process Discovery could be appropriate to create process models for the automation with RPA. In general, event logs are used in the context of Process Discovery methods in order to represent the occurred process activities in a control flow model. In this context, the corresponding data (e.g. specific excel cell) that allows the execution of activities with RPA is especially important and therefore must be represented in a process model [31]. In [31] the authors describe the concept of a multi-perspective Process Discovery method that integrates the data flow and the control flow in a single process model. Furthermore, instead of the well-known event logs, the UI logs are used in the context of RPA, which contain granular user interactions, e.g. clickstreams, keystroke protocols [32]. Several approaches of UI loggers already exist in the literature and allow logging user interactions in e.g. Microsoft Excel or Internet browsers [33,35]. The mentioned loggers produce logs that are applicable for further analysis with existing PM techniques.

In [11,30] the authors describe more holistic approaches. The introduced software tools include UI loggers and allow the automatic translation of recorded sequences into RPA flowcharts with all necessary data assignments. In [11] the translated sequence is selected based on the shortest duration in the log and Process Discovery is only part of the graphical representation of the high-level process. In [30] the tool allows the determination of automatable actions in the UI log and the subsequent translation in an RPA flowchart. Finally, in [23] the authors propose the automation of RPA itself, describing a self-learning approach that requires minimal human intervention.

Another related area is that of automated planning [17]. These techniques allow learning of operators that can model different instantiations of process models through automated planning, as an alternative to PM. With these models, very large-scale compositions can be generated and used to capture an RPA flowchart [19]. In the presented approach, each non-determinism accounts for a different instantiation.

Overall, the SLR showed that although theoretical research exists on combining RPA and PM, there is a lack of realization. Before automating a process with RPA, the analysis of the process is inevitable, which is why application of PM techniques seems to be promising. Therefore, this paper reviews whether it is possible to generate RPA flowcharts from UI logs based on Process Mining and which Process Discovery methods are most appropriate.

## 3   Existing Process Discovery Methods for RPA

In this chapter, existing approaches of Process Discovery methods and the corresponding algorithms are assessed and compared regarding their suitability for use in the context of RPA. It will be verified whether the existing algorithms are suitable to create process models based on UI logs. These process models should subsequently be able to be used for translation into an RPA flowchart.

In the literature, there is a wide variety of Process Discovery algorithms that can be used for discovering process models from event logs. Meanwhile, there are more than 40 different algorithms that tackle the discovery problem using different techniques. As the algorithms differ, so do the results, i.e. in the modeling language and thus in the resulting process models. For example, there are algorithms for generating Petri nets, BPMN models, directly-follows graphs, process trees, or Declare models. Some of the well known algorithms might be the *alpha-algorithm* [8], *Heuristics Miner* [40], *Inductive Miner* [27], or the *Evolutionary Tree Miner* [16]. The following section briefly summarizes the characteristics of the algorithms of the mentioned Process Discovery methods, including some evolutions.

The algorithm in [42] is an evolution of the original $\alpha$-algorithm, which is one of the first Process Discovery algorithms in the literature [8]. The *alpha++* ($\alpha$++) algorithm allows the determination of non-free-choice constructs in process models.

In [41] the authors describe an evolution of the original heuristics miner [40]. The goal of the *Flexible Heuristics Miner* (FHM) is to fully utilize the advantages of the basic idea of the original algorithm. For this purpose, causal nets for the representation of process models are defined for the first time.

The *Inductive Miner - Infrequent* (IMi) [28] is an algorithm that generates process trees and allows the processing of infrequent behavior in the event logs.

The IMi ensures that the model is represented correctly and with high fitness, i.e. the process model should represent the behavior in the event log.

The *Evolutionary Tree Miner* (ETM) is based on a genetic approach [16]. The algorithm generates an initial population of process trees that represent the behavior in the log. In further steps, the initial population is continuously adapted until defined criteria in the four quality dimensions fitness, precision, generalization and complexity are met.

A closer look at the aforementioned Process Discovery algorithms reveals that they all discover only the control flow in the generated process models. In this context, control flow describes the order of the different activities of an event log [3]. The same applies to the other existing algorithms in the literature. Based on the approach of the SLR, no algorithms were identified that reflect the data and data transformation in the model or considered these data when generating the models. Since the existing algorithms only consider the control flow in the process models based on activity ID, they do not distinguish between same activities with different data assignments. Furthermore, as the data is not included when generating the models, it is not possible to assign the data to the activities afterwords. The models generated in this way are therefore not suitable for translation into an RPA flowchart that already contains the data required for automation. Thus, to enable the automated translation of a process model into RPA flowcharts, adjustments must be made.

## 4   Prototypical Approach and Architecture

As described in the last chapter, the current Process Discovery algorithms do not consider data assignments when generating process models. Thus, without modifications they are not suitable for generating process models that can be used to create RPA flowcharts. Since the assigned data is essential for execution within RPA, the prototype is intended to implement a solution that enables the generation of process models with the necessary data assignments. Ultimately, the creation of RPA flowcharts is desired that already contain the required data for automation. Generally various solutions are conceivable for achieving this goal. On the one hand, existing Process Discovery algorithms could be adapted or new ones developed. The development of algorithms that integrate the data in the generation of process models seems promising, e.g. multi-perspective process mining [31]. These enable the mapping of the data flow in addition to the control flow. On the other hand, the adaption of the UI logs used as input for Process Discovery could also be goal-oriented. Existing algorithms distinguish activities mainly based on activity ID. Thus, to distinguish same activities with different data, the data could be noted in the activity ID. As an example, an activity *copy* could be extended by *cell A2* and could be clearly distinguished from the activity *copy cell A3*. Hereby, existing algorithms should distinguish the different activities based on the extended activity ID and map them to different nodes of the process model.

Comparing the options, the development of new Process Discovery methods is a sustainable solution. Here, the requirements for process models in the context of RPA can already be considered in the design of the method and the corresponding algorithm. Due to the scope of the paper at hand, this is not possible and is a future goal. However, the considerations for the adaption of the UI logs already seem to enable the required process models for a prototypical implementation. Additionally, the application of existing analysis tools to the discovered process models is possible and results of conducted evaluations of existing algorithms may be considered for the selection of the algorithm. Therefore, in the following sections, a prototype is designed and implemented to automatically generate RPA flowcharts based on Process Discovery.

## 4.1  Design and Implementation of the Prototype

For the presented prototype an adaption of the UI logs is chosen. That means that the UI logs should be adapted in such a way to enable the generation of process models with data assignments. To generate the process model, Process Discovery is applied on the adapted UI logs. Finally, based on the discovered process model, a single path of the model can be chosen and translated into an RPA flowchart. Consequently, the prototype is composed of three different parts, preceded by an existing UI logger and followed by an existing RPA tool. Its overall design is presented in Fig. 2.

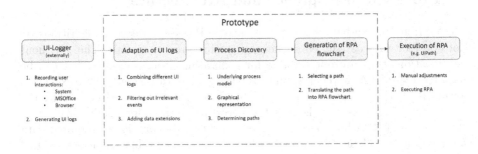

**Fig. 2.** Prototype structure

For recording user interactions the UI logger presented and implemented in *Smart-RPA* [11] is used. It allows the detailed recording of all relevant user interactions and especially the data necessary for the generation of an RPA flowchart, e.g. Excel cell, browser URL. The same paper is also the basis for the translation of events into RPA activities, but here adjustments have been made to build a prototype that enables the manual selection of a variant for automation based on the analysis of a given process model. This process model should include all data of a given UI log, as opposed to simply selecting the most frequent or shortest variant.

The prototype handles user interactions in the operating system, Microsoft Excel and Microsoft Edge. These already allow a variety of different actions and build the basis for the integration of further applications. In a first step, the UI logs have to be adjusted. The input of the algorithm consists of CSV files generated with the described UI logger. Each CSV file captures one instance of the process, since different instances cannot be distinguished in a single file. These are imported into a data frame and the column labels are matched to the XES standard, e.g. *time:timestamp*. Additionally, a *case:concept:name* column is created to represent the case ID. This is repeated for each CSV file and finally a combined data frame is created with all user interactions. The combined file captures all recorded instances of a process and is thus comparable to an event log. To enable the generation of process models with data assignments using existing Process Discovery algorithms, the column *concept:name* is extended with the relevant data from the other columns. Since Process Discovery algorithms generate process models based on this column, the data is included in the process model. Thus, it is possible to link it to a specific activity in the UI log. Since the goal is to distinguish between activities with the same activity name, the data in which the activities differ is added. For example, Excel activities can be distinguished by worksheet and cell, or browser activities by URL and element ID. The final data frame contains passed to the second component, which contains the Process Discovery. The compliance with the XES standard enables the application of traditional Process Discovery algorithms. According to this description, the algorithm presented in Fig. 3 results. Since the implementation of the prototype is done in Python, the implementation of the Process Discovery is done with PM4PY [20], which is a Process Mining library that provides Process Mining algorithms. From this library, the $\alpha$–miner is chosen for Process Discovery in the prototype. The algorithm generates process models that are mainly low in complexity and therefore particularly suitable for graphical representation.

**Algorithm 1:** Adaption of UI logs

    **Procedure** LogProcessing(*csv_list*):

        **for** *CSV-Logs* in *csv_list*:

            *df* ← Rename columns to match XES standard
            *df* ← Create *column case:concept:name*
            *df* ← Generate dataframe with user interactions of csv

        **end**
        *combined_df* ← Combine all user interactions in one dataframe
        *combined_df* ← Create column *frequency*

        **for** *rows* in *combined_df*:

            **if** *activity* in *activity_list*:

                *combined_df* ← Extend *concept:name* with data of relevant columns

**Fig. 3.** Adaption of UI logs

Even though incorrect models may arise with it, the $\alpha$–miner is chosen as it provides a simple realization for the first implementation that should provide appropriate results. The generated process model serves as input for the generation of an RPA flowchart. Additionally, it is input for the graphical representation, which is a support for the selection of an execution path. To identify all paths in the process model, an algorithm is used that creates paths in a loop until each transition in the process model has been involved at least once. These paths are the basis for selection by the user. In case of only two different paths, they can both be transferred in an RPA flowchart together with an if-else relation.

The identified paths are passed to the third component of the prototype. Here, the frequency of the paths is determined by the minimum frequency of the involved nodes. This information, in addition to the graphical representation of the process, is another support for the user. After selecting a path, the activities of the selected path are translated into software code. The XAML format (Extensible Application Markup Language) is used, which can be read by common RPA tools. The translation of the activities is possible by adapting the UI logs, as this allows the representation of data in the process models and the distinction between activities. Thus, the data is correctly assigned to the activities in the flowchart and it can be executed in an RPA tool.

## 5    Evaluation of the Prototype

In this section, the prototype is tested. For this purpose, an exemplary use case and the resulting outcome of the prototype are presented first. Afterwards, the strengths and weaknesses of the prototype are discussed.

### 5.1    An Use Case in the Prototype

To evaluate the prototype, a test case is used that contains activities from the applications Microsoft Excel, Microsoft Edge and the operating system. In addition, several paths are included to test the identification of them. The test case contains three different paths and thus 2 different decisions in the process model. For this, the data of 3 different persons are to be transferred from an Excel file into a web form. A total of 15 different executions of the activities are recorded with the UI logger, the frequencies of the paths are 10, 4 and 1. The first component of the prototype correctly adapts the 15 UI logs and combines them in a CSV file. In this file, the XES standard is applied and the frequency column is created. Furthermore, the contents of the column *concept:name* are correctly extended with the respective data of the other columns, e.g. the activity *copy-Excel* with the cell, file and data sheet. A section of the combined CSV file is given in Fig. 4. Since the combined UI log contains the required adaptions, the next step is to start Process Discovery. A section of the generated process model is given in Fig. 5. The figure shows that 3 different paths and 2 decisions have been included in the process model. The intended information is visualized in the transitions. The label is composed of the node ID, the activity name and

| case:concept:name | time:timestamp | org:resource | category | application | concept:name |
|---|---|---|---|---|---|
| 413171032005000 | 2021-04-13T17:10:32.005 | fabia | MicrosoftOffice | Microsoft Excel | openWorkbook Daten_Formular.xlsx |
| 413171032005000 | 2021-04-13T17:10:39.504 | fabia | Browser | Edge | open Browser |
| 413171032005000 | 2021-04-13T17:11:02.410 | fabia | Browser | Edge | autoBookmark https://www.google.de/ |
| 413171032005000 | 2021-04-13T17:11:09.375 | fabia | Browser | Edge | autoBookmark https://www.umfrageonline.com/s/e6e569b |
| 413171032005000 | 2021-04-13T17:11:14.130 | fabia | MicrosoftOffice | Microsoft Excel | getCell A3 Daten_Formular.xlsx Personendaten |
| 413171032005000 | 2021-04-13T17:11:16.571 | fabia | Browser | Edge | clickTextField https://www.umfrageonline.com/s/e6e569b element-11-23828958 |
| 413171032005000 | 2021-04-13T17:11:16.717 | fabia | MicrosoftOffice | Microsoft Excel | pasteExcel A3 |
| 413171032005000 | 2021-04-13T17:11:18.046 | fabia | Browser | Edge | changeField_pasteExcel https://www.umfrageonline.com/s/e6e569b element-11-23828958 |
| 413171032005000 | 2021-04-13T17:11:26.270 | fabia | Browser | Edge | close Browser |
| 413171032005000 | 2021-04-13T17:11:28.420 | fabia | MicrosoftOffice | Microsoft Excel | closeWorkbook Daten_Formular.xlsx |

**Fig. 4.** Combined CSV file generated by the prototype

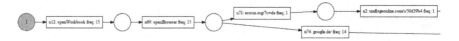

**Fig. 5.** Section of a process model in the prototype

the frequency. Especially the data relevant for RPA are represented and allow the translation into an RPA flowchart. In the given case, the most frequent one is selected for the generation of the RPA flowchart. With the third component of the prototype, the activities are translated into an XAML file. This file is then imported with the RPA tool *UiPath*. The tool translates the XAML file and represents it in a flowchart. A section of the generated flowchart is given in Fig. 6. The figure shows that the required data is correctly inserted into the respective activities. Finally, the flowchart is executed and the software robot transfers the personal data to the web form as intended.

### 5.2    Strengths and Improvement Potentials

The use case of the prototype has shown that the desired functions are achieved. In particular, it replaces the current trial-and-error approach in the application of RPA. The benefit results in the automated generation of a process model in which the data necessary for RPA is included. This process model simplifies the analysis of the processes to be automated. In addition, the current approach is improved by the automated generation of RPA flowcharts based on a process model. There is no need to manually transfer user interactions from UI logs to the flowchart and the analyst can focus on monitoring the execution and making any adjustments. In general, the prototype at this stage works particularly well for rather simple processes that contain clearly definable paths without loops.

Although the prototype implements the desired functions, some improvements are still necessary. Particularly, the complexity of the processes is currently a limiting factor. In the test, a process model with loops was created unintentionally. Due to the current algorithmic implementation, this can not be processed in a reasonable way at this stage of the prototype. Another limiting

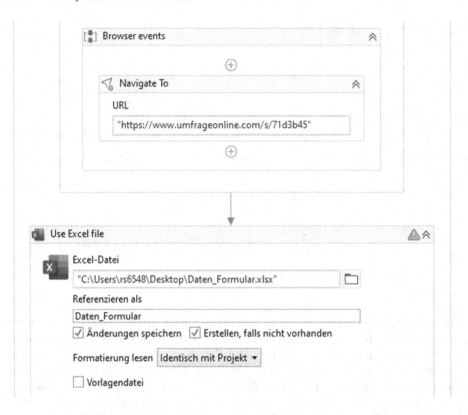

**Fig. 6.** Flowchart based on the generated XAML file

factor is the fact that process models might get very large. Since the data is notated directly in the activities of the process model, activities with different data are no longer mapped in one node of the model. This can lead to an explosion of the model, e.g. in the exemplary case there are 8 different activities. Due to the assignment of the explicit data to the activities, the paths in the process model have a minimum of 37 nodes. Finally, there is also potential for improvement in the graphical representation of the process models. For example, it would be useful to graphically highlight the most frequent path in the process model.

## 6   Conclusion

The goal of the paper at hand was the automatic creation of RPA flowcharts with Process Discovery. For this purpose, the capability of existing Process Discovery algorithms for the generation of RPA flowcharts has been analyzed. In addition, a prototype has been developed that enables this generation and thus overcomes existing shortcomings. Based on the results of the research, the prototypical implementation of a process discovery algorithm in the context of RPA could be realized. The prototype fulfilled the desired functions in the tests and enabled the automatic generation of RPA flowcharts with required data assignments. The results also show that Process Discovery methods need to be adapted for use in the context of RPA. As the prototype showed, however, this is already possible by simply adapting the UI logs used, with all the accompanying advantages and disadvantages. The advantage is the automated generation of a process model in which the data required for automation with RPA is included. This model includes all variants of the examined process and thus enables the assessment and selection of the most suitable variant for automation.

While the prototype allowed the implementation of the functions desired, there is still room and need for improvement. Since the detection of paths in the process model is done with a rather simple algorithm, the complexity of the processes is a limiting factor. Here, the functionality could be extended to loops for example. Another interesting study would be whether the analysis of all variants of a process with Process Discovery brings great practical advantages in the context of RPA. In the prototype, two different paths can be transferred in an RPA flowchart with an if-else relation.Here, a comparison of the selection based on the analysis of all variants with the selection of the shortest or most frequent variant in the UI log and a transfer of more than two variants would be desirable.

# References

1. van der Aalst, W.: Process discovery: capturing the invisible. IEEE Comput. Intell. Mag. **5**(1), 28–41 (2010)
2. van der Aalst, W.M.P.: Introduction. In: Process Mining, pp. 1–25. Springer, Heidelberg. https://doi.org/10.1007/978-3-642-19345-3_1
3. van der Aalst, W.: Process mining: overview and opportunities. Assoc. Comput. Mach. **3**(2), 1–17 (2012)
4. van der Aalst, W.: Process Mining: Data Science in Action, vol. 2. Springer, Heidelberg (2016). https://doi.org/10.1007/978-3-662-49851-4
5. van der Aalst, W., et al.: Process mining Manifesto. In: Daniel, F., Barkaoui, K., Dustdar, S. (eds.) BPM 2011. LNBIP, vol. 99, pp. 169–194. Springer, Heidelberg (2012). https://doi.org/10.1007/978-3-642-28108-2_19
6. van der Aalst, W., Bichler, M., Heinzl, A.: Robotic process automation. Bus. Inf. Syst. Eng. **60**(4), 269–272 (2018). https://doi.org/10.1007/s12599-018-0542-4
7. van der Aalst, W., et al.: Business process mining: an industrial application. Inf. Syst. **32**(5), 713–732 (2007)
8. van der Aalst, W., Weijters, T., Maruster, L.: Workflow mining: discovering process models from event logs. IEEE Trans. Knowl. Data Eng. **16**(9), 1128–1142 (2004)
9. Agaton, B., Swedberg, G.: Evaluating and Developing Methods to Assess Business Process Suitability for Robotic Process Automation. Master's thesis, Chalmers University of Technology and University of Gothenburg (2018)
10. Agostinelli, S.: Synthesis of strategies for robotic process automation. In: Symposium on Advanced Database System, vol. 2400, pp. 5–8. CEUR-WS.org (2019)
11. Agostinelli, S., Lupia, M., Marrella, A., Mecella, M.: Automated generation of executable RPA scripts from user interface logs. In: Asatiani, A., et al. (eds.) BPM 2020. LNBIP, vol. 393, pp. 116–131. Springer, Cham (2020). https://doi.org/10.1007/978-3-030-58779-6_8

12. Agostinelli, S., Marrella, A., Mecella, M.: Towards intelligent robotic process automation for BPMers. In: Proceedings of the AAAI Workshop on Intelligent Process Automation (2020)
13. Aguirre, S., Rodriguez, A.: Automation of a business process using Robotic Process Automation (RPA): a case study. In: Figueroa-García, J.C., López-Santana, E.R., Villa-Ramírez, J.L., Ferro-Escobar, R. (eds.) WEA 2017. CCIS, vol. 742, pp. 65–71. Springer, Cham (2017). https://doi.org/10.1007/978-3-319-66963-2_7
14. Barnett, G.: Robotic Process Automation: Adding to the Process Transformation Toolkit (2015). https://www.blueprism.com/uploads/resources/white-papers/RPA-Adding-to-the-process-automation-toolkit.pdf
15. Bosco, A., Augusto, A., Dumas, M., La Rosa, M., Fortino, G.: Discovering automatable routines from user interaction logs. In: Hildebrandt, T., van Dongen, B.F., Röglinger, M., Mendling, J. (eds.) BPM 2019. LNBIP, vol. 360, pp. 144–162. Springer, Cham (2019). https://doi.org/10.1007/978-3-030-26643-1_9
16. Buijs, J.C., Van Dongen, B.F., van der Aalst, W.: Quality dimensions in process discovery: the importance of fitness, precision, generalization and simplicity. Int. J. Coop. Inf. Syst. **23**(01), 1440001/1-39 (2014)
17. Callanan, E., De Venezia, R., Armstrong, V., Paredes, A., Chakraborti, T., Muise, C.: MACQ: a holistic view of model acquisition techniques. In: ICAPS 2022 Workshop on Knowledge Engineering for Planning and Scheduling (2022)
18. Cernat, M., Staicu, A.N., Stefanescu, A.: Towards automated testing of RPA implementations. In: Proceedings of the 11th ACM SIGSOFT International Workshop on Automating TEST Case Design, Selection, and Evaluation. Association for Computing Machinery, New York (2020)
19. Chakraborti, T., Agarwal, S., Khazaeni, Y., Rizk, Y., Isahagian, V.: D3BA: a tool for optimizing business processes using non-deterministic planning. In: Del Río Ortega, A., Leopold, H., Santoro, F.M. (eds.) BPM 2020. LNBIP, vol. 397, pp. 181–193. Springer, Cham (2020). https://doi.org/10.1007/978-3-030-66498-5_14
20. FIT: PM4PY (2022). https://pm4py.fit.fraunhofer.de/
21. Fleischmann, A., Oppl, S., Schmidt, W., Stary, C.: Ganzheitliche Digitalisierung von Prozessen. Springer Nature, Wiesbaden (2018). https://doi.org/10.1007/978-3-658-22648-0
22. Fung, H.P.: Criteria, use cases and effects of information technology process automation (ITPA). Adv. Robot. Autom. **3**, 1–10 (2014)
23. Gao, J., van Zelst, S.J., Lu, X., van der Aalst, W.M.P.: Automated robotic process automation: a self-learning approach. In: Panetto, H., Debruyne, C., Hepp, M., Lewis, D., Ardagna, C.A., Meersman, R. (eds.) OTM 2019. LNCS, vol. 11877, pp. 95–112. Springer, Cham (2019). https://doi.org/10.1007/978-3-030-33246-4_6
24. Institute for Robotic Process Automation: Carnegie Mellon University: Introduction to Robotic Process Automation (2015). https://irpaai.com/wp-content/uploads/2015/05/Robotic-Process-Automation-June2015.pdf
25. Jimenez-Ramirez, A., Reijers, H.A., Barba, I., Del Valle, C.: A method to improve the early stages of the robotic process automation lifecycle Andres. In: Advanced Information Systems Engineering (CAiSE), vol. 1, pp. 659–674. Springer, Cham (2019). https://doi.org/10.1007/978-3-030-21290-2_28
26. Lacity, M.C., Willcocks, L.P.: A new approach to automating services. MIT Sloan Manage. Rev. **58**(1), 41–49 (2016)
27. Leemans, S.J.J., Fahland, D., van der Aalst, W.M.P.: Discovering block-structured process models from event logs - a constructive approach. In: Colom, J.-M., Desel, J. (eds.) PETRI NETS 2013. LNCS, vol. 7927, pp. 311–329. Springer, Heidelberg (2013). https://doi.org/10.1007/978-3-642-38697-8_17

28. Leemans, S.J.J., Fahland, D., van der Aalst, W.M.P.: Discovering block-structured process models from event logs containing infrequent behaviour. In: Lohmann, N., Song, M., Wohed, P. (eds.) BPM 2013. LNBIP, vol. 171, pp. 66–78. Springer, Cham (2014). https://doi.org/10.1007/978-3-319-06257-0_6

29. Leno, V., Augusto, A., Dumas, M., La Rosa, M., Maggi, F.M., Polyvyanyy, A.: Identifying candidate routines for robotic process automation from unsegmented UI logs. In: ICPM, pp. 153–160 (2020)

30. Leno, V., Deviatykh, S., Polyvyanyy, A., La Rosa, M., Dumas, M., Maggi, F.M.: Robidium: automated synthesis of robotic process automation scripts from UI logs. In: Proceedings of the Best Dissertation Award, Doctoral Consortium, and Demonstration and Resources Track at BPM, vol. 2673, pp. 102–106. CEUR-WS.org (2020)

31. Leno, V., Dumas, M., Maggi, F.M., La Rosa, M.: Multi-perspective process model discovery for robotic process automation. Adv. Inf. Syst. Eng. (CAiSE) **2114**, 37–45 (2018)

32. Leno, V., Polyvyanyy, A., Dumas, M., La Rosa, M., Maggi, F.M.: Robotic process mining: vision and challenges. Bus. Inf. Syst. Eng. **63**(3), 301–314 (2020). https://doi.org/10.1007/s12599-020-00641-4

33. Leno, V., Polyvyanyy, A., La Rosa, M., Dumas, M., Maggi, F.M.: Action logger: enabling process mining for robotic process automation. In: Proceedings of the Dissertation Award, Doctoral Consortium, and Demonstration Track at BPM, vol. 2420, pp. 124–128. CEUR-WS.org (2019)

34. Lhuer, X., Willcocks, L.: The next acronym you need to know about: RPA (robotic process automation) (2016). https://www.mckinsey.de/business-functions/mckinsey-digital/our-insights/the-next-acronym-you-need-to-know-about-rpa

35. Linn, C., Zimmermann, P., Werth, D.: Desktop activity mining - a new level of detail in mining business processes. In: INFORMATIK (Workshops). pp. 245–258. Köllen Druck+Verlag GmbH, Bonn (2018)

36. Slaby, J.R.: Robotic automation emerges as a threat to traditional low-cost outsourcing (2012). https://neoops.com/wp-content/uploads/2014/02/Robotic-Automation-A-Threat-To-Low-Cost-Outsourcing_HfS.pdf

37. Sutherland, C.: Framing a constitution for Robotistan (2013). https://neoops.com/wp-content/uploads/2014/03/RS_1310-Framing-a-constitution-for-Robotistan.pdf

38. UiPath: Handbook UiPath Studio (2022). https://docs.uipath.com/studio/lang-de/docs/flowcharts

39. Verbeek, H., van der Aalst, W., Munoz-Gama, J.: Divide and conquer: a tool framework for supporting decomposed discovery in process mining. Comput. J. **60**(11), 1649–1674 (2017)

40. Weijters, A.J., van der Aalst, W., De Medeiros, A.K.: Process mining with the heuristics miner-algorithm. Technische Universiteit Eindhoven, Technical report, WP 166 (2006)

41. Weijters, A.J., Ribeiro, J.T.: Flexible Heuristics Miner (FHM). In: IEEE Symposium on CIDM, pp. 310–317 (2011)

42. Wen, L., van der Aalst, W., Wang, J., Sun, J.: Mining process models with non-free-choice constructs. Data Min. Knowl. Disc. **15**(2), 145–180 (2007). https://doi.org/10.1007/s10618-007-0065-y

# Can You Teach Robotic Process Automation Bots New Tricks?

Yara Rizk[✉], Praveen Venkateswaran, Vatche Isahagian, Vinod Muthusamy, and Kartik Talamadupula

IBM Research AI, Cambridge, MA, USA
yara.rizk@ibm.com

**Abstract.** Over the past decade, robotic process automation (RPA) has emerged as a lightweight paradigm for automation in business enterprises, making automation more accessible to non-techie business users. In the industry, RPA vendors have not only provided out-of-the-box RPA bots to automate manual tasks on legacy software; they have also provided users a recorder to create their own bots for specialized tasks. However, if these recorders do not create generalizable bots, users risk facing a "bot sprawl" and governance problem. Building generalizable bots currently requires intervention from IT departments which are typically oversubscribed given their limited resources. Furthermore, the generalization process is typically long and tedious; it does not scale to cover the expansive needs of business users. We thus need a tool that can empower business users to act as citizen developers and build generalized bots themselves. In this work, we argue that the next generation of RPA bots must leverage artificial intelligence to learn from user interactions (through natural language or other modalities intuitive to citizen developers) and generalize to unseen settings. To achieve this, we first survey and assess the current state of the art in the RPA field for enabling citizen developers; and identify several key research challenges at the intersection of AI, RPA, and interactive task learning that must be addressed to realize the vision of RPA bots that continually learn new automation solutions from user interactions.

**Keywords:** Robotic process automation · Artificial intelligence · Learning · Teaching by instruction

## 1 Introduction

The prevalence of artificial intelligence and automation in our daily lives – from smart kitchen appliances to digital personal assistants, and enabled by computing advancements – has set the expectation for a similar level of automation in our professional lives. If people no longer waste time keeping track of their grocery list, why should they still waste their professional time on manual, repetitive tasks (such as data entry) that take them away from doing the things they

© Springer Nature Switzerland AG 2022
A. Marrella et al. (Eds.): BPM 2022, LNBIP 459, pp. 246–259, 2022.
https://doi.org/10.1007/978-3-031-16168-1_16

actually love in their job? Furthermore, the COVID pandemic has significantly accelerated the pace of the digital transformation of business enterprises.

As businesses have looked to adopt computing advancements to modernize their processes, one of the technologies driving their digital transformation is robotic process automation (RPA) [19,28,29]. RPA focuses on automating repetitive, error-prone professional tasks without requiring the overhaul of legacy software [7]. By automating the mouse click on user interfaces, RPAs can perform tasks on behalf of employees: tasks like copying data from one software tool to another and transforming data from one format to another among others, thus freeing employees up to focus on more critical and engaging tasks.

All RPA vendors provide a bot store selling out-of-the-box bots that perform some of the most common tasks. Furthermore, RPA bots have been embedded into conversation assistants to provide business users with a more natural and accessible interaction experience. However, it is not possible to provide comprehensive solutions out of the box from day one. Therefore, these tools need to evolve with time to expand their capabilities. Hence, most vendors also provide users the ability to create their custom bots using low-code approaches that do not require any programming knowledge [13]. One such tool is what is generally referred to as a recorder that allows users to record themselves performing a task they want to automate [4], such as copying data from one place to another. The recorder generates an RPA script that can then be run to autonomously perform the task. In the business world, many companies have proprietary software and procedures that require specialized bots.

While the ability to create custom bots in companies by RPA users is crucial in businesses, the typical scope of execution of these bots still focuses on an individual's task needs as opposed to the team's. However, if these bots are not generalizable enough, companies will face a governance problem. In other words, creating a bot for every new narrow task will cause a bot sprawl problem and will make it very difficult for companies to keep track of, maintain, and reuse their bots. Instead, they will keep creating new bots without maximizing the return on investment of creating bots; i.e., the bots will not be used enough times to justify the overhead of creating them.

Currently, building generalizable bots requires intervention from IT departments that are typically oversubscribed and have limited resources. Furthermore, the generalization process is typically long and tedious and thus, does not scale to cover the expansive needs of business users. It is therefore imperative to empower business users to act as citizen developers to build generalized bots themselves. But this will require building tools that are intuitive to citizen developers (e.g. using natural language), which simplify and abstract away the complexities of traditional development and debugging. Providing such tools requires balancing the trade-offs between their simplicity and the ambiguity and imprecision that results from providing such intuitive tools (e.g., natural language interface is familiar to users but is not precise).

While the issues outlined above represent a set of complex problems, we note that variants of these problems have been an active topic of research in other computing and AI disciplines. This includes conversational digital assistants that need to interact with multiple users whose utterances may be noisy [22], software engineering where the goal is to provide user friendly visual languages to enable good specifications of desired behavior, and AI planning and robotics where creating specialized robots for narrow tasks is not feasible, given the physical nature of the bots. In particular to the robotics field, different approaches have been suggested that focus on teaching robots how to perform new tasks as they operated in the world performing tasks they already knew. Topics like transfer learning, learning from instruction, learning from demonstration, self-learning, active learning and others have been maturing in the fields of artificial intelligence, machine learning and deep learning for the application of robotics. These algorithms could also prove beneficial for RPAs.

We posit that the next generation of RPA bots used in enterprise settings will be taught *iteratively* through *intuitive* and *multi-modal* interactions dependent on the type of task and skill of the teacher(s). These bots should be characterized by their ability to continually learn, customizability by end users with minimal coding expertise, evolution with changing business processes, and generalizability to new process automation tasks by understanding their context. We also discuss some of the challenges of realizing this vision, focusing on challenges at the intersection of machine learning, process management, and interactive task learning.

The remainder of this paper is organized as follows. In the next subsection (c.f. Sect. 1.1), we present a running example which we will use throughout the paper. In Sect. 2, we define a few key terms to set the context for our discussion. In Sect. 3, we highlight existing RPA work. In Sect. 4 and 5, we present our vision of evolving from traditional approaches to build the next generation of RPA bots that can be taught new skills, and outline the research challenges to achieve this vision. We present our conclusions in Sect. 6.

## 1.1   Running Example

Consider the following motivating use case involving an enterprise employee from the sales department seeking business travel pre-approval, which will need to be approved by their manager before travel. In a typical enterprise setting the employee would need to submit the reason for travel along with an estimated budget. While there may be multiple enterprise applications that they could use, for the purposes of this paper we will assume that the employee will need to fill out an excel sheet of the itemized estimated expenses. To fill out the excel sheet, the employee needs to pick a mode of transportation (flight, train), identify the source (Boston, MA) and destination cities (New York, NY), travel dates, hotel reservations, taxi or car rental. There may also be specific types of policies that the employee needs to adhere to, such as company-wide policies (e.g. applying company agreement discount for the car rental, choosing hotels from the list of approved ones) governmental regulations (e.g. visa restrictions for international travel), and customer preferences (e.g. requiring a proof of COVID vaccination while visiting). Furthermore, the employee would need to submit this pre-approval for each sales trip that they plan to take.

To ease the burden of doing this repetitive mundane task, the employee can go through the hassle of building an RPA which will search for flights – because it is the employee's preferred mode of transportation – from Boston, MA, parse the result and populate the corresponding cell in the excel sheet, then proceed to repeat similar steps for both hotel reservations and car rental.

While such an RPA would perform adequately for some trip pre-approvals, there are certain gaps and challenges that prohibit this from being a wide ranging solution for all the travel needs of the employee as well as others in the organization. Consider the case when the employee decides to take a personal trip. While the RPA can be used to search for flights, hotels and cars. The employee is no longer restricted to book only the company approved ones, and may not be allowed to apply the company discount when renting a car. Thus, they would have to perform manual adjustments.

More critical cases that may need significant changes to the RPA or can result in RPA execution failure occurs when say the excel sheet formatting is updated due to changes in company or governmental policies (e.g. pandemic), or when an employee in another branch of the same enterprise company (e.g. London, England) starts using the same RPA to get a travel pre-approval. Since the RPA's home city is set to Boston, MA, the RPA may report an incorrect flight estimate. It will also, as programmed, apply the policies of that of the US branch, which may be different from England's branch.

While these may not be crucial changes, this example serves to highlight the challenges in building and deploying RPAs that are focused on solving one particular use case and are not able to handle relatively general cases.

## 2  Background: Intelligence and Learning

Creating truly intelligent machines, generally referred to as artificial general intelligence [11], has been the ultimate goal of artificial intelligence researchers; its objective is to endow machines with human-like intelligence. However, researchers still debate the exact definition of intelligence in humans (in addition to living beings more generally and hence machines). With theories ranging from the g-factor and the Cattell–Horn–Carroll theory [3] to Gardner's multiple intelligences [10], the one common theme in all is the ability *to learn*. The Merriam-Webster dictionary defines intelligence as "the ability to learn or understand or to deal with new or trying situations" and "the ability to apply knowledge to manipulate one's environment or to think abstractly as measured by objective criteria". The dictionary also defines learning as the ability "to gain knowledge or understanding of or skill in by study, instruction, or experience".

In [26], agents learn if they can update and expand their internal model of the environment (which influences their actions) based on the experience they gain from operating in the environment. This does not mean that the agent cannot have prior knowledge obtained from its designer, but it is expected to build on this prior knowledge by incorporating its understanding of the world it is perceiving. It also means that agents should not just *imitate* or *memorize* what an instructor taught the system (by repeating the exact same steps independent to context); in other words, for the agent to have truly learnt from an instructor, it should successfully complete tasks in different environments than when it learned the task.

Multiple methods can be adopted to acquire new knowledge. One approach to learning is to simply observe other agents (e.g., humans) performing the task through user logs. Another is to obtain instructions from end users dynamically; this allows the system to ask clarifying questions (if it identifies ambiguous instructions). Yet another approach is learning from demonstrations where end users perform the task and the system records a video (or something similar). Finally, learning from examples entails the system receiving a bunch of input/output examples from users and extrapolating the rules. We use the term "teaching" when the system is explicitly receiving instructions from an end user and we use the term "learning" when the system indirectly obtains the knowledge of performing new tasks.

Focusing on RPAs, this means that RPAs must learn from experience or through instruction to exhibit intelligence. While [33] concluded that RPAs are not intelligent and [5] discussed multiple requirements to transform RPAs from robotic to intelligent process automation bots, in what follows, we will focus on the learning aspect of intelligence and discuss how RPAs become more "intelligent" by providing them with the tools to learn (and generalize to) new tasks.

## 3    RPA State-of-the-Art

Many techniques to create RPAs have been adopted in the literature and in industry since the inception of RPAs. These techniques require varying degrees of human involvement, some in a direct and others in an indirect manner. The most common approach is the manual creation of RPA scripts by human experts (IT personnel or subject matter experts or both in collaboration). However, many approaches have also been developed to automate the generation of scripts. Some of these automated approaches directly place the human in the loop; these generate scripts from a human's demonstrations or instructions [21]. Others indirectly leverage humans by generating scripts from user logs performing the tasks that must be automated [1] or process instruction manuals [20]. Almost all automated approaches attempt to leverage artificial intelligence algorithms with various levels of sophistication [16]. In this section we review two popular approaches leveraged by existing solutions.

### 3.1    Recorders

Beyond out-of-the-box bots in RPA stores, many vendors, including IBM RPA, Automation Anywhere, UIPath and Blueprism, provide studios to build your own RPA bots by writing scripts (which may have learning curves with varying degrees of difficulty). They also provide a feature in these studios known as *a recorder*. These recorders allow business users to teach RPA bots by demonstration.

Users record their actions while performing a task on their desktop and the recorder automatically generates a script that can be run to perform the task. Generally speaking, these recorders possess a set of atomic actions that they recognize from recordings. The functions for these actions are already part of the studio's library and hence all that must be done is combine these functions together into a new script and plug in their inputs that are identified from the recording. For the running example in Sect. 1.1 where the employee needs to fill out an Excel spreadsheet of travel expenses, an example of an atomic function is copying the contents of a cell in Excel into another cell. Of course, the bot must have an understanding of Excel's domain which consists of the concept of cell, among others, and know how to refer to a cell. This allows it to identify the source and target cells and pass them as arguments to the function that performs the action of copying.

On the one hand, recorders can produce a time series data stream of atomic actions and convert the time series to an automation script. On the other hand, they could produce a sequence of image frames that can be analyzed by image processing technology (including segmentation and understanding) to identify actions being performed and map them to a sequence of atomic functions. Each approach presents different challenges and impact on the effectiveness of the recorders in generating RPA scripts. The different recorders also require different learning curves based on the user's existing knowledge and abilities, thereby presenting varied levels of usability. The choice of recorder and the expertise of the user will impact the performance of the RPA scripts being generated as well as its generalizability across different users.

## 3.2   Conversational RPAs

Depending on automation tasks and the users' comfort levels, invoking RPA bots may not always be easy. Hence, commercial vendors also offer digital conversational assistants [25,32] to allow users to interact with RPA bots in natural language. Various vendors and approaches in the literature provide varying degrees of sophistication as far as natural language comprehension is concerned.

UIPath leverages the Google Dialogflow chatbot authoring tool to allow RPA bot creators to endow their bots with conversational abilities[1]; in [14] RPA bots are integrated with RASA, a natural language understanding and dialog open source library and [23] uses regular expression matching. Yet another approach provides more flexibility in the choice of natural language processing sophistication to integrate with RPAs; in [25], RPA bots could be combined with regular expression or Watson Assistant or other types of natural language understanding components as a result of the modular architecture presented.

---

[1] https://docs.uipath.com/chatbots/docs/connect-to-google-dialogflow.

The building blocks comprise of skills that are atomic functions to *understand* user inputs, *act* upon them, and finally *respond* back appropriately. Understand skills are functions driven by language models that serve to interpret the users' natural language utterances to determine the underlying intent, identify key entities, and determine the act skills to subsequently execute. Act skills are RPAs that execute the user's intent and automate business process tasks to produce the expected outcome. These skills are of two types – those that have a lasting change on the state of the world (e.g.) sending emails or paying an invoice, and those that do not (e.g.) reading emails or checking account balance. Respond skills are RPAs that produce a human-consumable response from the act skill's output. This could be a natural language utterance, a visual representation, etc. These skills could range from simple template responses to more sophisticated text generation deep learning models.

In the running example in Sect. 1.1, the employee could provide a natural language input (e.g.) "seeking travel pre-approval", to a conversational assistant. It would interpret the input and trigger the appropriate RPAs to automate filling out relevant information such as the location, dates, etc. The assistant would then return with a natural language response indicating whether the approval had been granted or not.

## 4   The Vision

The traditional approach to developing RPA bots, shown on the left in Fig. 1, has separate phases for development and usage. Failures and issues encountered by users are logged over time and eventually get acted upon by developers to build and deploy new and improved versions of the bots. However, this process can have long turnaround times, and often results in developers prioritizing the resolution of certain issues over others due to lack of resources, causing frustration to the users.

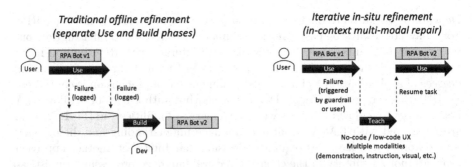

**Fig. 1.** Multi-modal teaching: RPA Bots should evolve by enabling users to teach using a combination of modalities in the context of where the bots fail.

In contrast, we envision the next generation of RPA bots where the development and refinement of automation capabilities are done in an iterative manner, as shown on the right in Fig. 1. In this approach, users are empowered to teach the RPA bots new capabilities to address any shortcomings. There are three aspects that differentiate this approach from the traditional way to develop RPA bots. First, we define a teaching mode that the system enters when automation failures interrupt the user's task. These failures could occur due to guardrails defined by the developer or by the business, which limit the execution bounds for the RPA, or because of the user explicitly indicating their dissatisfaction with the actions taken. Second, users address specific failures in the context of the task they were attempting to perform. Since teaching happens in the context of the failures, it enables the user to quickly address the issue. We believe this is an easier way for non-developers to teach and refine the RPAs. Third, since the same user is responsible for teaching the bot how to handle the failures, it would empower users to evolve and customize them according to their needs.

This iterative approach can have two variants: *Mode 1* where the user resumes their task after repairing the failure as depicted in Fig. 1; and *Mode 2* where the user restarts their task with the new version of the automation. The latter may be appropriate if there have been no side effects so far, or if the automation can roll back or compensate the steps already taken.

The process of teaching these bots needs to be intuitive to citizen developers with limited programming knowledge by supporting multiple modalities and users' teaching abilities. Certain automations may need to be taught through demonstration and recordings, while others may be more easily taught through natural language instructions. In addition, some users may be more articulate and knowledgeable than others in teaching their automation needs, and the bots should be able to identify information gaps and communicate with the users.

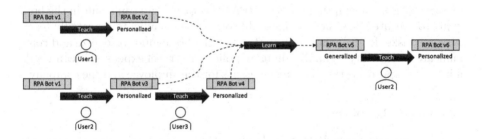

**Fig. 2.** Collaborative teaching: The evolution of a bot is not simply a one-time teaching step by a single user, but instead consists of iterative phases of users teaching bots to personalize them for their needs, and the system learning to generalize from the teachings.

The iterative teaching approach needs to consider the lifecycle of how multiple users collaborate, either directly or indirectly, to teach and improve the bot over time. As shown in Fig. 2, users may independently teach and personalize a bot resulting in a graph of versioned bots. In addition, the system can learn from these individual teachings to generate a generalized bot that is more robust than any of the personalized bots. Users may then choose to use this generalized bot and further personalize it, continuing the cycle of iterative refinement. There is an open problem on how the system can learn from individual teachings, and many of the challenges in Sect. 5 are relevant to solving this problem.

In the individual and collaborative teaching and learning phases, it is important that the bots adhere to certain requirements, such as those derived from industry regulations or company policies. For example, there may be a requirement that a bot only book international flights if the user is vaccinated, or check that users only reserve hotels from a company-approved list. Both bots personalized by users and those generalized by the system need to satisfy these requirements. In some cases, these guardrails may be checked at the time of teaching or learning, similar to static checking of programs by a compiler. In other cases, it is only at runtime that guardrail violations can be detected. The problems of generating guardrails from requirements and checking for violations are open challenges.

## 5   Challenges and Opportunities

### 5.1   Metacognition

Arguably one of the most important aspects of learning is metacognition [6]: understanding what one does not know. Also referred to as introspection, [12,31] argued for its importance in machine classification tasks. In the context of RPAs, it is important for these bots to have an awareness, and possibly understanding, of what they do and do not know, especially in systems that learn from human demonstration or instruction. In such situations, it would be very difficult, and even frustrating, for users attempting to teach RPAs new tasks, and can result in the bots trying to execute tasks they are incapable of, or users having to repeatedly teach the same tasks. In the example in Sect. 1.1, if the bot cannot recognize and communicate its inability to apply company policies to travel expenses, it can result in incorrect approvals thereby causing issues for the employee and the company.

### 5.2   Generalizability

To truly learn new tasks, RPA bots should not simply memorize steps by converting them to scripts but should be capable of combining the newly acquired steps to previous knowledge and extrapolate to new similar tasks. In other words, as the bots learn new tasks, they should require fewer samples to learn. They should also extrapolate to new tasks more easily as their experience increases. This is a challenging problem because it is difficult to identify what parameters will vary across tasks and by how much and which parameters are fixed. In some cases, this requires understanding the underlying intentions of the action and not just how to perform an action. For instance, in the example in Sect. 1.1, say the employee creates an RPA to submit their travel request from Boston to New York. The bot should be able to identify that Boston and New York are parameters that can vary, and thereby generalize the automation for other employees who may wish to input other locations, without them having to recreate it. The machine learning literature has looked at the generalizability problem in multiple domains; transfer learning [30] is one sub-field of machine learning that was developed to study how models can transfer knowledge they have learned for one task to a new but related task. RPA bots may also look to leverage transfer learning approaches to generalize their capabilities.

## 5.3  Catastrophic Forgetting

Another concern in machine learning algorithms that learn over time, especially neural network models, is catastrophic forgetting [9]; bots may forget what they had already learnt as they learn new tasks. This phenomenon is problematic because it means that bots are replacing previously acquired knowledge instead of combining and augmenting their models. Users may need to reteach bots tasks they had previously taught them even though these tasks had not become obsolete. This issue would also hinder the solution to bot sprawl and governance since new bots would need to be created to learn new tasks as opposed to teaching existing ones. Furthermore, it would create a frustrating experience for business users who may opt not to use such tools at all. For example, if a users instructs a bot to book a trip using a plane and another user instructs the same bot to book a train instead, we expect the bot to (at the very least) ask the user the next time around whether to use a plane or a train instead of replacing the plane with a train every single time.

## 5.4  Citizen Developers

RPAs are generally adopted by business users who do not have much knowledge of programming or machine learning. Since it is these users who will also need to teach these bots new tasks, the user experience for teaching bots must be robust enough to handle failures and recover from them gracefully enough to allow users to understand and resolve issues. Further, these systems should interact with users in natural language, the most natural medium for business users. However, it should also support multi-modality [17] since natural language is inherently noisy and may not be the most optimal modality. So a mix of teaching by instruction and demonstration may be necessary. Unfortunately, there exists a tradeoff between the learning curve of a tool and this tool's ability to create complex automation due to the limitations of the current technology. In other words, if we want to reduce the learning curve for citizen developers by using natural language instructions to program RPA bots, we will need to restrict the scope to create simple bots only because the limitations of natural language processing technology limits the scope of phrases it can parse and understand [18].

## 5.5  Automation Lifecycle

Enabling citizen developers to teach bots to address their manual, repetitive tasks creates a new set of challenges about the overall automation lifecycle.

The first of these challenges is dealing with changes in the specification of the underlying system that the bot is targeting (e.g., changes in the destination travel reservation APIs to include additional required fields, or changes in the format of the pre-approval Excel sheet to include other required fields). The second challenge is dealing with changes to the environment where the bot is deployed that affects its state of execution (e.g., in case of the digital assistant, the natural language understanding AI model can change as result of new training data). Also, while enabling citizen developers to build their own bots, it would be naive to expect them to teach a bot that not only addresses their automation needs (happy execution path) but also addresses the side effects of errors in that execution (e.g., what to do when there are exceptions, branching based on certain output status). Thus, that bot needs to evolve to address changes in the outside environment, accommodate changes to its environment, and most importantly provide a mechanism and tools, which includes the user in the loop, to address situations that the citizen developer did not consider. Finally, these bots need to also be maintained throughout their lifecycle as the software that they interface with changes and the technology they leverage evolves.

## 5.6   Interpretability

As users interact with RPA bots to teach them new capabilities, it becomes even more important for these bots' behavior to be transparent and interpretable. This allows users to build a more accurate mental model of the bots' inner workings and adapt their instructions accordingly, reducing the amount of guessing that ultimately leads to a frustrating user experience. However, since the technology that enables RPA bots to learn relies on AI, often complex black box models like deep neural networks, explainability may not be intrinsic or easy to implement [27].

## 5.7   Interactive and Informative Experience

Teaching RPA bots, either by demonstration or instructions, can result in them behaving differently from user expectations which can prove costly. For instance, the bots may misinterpret the taught information and perform incorrect actions, or may just not learn at all. This makes it important for the bots to provide informative responses during the teaching process reflecting their understanding at every step. In addition, the responses need to be user-friendly and interpretable, and the bots should also be able to elicit user feedback in an interactive manner to determine when their expectations are not being met. Authors in [2,24] discuss the importance of defining mechanisms for bots to "fail gracefully", but the vast space of multi-modal user inputs presents a challenge to identify the appropriate response.

## 5.8   Learning New Skills

Initial research efforts towards teaching assistants [2,22] have limited their scope to specific use-cases. However, in real-world settings, RPA bots need to be able to execute automation skills that vary widely in their complexity, scope, usage, etc., and interact with users who differ in their expertise and teaching abilities. Learning new skills may involve incrementally updating existing ones, training a model through user-provided input/output examples, or even synthesising new code from specifications, making it important for bots to be able to identify what and how to learn. In addition, large teams of users may teach overlapping skills, or with conflicting requirements, thereby requiring bots to have the ability to disambiguate. Analysis of user interaction logs to identify similarities and differences, transferring relevant knowledge across skills, and continual learning with changing business and user needs, are important capabilities for bots to possess to enable intuitive teaching models.

### 5.9   Guardrails

Any time we empower end users and/or the RPA bot itself to evolve throughout the bot's lifecycle, we must place guardrails to ensure that the behavior does not cause harm or break any rules [8,15]. Microsoft's chatbot disaster (where users taught the chatbot to become racist within 24 h) [34], while one of the worst case scenarios that could unravel, is not the only reason why safeguards are important. Less dramatic reasons like company policy or domain best practices may need to be encoded into such the bots; the users should not be able to use teaching frameworks to circumvent such policies. Similarly, the bot should not generalize to a state that may disregard these policies or best practices.

## 6   Conclusions

RPAs have been the crucial driving force in the digital transformation of enterprises, and are considered by many the seminal innovation in helping lower the bar of development for business users. Most of this effort, however, is just in its infancy. Our vision aims to lower the bar even further by advocating for the next generation of RPA bots. By tapping into recent advancements in Artificial Intelligence, RPA bots should provide users with an intuitive, interactive, and multi-modal framework to teach RPAs new capabilities, that can be learned and refined in an iterative manner. In addition, to achieve wide ranging use, these bots would also need to have the ability to generalize the skills that they learn across different automation tasks and users with varied needs and expertise. This would empower business users to become citizen developers and leverage RPAs to work for their personal and organizational automation requirements. In this work, we outline several challenges that span across machine learning, process automation, and interactive task learning that need to be addressed towards making this vision a reality.

## References

1. Agostinelli, S., Lupia, M., Marrella, A., Mecella, M.: Automated generation of executable RPA scripts from user interface logs. In: Asatiani, A., et al. (eds.) BPM 2020. LNBIP, vol. 393, pp. 116–131. Springer, Cham (2020). https://doi.org/10. 1007/978-3-030-58779-6_8

2. Azaria, A., Srivastava, S., Krishnamurthy, J., Labutov, I., Mitchell, T.M.: An agent for learning new natural language commands. Auton. Agents Multi-Agent Syst. **34**(1), 1–27 (2019). https://doi.org/10.1007/s10458-019-09425-x

3. Carroll, J.B., et al.: Human Cognitive Abilities: A Survey of Factor-Analytic Studies, vol. 1. Cambridge University Press, Cambridge (1993)

4. Cewe, C., Koch, D., Mertens, R.: Minimal effort requirements engineering for robotic process automation with test driven development and screen recording. In: Teniente, E., Weidlich, M. (eds.) BPM 2017. LNBIP, vol. 308, pp. 642–648. Springer, Cham (2018). https://doi.org/10.1007/978-3-319-74030-0_51

5. Chakraborti, T., et al.: From robotic process automation to intelligent process automation. In: Asatiani, A., et al. (eds.) BPM 2020. LNBIP, vol. 393, pp. 215–228. Springer, Cham (2020). https://doi.org/10.1007/978-3-030-58779-6_15

6. Cox, M.T.: Metacognition in computation: a selected research review. Artif. Intell. **169**(2), 104–141 (2005)

7. Czarnecki, C., Fettke, P.: Robotic process automation. In: Robotic Process Automation, pp. 3–24. De Gruyter Oldenbourg (2021)

8. Eaneff, S., Obermeyer, Z., Butte, A.J.: The case for algorithmic stewardship for artificial intelligence and machine learning technologies. Jama **324**(14), 1397–1398 (2020)

9. French, R.M.: Catastrophic forgetting in connectionist networks. Trends Cogn. Sci. **3**(4), 128–135 (1999)

10. Gardner, H., et al.: Multiple Intelligences, vol. 5. Minnesota Center for Arts Education (1992)

11. Goertzel, B.: Artificial general intelligence: concept, state of the art, and future prospects. J. Artif. Gen. Intell. **5**(1), 1 (2014)

12. Grimmett, H., Paul, R., Triebel, R., Posner, I.: Knowing when we don't know: introspective classification for mission-critical decision making. In: 2013 IEEE International Conference on Robotics and Automation, pp. 4531–4538. IEEE (2013)

13. Hirzel, M.: Low-code programming models. arXiv preprint arXiv:2205.02282 (2022)

14. Hung, P.D., Trang, D.T., Khai, T.: Integrating chatbot and RPA into enterprise applications based on open, flexible and extensible platforms. In: Luo, Y. (ed.) CDVE 2021. LNCS, vol. 12983, pp. 183–194. Springer, Cham (2021). https://doi.org/10.1007/978-3-030-88207-5_18

15. Hwang, T.J., Kesselheim, A.S., Vokinger, K.N.: Lifecycle regulation of artificial intelligence-and machine learning-based software devices in medicine. Jama **322**(23), 2285–2286 (2019)

16. Jha, N., Prashar, D., Nagpal, A.: Combining artificial intelligence with robotic process automation—an intelligent automation approach. In: Ahmed, K.R., Hassanien, A.E. (eds.) Deep Learning and Big Data for Intelligent Transportation. SCI, vol. 945, pp. 245–264. Springer, Cham (2021). https://doi.org/10.1007/978-3-030-65661-4_12

17. Kephart, J.O.: Multi-modal agents for business intelligence. In: Proceedings of the 20th International Conference on Autonomous Agents and MultiAgent Systems, pp. 17–22 (2021)

18. Khurana, D., Koli, A., Khatter, K., Singh, S.: Natural language processing: State of the art, current trends and challenges. arXiv preprint arXiv:1708.05148 (2017)

19. Kirchmer, M., Franz, P.: Value-driven Robotic Process Automation (RPA). In: Shishkov, B. (ed.) BMSD 2019. LNBIP, vol. 356, pp. 31–46. Springer, Cham (2019). https://doi.org/10.1007/978-3-030-24854-3_3

20. Leopold, H., van der Aa, H., Reijers, H.A.: Identifying candidate tasks for robotic process automation in textual process descriptions. In: Gulden, J., Reinhartz-Berger, I., Schmidt, R., Guerreiro, S., Guédria, W., Bera, P. (eds.) BPMDS/EMMSAD -2018. LNBIP, vol. 318, pp. 67–81. Springer, Cham (2018). https://doi.org/10.1007/978-3-319-91704-7_5
21. Li, T.J.J., Mitchell, T., Myers, B.: Interactive task learning from GUI-grounded natural language instructions and demonstrations. In: Proceedings of the 58th Annual Meeting of the Association for Computational Linguistics: System Demonstrations, pp. 215–223 (2020)
22. Li, T.J.J., Radensky, M., Jia, J., Singarajah, K., Mitchell, T.M., Myers, B.A.: PUMICE: a multi-modal agent that learns concepts and conditionals from natural language and demonstrations. In: Proceedings of the 32nd Annual ACM Symposium on User Interface Software and Technology, pp. 577–589 (2019)
23. Moiseeva, A., Trautmann, D., Heimann, M., Schütze, H.: Multipurpose intelligent process automation via conversational assistant. arXiv preprint arXiv:2001.02284 (2020)
24. Ortiz, C.L.: Holistic conversational assistants. AI Mag. 39(1), 88–90 (2018)
25. Rizk, Y., et al.: A conversational digital assistant for intelligent process automation. In: Asatiani, A., et al. (eds.) BPM 2020. LNBIP, vol. 393, pp. 85–100. Springer, Cham (2020). https://doi.org/10.1007/978-3-030-58779-6_6
26. Russell, S., Norvig, P.: Artificial intelligence: a modern approach (2020)
27. Samek, W., Wiegand, T., Müller, K.R.: Explainable artificial intelligence: Understanding, visualizing and interpreting deep learning models. arXiv preprint arXiv:1708.08296 (2017)
28. Schmitz, M., Dietze, C., Czarnecki, C.: Enabling digital transformation through robotic process automation at Deutsche Telekom. In: Urbach, N., Röglinger, M. (eds.) Digitalization Cases. MP, pp. 15–33. Springer, Cham (2019). https://doi.org/10.1007/978-3-319-95273-4_2
29. Siderska, J.: Robotic process automation-a driver of digital transformation? Eng. Manage. Prod. Serv. 12(2), 21–31 (2020)
30. Tan, C., Sun, F., Kong, T., Zhang, W., Yang, C., Liu, C.: A survey on deep transfer learning. In: Kůrková, V., Manolopoulos, Y., Hammer, B., Iliadis, L., Maglogiannis, I. (eds.) ICANN 2018, Part III. LNCS, vol. 11141, pp. 270–279. Springer, Cham (2018). https://doi.org/10.1007/978-3-030-01424-7_27
31. Triebel, R., Grimmett, H., Paul, R., Posner, I.: Driven learning for driving: how introspection improves semantic mapping. In: Inaba, M., Corke, P. (eds.) Robotics Research. STAR, vol. 114, pp. 449–465. Springer, Cham (2016). https://doi.org/10.1007/978-3-319-28872-7_26
32. UIPath: RPA + AI-powered chatbots? Now you're talking (2022). https://www.uipath.com/product/chatbots-automation
33. Viehhauser, J.: Is robotic process automation becoming intelligent? Early evidence of influences of artificial intelligence on robotic process automation. In: Asatiani, A., et al. (eds.) BPM 2020. LNBIP, vol. 393, pp. 101–115. Springer, Cham (2020). https://doi.org/10.1007/978-3-030-58779-6_7
34. Wolf, M.J., Miller, K.W., Grodzinsky, F.S.: Why we should have seen that coming: comments on Microsoft's tay "experiment," and wider implications. ORBIT J. 1(2), 1–12 (2017)

# API as Method for Improving Robotic Process Automation

Petr Průcha[(✉)] and Jan Skrbek

Technical University of Liberec, Liberec, Czechia
`petr.prucha@tul.cz`

**Abstract.** Robotic process automation has been maturing with significant speed and organizations that started using RPA during the past years now automate the "low hanging fruit" with the best return on investment. The organizations strive to utilize RPA to get a higher value. However, the deployment of RPA on multiple processes is complex and demanding in terms of costs and resources. In this research, we focused on the reduction of the RPA bot duration by using the applications' API in the RPA automation. We compared the duration of RPA bots on three processes. The three processes were automated with RPA technology via GUI and also via API. The results indisputably show that using API in RPA automation has a positive impact on its duration. RPA bot using API was, in some cases, ten times faster than RPA bot using GUI. The average duration change of using API against GUI was in the interval from 84.01% to 91.87%. This change shows the enormous acceleration of RPA bot and the impactful benefits of the use of API in RPA automation of processes. We may thus conclude that API is a good complement to RPA automation. RPA developers and RPA architects should use API when it is possible for better utilization of RPA robots.

**Keywords:** Robotic process automation · Application programming interface · RPA standardization

## 1 Introduction

From the beginning of humankind, people have been trying to make complex tasks easier to increase value without having to work that hard. Thanks to this evolutionary feature, we have reached a point where we have RPA bots that do routine and monotonous work that used to be done by humans. During the last years, rather than an emerging technology, RPA has been turning into a mainstream technology and the wave of hype around RPA has been slowly subsiding. In the near future, we may expect that fewer companies will be willing to pay the extra money for a mainstream technology that, in some cases, is incomplete, buggy, and hard to scale up. More frequently, companies and organizations are interested in the total cost of RPA, the Full-time equivalent (FTE) saved, and the overall Return on Investment (ROI). Therefore, the management of companies is pushing the Center of Excellence (CoE) department to maximize efficiency, deploy RPA bots as much as possible and maximize the utilization of licenses and robots [15, 26].

© Springer Nature Switzerland AG 2022
A. Marrella et al. (Eds.): BPM 2022, LNBIP 459, pp. 260–273, 2022.
https://doi.org/10.1007/978-3-031-16168-1_17

On the other hand, when RPA bots reach a certain critical mass, they often require a lot of additional work from RPA developers and the CoE maintenance team to produce any value to the organization. In some cases, it can be a never-ending cycle of maintenance of RPA bots [17]. Governance of deployed robots does not bring any additional value to the organizations and consumes limited resources of CoE. Due to spending resources on the maintenance of the robotic process, developers do not have enough time to deploy new robots into production and improve the utilization of the RPA ecosystem. Originally, RPA robots were primarily used for the automation of legacy systems, but with the development of RPA technology, they are increasingly used for all kinds of automation. The automation maintenance of legacy systems is less demanding than modern systems. Legacy systems and their automation are usually more stable due to the fact that there are no updates that would change the UI of the app or change the operation of the app. As opposed to legacy systems, modern systems/apps are developed agilely and are updated regularly in 14 day intervals (depending on the length of the development sprint). Every new update may break any automation using the app; the more RPA robots deployed, the more time spent on maintenance.

In 2000, Richard Fielding [14], in his dissertation thesis, laid the foundation for REST API. This invention started a new type of back-end automation. REST API transformed the IT world and led to the foundation of the new data-based economy [18]. API automation is heavily used, especially for high volumes of processes, but finds application also in less busy processes with few uses per day or even per month. Back-end automation is considered reliable. However, the applications have to support the API. The support of APIs is especially problematic in legacy software that was coded in previous millennia, where much of today's applications were written in programming languages that are now considered obsolete, such as FORTRAN or COBOL [1, 3, 28]. Even today's applications, written in modern programming languages, often do not support APIs, which leads to the impossibility of back-end automation. If the application does not have API, the only possibility for implementing automation is on the front-end, which is the domain of RPA. Another reason why back-end automation is used just in specific cases is due to the fact that back-end automation requires much deeper programming knowledge than RPA bots. This can be problematic in a world with a shortage of experienced software engineers. Creating a simple API call is relatively easy and similar to classic programming languages and RPA, but the subsequent processing of information is more demanding than it is for RPA bots, which can be programmed by a so-called citizen developer. Research published by Van der Aalst et al. [1] describes when it is more suitable to use API versus when RPA is more appropriate. However, many processes require connecting more than one application. If we automate a process through multiple applications, some of these applications may allow API use, and some may not.

In this paper, we investigate the possibilities of API use in automation via RPA, which could help to solve the stability issues of RPA bots. We focus on researching the potential benefits of API in RPA automation for the duration of a process. The current state of the art suggests that API can be beneficial for duration of RPA bots but does not provide enough evidence for this suggestion.

The paper is structured as follows; first, we investigate work related to our research, then we introduce a case-study example for testing the potential benefits of API in RPA automation. Later we introduce the methodology of how we conducted the research and the results of our research. To conclude, we discuss the results in the context of other studies on this topic.

## 2 Process Automation

The automation of a process comprises numerous approaches and is not limited to the robotic process automation and API automation presented in the introduction. Also, a relatively frequent type of process automation is workflow automation [24]. Workflow automation is usually limited to platforms such as CRM or ERP [9]. Extending the workflow automation out of the platform is essential for opening the application interface for API automation or for using RPA on the graphical user interface (GUI) of the software. In Table 1, the advantages and disadvantages of the aforementioned types of process automation are presented.

**Table 1.** Overview of process automation

| Process automation | Advantages | Disadvantages |
|---|---|---|
| RPA automation | The simplicity of automation, changing the code of the application is not required, the possibility to automate various applications and connect them | Not as reliable and stable as API, it depends on third-party software |
| API automation | The stability and reliability of automation, the low operation cost of automation | The complexity of automation software without prepared API |
| Workflow automation | The simplicity of automation, the low operation cost of automation | Limited to a particular platform |

Workflow automation is not suitable for the automation of complex processes due to limitation to a particular platform [9, 24]. RPA and API automation are versatile and it is possible to use them for all kinds of automation [9]. RPA and API automation usually have different uses for automation. For RPA, these include lightweight automation with many applications, legacy systems, or when a hardcoded solution is not rentable [1, 3]. API automation is more typically used for automation processes that are high in volume or an application with programmed API [1, 3]. RPA automation is considered as an example of front-end automation and API automation as an example of back-end automation. Front-end and back-end automation are not necessarily two separate approaches. Web-scraping or software testing are good examples of connecting back-end and front-end automation. The RPA technology from the main RPA vendors allows the use of API calls in RPA automation. It is also possible to hardcode some code into a process automated by RPA.

# 3 Related Work

A comparison of API and RPA already started to appear in the first studies on RPA technology, where it was necessary to point out suitable use cases and explore where the technology has an advantage compared to other technologies. With the growing popularity of RPA, the comparison of RPA and back-end automation has been declining. More articles focus on the use, expansion, and enhancements of RPA robots [13, 27]. A frequently used example is intelligent process automation (IPA[1]), combining robotic process automation with artificial intelligence [7, 19, 21]. The combination of RPA with Business Process Management (BPM) represents the scope extension of RPA. In this case, the process automation is used as a backup of a working process. It has the advantage since the process for automation has to be mapped and documented. The process for RPA bots has to be deterministic, so it also benefits from the consistency of results [11]. There are also studies on the use of RPA for software testing and the use of RPA in test-driven development [6]. Another example of improving and extending the use of RPA is the use of process logs and RPA logs to compare manual and robotic performance and detect potential process errors [10].

## 3.1 Improvement of RPA

In this study, we focused on the improvement and optimization of RPA bots. By improvement and optimization, we mean increasing efficiency or performance. Coudhary and Karmel [8], in their research, focused on the optimization of RPA bots by more effective implementation of the infrastructure that RPA bots use. They achieve the improvement by automatic scheduling activities that take place on the infrastructure. Similar research was conducted to increase the efficiency of bots by scheduling. RPA bots were scheduled to complete all tasks in a queue with limited resources as quickly as possible. The authors use linear integer programming to solve this problem [23]. This problem is very complex and can be compared to the familiar "traveling salesman problem[2]" consisting in finding ways to best allocate and utilize limited resources. Better utilization of limited resources and efficiency improves the ROI of all RPA automation. An example of RPA improvements is the framework for optimizing RPA robots in auditing. Eulerich et al. [12] have developed a set of procedures for creating more efficient RPA robots in the field of auditing; this has also led to the design and creation of architecture for RPA robots. So far, there is no global standard that developers should follow. Most often, developers use standards from RPA vendors that are not always optimal. This issue is mentioned in the comprehensive study of RPA, and one of the future challenges is to create a standard or at least the best case practices that developers will adhere to [25].

## 3.2 API in RPA

Most of the research papers about RPA and API or back-end automation mention RPA and API in relation to the comparison of these two technologies. Existing studies describe

---

[1] IPA is an abbreviation made from RPA + AI = IPA.

[2] Also known as "Traveling agent problem", "travelling salesperson problem" or TSP.

when it is beneficial to use back-end automation and when it is preferable to use RPA [1, 2]. In addition, there are studies that use third-party API applications to enhance or extend the use of RPA such as the Google Vision API or the IBM Watson connection API [5, 20]. In previous research [22], in-depth interviews with RPA experts revealed that the use of RPA and API can be very beneficial for speeding up the process automation and increasing the stability of the automated ecosystem.

Based on the research objective, we formulate the following hypothesis:

*H0: API does not affect the duration of the process, automated by RPA bot.*
*H1: API does affect the duration of the process, automated by RPA bot.*

## 4  Case Examples

A common RPA use case is the automation of the processes with sufficient maturity, where the changes in the process are not typical, and adequate volume of transactions to process. This use case is believed to have the most significant ROI. However, over time, RPA automation has also proven to have other benefits as increasing the quality of services, providing transparency to the process and improving the satisfaction of employees [25]. This research paper and case examples examine these potential benefits. Specifically, we test the effects of API in RPA automation on the duration of the processes. We conduct the tests using three business processes. These business processes are specific in that it is possible to perform the given activities using the API calls as well as the graphical user interface (GUI) employed by the user. These three processes are presented in Figs. 1, 2 and 3. While the three processes have been chosen to simulate simple tasks that can be performed in most companies, they carry a certain level of complexity. The flowcharts in Figs. 1, 2 and 3 represent the flow performed by the RPA robot via GUI. A more detailed description of how we automate and measure these processes is provided in the methodology section.

*Process number one (P1)* represents the automation of the social media page administration (Fig. 1). It simulates the process where a social media manager posts a specific text or image to the Facebook page. P1 is the simplest case example of process automation. **The automation for RPA automation of the GUI** sequence is the following: RPA robots open the web browser, go to the given URL address of the Facebook page, log into FB, click on the button to create a post, write a prepared text, then click the post button, wait 3 s, click on the arrow button and finally on the log out button. The **API automation sequence** consists of RPA robot reading parameters such as URL, the API-key and message of the API call and, via the REST API POST method, sends it. The program receives the standard message with code about the progress.

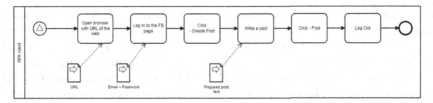

**Fig. 1.** P1 - management of social media

*Process number two (P2)* illustrates the automation of predefined key performance indicator (KPI) reporting from an online database to the process owner or manager of the KPI (Fig. 2). It simulates the process of reporting the net promoter score (NPS) from a customer survey. The manager receives the NPS via email. P2 is constructed from the following steps. Firstly, P2 reads the results from the online database, the RPA robot calculates the NPS, and then emails it to the designated administrator. **The automation sequence for RPA automation of GUI** consists of the following steps: RPA robots open the web browser, go to the given URL address of the Airtable database, read the number of promoters and detractors displayed on the web, click the user account button, click on log out, RPA calculates the results with the formula for NPS and RPA use inbuild method for sending email via Outlook. The **API automation sequence is**: the RPA robot reads parameters such as the URL of the database and the API key of the API call and, via the REST API GET method, sends it. The API call receives JSON of records in the database. The RPA robot parses the values from JSON to get the number of promoters and detractors. The last two parts of the process are the same as GUI RPA automation. The RPA robot calculates the NPS and uses the inbuilt Outlook method for sending the email.

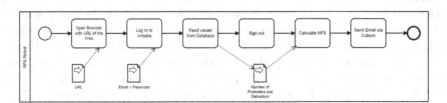

**Fig. 2.** P2 - reading values from database

*Process number three (P3)* is the automation of managing information about customers, leads, and users. This automation addresses the problem of legislative requirements regarding the management of information about customers, leads, and users. It simulates the process of deleting a customer from a CRM system because of legislative requirements or based on customer request. P3 deletes all records from CRM and sends an email to the user of CRM confirming that all records were deleted from the CRM system. P3 automation is made up of two parts, one deletes the records in CRM, and the second sends the email. The **automation sequence for RPA automation of GUI** includes RPA robots opening the web browser, going to the given URL address of Hubspot CRM, logging into CRM, searching for the user that should be deleted, selecting the user, clicking the delete button, writing 1 for confirmation in the pop-up window, clicking the delete button again, clicking on the account menu button, and clicking sign out. Finally, the RPA robot sends an email via the STMP server to the user of CRM providing confirmation of the deletion. During the **API automation sequence**, the RPA robot reads the parameters such as the URL, API-key, ID user for deletion of the API call and, via REST API DELETE method, sends it. The program receives the standard message with code about the progress. If the deletion was successful, the RPA robot sends the email with confirmation to the user of CRM via the STMP method.

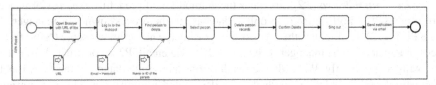

**Fig. 3.** P3 - deleting person from CRM

## 5  Experimental Design

The aim of this paper is to analyze the potential benefits of using API calls in RPA automation in terms of the duration of the process, automated by RPA bots. We designed and automated three processes (P1, P2, and P3) introduced in the previous chapter. To examine and validate the benefits of API calls in RPA automation, we used the three most dominant RPA vendors selected by Gartner: UiPath (UP), Automation Anywhere (AA), and Blue Prism (BP) [16]. For the purposes of this experiment, we used the demo licences from each vendor. We did not use any paid feature offered by the vendors. In addition to the demo software, we also used Python. For each vendor's platform, we built a version of the process automation using API and a version using GUI without API. We made a validation program in Python to discern the time for the raw API call. In total, we made 21 automations, 18 RPA automations, and three Python automations. For each case example process on every platform for API and GUI six processes were used. We used the same API calls with the same keys and URL for all processes. Also, the login credentials were identical for GUI web apps. Because API calls are direct to the greatest extent possible, we also tried to make the design of front-end automation as fast as possible. In the GUI process, we directed the automation to the page where all the actions are made. In Table 2, there is an overview table where it is possible to see which processes were automated and which technology was used.

**Table 2.** Overview table of automation

|  | Automation anywhere | Blue prism | UiPath | Python |
|---|---|---|---|---|
| API | P1, P2, P3 | P1, P2, P3 | P1, P2, P3 | P1, P2, P3 |
| GUI | P1, P2, P3 | P1, P2, P3 | P1, P2, P3 | - |

For all the RPA processes, we enabled the logging of RPA robots in order to log activities in the processes. This way, we could later analyze the run-time of the processes and duration of the activities. In Python, for tracking time, we used the datetime module. We used time as the metric due to transparent comparison and the possibility of transferring the time to the price of bot per hour. With time the analysis of the cost and total ROI is straightforward.

We ran our automation for each process ten times. The process had to be completed successfully, otherwise it was excluded from the time samples. Before running the process on another platform, the computer was restarted to free operational memory. The experiment was run on a computer with Windows 10, Intel Core i7-4712MQ and 8 GB of RAM. The versions of RPA platforms were Automation Anywhere 360 v.24 - Build 12350 (Community Edition), Blue Prism 7.0.1 and UiPath 2021.10. The version of Python used was 3.9.1.

From the RPA robot's event logs, we used process mining techniques on RPA processes to get the duration of all processes. All duration of activities is in appendix. For a better understanding and representation of the results, we calculated the arithmetical mean of measured values displayed in Table 3. For validation of our hypothesis that API has an impact on the duration of the process, we used analysis of variance (ANOVA).

We aggregated the processes by the name of the case example and tested if there is a difference in duration between the GUI RPA automation and RPA automation with the use of API on the same process, the results are in Table 5.

## 6  Results

Table 3 contains the results of the average bot duration of the test processes. Values are displayed in seconds, rounded to two decimal places. We can see that bot duration with API was faster than bot using GUI in all processes. At first glance, we can see that the differences are enormous. In some cases, API is more than 10x faster than RPA bot using GUI. Table 4 describes the results of the analysis of variance for the measured values. The results of the P-Value clearly show that the difference is statistically significant, and we can reject the null hypothesis that API does not have any impact on the duration of the process, automated by RPA bot. We accept the alternative hypothesis that API does affect the duration of the process, automated by RPA bot. Based on the P-value, we can reject the null hypothesis on every level of statistical significance (0.1, 0.05, and 0.01). We may thus conclude that the use of API in RPA automation positively influences the duration of the process and leads to a significant decrease. The results clearly show that API is beneficial and reduces the run-time of RPA bot compared to RPA bot using GUI.

**Table 3.** Mean of the duration of processes

| Technology | P1 GUI (s) | P1 API (s) | P2 GUI (s) | P2 API (s) | P3 GUI (s) | P3 API (s) |
|---|---|---|---|---|---|---|
| AA | 27.3 | 2.1 | 21.7 | 3.8 | 30 | 2.1 |
| BP | 31.41 | 2.11 | 25.21 | 2.27 | 29.58 | 1.47 |
| UP | 21.85 | 3.61 | 16.55 | 4.07 | 16.8 | 2.65 |
| Python | – | 1.7 | – | 1.78 | – | 1.05 |

**Table 4.** Results of statistical significance of ANOVA

| Name of the process | ANOVA coefficient (F) | P-value |
|---|---|---|
| P1 | 8993.04 | 9.64e−62 |
| P2 | 792.88 | 5.9e−35 |
| P3 | 6127.28 | 2.83e−57 |

In Table 5, we have an arithmetical average of duration for the processes automated via GUI and the processes automated via API. We aggregate values from all platforms for one process to obtain the average duration for each process with a certain method. In the fourth column, we can see the average difference calculated by subtracting the API duration from the GUI duration. In the last column, we derived the percentage change from the average, where we take the GUI method as the base and the API method as the new result. We arrived at the percentage change of a decrease in the interval of 84.02% to 91.87%.

**Table 5.** Difference between RPA automation using GUI vs API

| Name of the process | AVG GUI duration (s) | AVG API duration (s) | AVG difference (s) | AVG percentage change (%) |
|---|---|---|---|---|
| P1 | 26.85 | 2.6 | 24.25 | −90.32 |
| P2 | 21.15 | 3.38 | 17.77 | −84.02 |
| P2 | 25.46 | 2.07 | 23.39 | −91.87 |

## 7 Discussion

In this research, we focused on exploring the potential benefits of API on duration in RPA automation. Due to the fact that RPA automation is widely used to connect multiple applications, it is highly probable that one or more applications have an application programming interface. Based on our experiment with three case examples processes that were automated via the most commonly used RPA platforms, we confirm our hypothesis that API in RPA automation has a positive impact on the duration of the process, automated by RPA bot. Since RPA is associated with significant costs, the opportunity to speed up the automation so that the robot can produce more value in less time has a positive impact on the total return of investment into automation.

Another advantage of APIs in RPA automation mentioned by Prucha [21], that APIs can improve the stability of RPA bots, is probably true. When developing RPA automation on the GUI, we had to repair and reselect some selectors on the GUI before running the test, despite trying to select stable selectors. This assumption was not properly tested, but we can recommend exploring it in future research. This finding brings us back to design patterns and standard RPA architecture, as these findings could be considered key in the development of the standard [23]. The results clearly show that the API will be faster than RPA bot using GUI, and if the API can be used, the RPA developer should use it.

The time tracking has certain limitations, which is essential to mention. Although they are software applications, the individual runs of RPA robots vary in terms of the duration of the automation. Thus, the duration of the same activity may differ and is not completely constant. Usually, the discrepancy is slight and we consider, as others have, that the distribution of outputs will be narrower than the normal distribution or follow the pattern of normal distribution [3, 4]. Based on this fact, the property of outliers emerges, which influences measured results and the final average. In the appendix, we can see that the variance of RPA bot is relatively extensive in some cases. This is due to the fact that we work with and use multiple elements in automation, such as the graphic design, the web browser, the server, the computer where the robot is running, the memory of the computer, the distance of a local station from the server, if the application is cached in the memory, and more. To calculate the overall ROI, it is important to consider all elements which can influence the RPA automation. In addition to time tracking limitations, there are also minor limitations in logging activities. Logging activities are limited to raw API calls made via Python. Because of the nature of the code, there is a delay in program execution as the datetime library needs to be loaded first to get the current time. Also, the precision of time measurement in Automation Anywhere is a little questionable because, with the Automation Anywhere community edition license, we get time accuracy to the second, not to tenths or hundredths of a second. It is important to note that we do not compare RPA vendors' platforms with each other. The differences in duration between platforms are not comparable because on some platforms, we spend more time with optimalization to reach some level of stability at the expense of speed. Also, for some platforms we spend more time automating the process, which would lead to biased results if we were to compare RPA platforms.

It is also important to add the cost of time spent on analyzing and building the automation of the process. The time for building the RPA automation with API was variable and dependent on more factors. In general, the API process was the fastest to automate, but in reality, it depends on the API documentation of the application. The type of API call also contribute to the development time. It is the difference between the GET, POST and DELETE call. The API GET call is different due to the fact that the received data need to be processed, which prolongs the build time. One of the contributors to total build time is the preparation of the application to work with API calls. Most applications need at least the API key. Enabling the use of API by an administrator of the application is also frequent. An organization with strict cyber security can delay the development of the RPA automation. The build time of the RPA bots with GUI was less variable in time and probably the development of GUI automation took slightly more time. However, build time was not tracked. These assumptions lead to another potential research challenge to track the build time of API or GUI RPA automation, especially from the perspective of citizen developers.

## 8   Conclusion

The results are unambiguously in favour of the use of API in RPA automation. API, in some cases, is more than ten times faster and, in general, the most modest difference is an 84% reduction of the duration of API compared to RPA bot using GUI. The results provide explicit evidence of the benefits of API on the duration of the process, automated by RPA bot. The results of this research contribute to the future development of RPA technology, and it can be used to help build standard design patterns for RPA automation. As future research, it is not limited just to other benefits of API in RPA technology, but also offers opportunities for researching other methods or design patterns to improve the stability of RPA, rendering RPA maintenance-free or speeding up the whole RPA process.

**Acknowledgment.** This research was made possible thanks to the Technical University of Liberec and the SGS grant number: SGS-2022-1004. This research was conducted with the help of Pointee.

## Appendix

**Table 6.** Blue Prism individual runs of the process

| P1 GUI (s) | P1 API (s) | P2 GUI (s) | P2 API (s) | P3 GUI (s) | P3 API (s) |
|------------|------------|------------|------------|------------|------------|
| 31.5 | 1.8 | 21 | 0.73 | 30.3 | 1.1 |
| 31.2 | 1.9 | 26 | 0.42 | 30.1 | 1.3 |
| 32.7 | 1.9 | 25.5 | 0.8 | 30.1 | 1.3 |
| 31.2 | 1.8 | 26 | 0.8 | 30 | 5 |
| 31.2 | 1.9 | 25.8 | 1.85 | 30.2 | 1 |
| 31.3 | 2 | 24.9 | 0.7 | 30 | 1 |
| 31.3 | 1.8 | 25.7 | 0.387 | 30.2 | 1.1 |
| 31 | 1.8 | 25.8 | 0.8 | 27.45 | 1 |
| 31.2 | 1.7 | 24.8 | 2 | 30.3 | 0.9 |
| 31.5 | 4.5 | 26.6 | 14.21 | 27.1 | 1 |

**Table 7.** UiPath individual runs of the process

| P1 GUI (s) | P1 API (s) | P2 GUI (s) | P2 API (s) | P3 GUI (s) | P3 API (s) |
|---|---|---|---|---|---|
| 24.5 | 3.1 | 24.5 | 6.8 | 19.3 | 2.2 |
| 23 | 3.6 | 15 | 3.8 | 16.7 | 3 |
| 23 | 4.3 | 17.1 | 3.7 | 16.2 | 2.7 |
| 20 | 3.1 | 16.8 | 3.8 | 16.2 | 2.6 |
| 20 | 3.3 | 15.7 | 3.6 | 18.2 | 2.5 |
| 20.7 | 3.3 | 16.2 | 3.7 | 16.1 | 2.4 |
| 19.2 | 4 | 14.6 | 4.2 | 16.6 | 2.4 |
| 20.9 | 4 | 13.8 | 3.7 | 15.8 | 2.4 |
| 25 | 3.4 | 15.6 | 2.7 | 16.8 | 3.1 |
| 22.2 | 4 | 16.2 | 4.7 | 16.1 | 3.2 |

**Table 8.** Automation anywhere individual runs of the process

| P1 GUI (s) | P1 API (s) | P2 GUI (s) | P2 API (s) | P3 GUI (s) | P3 API (s) |
|---|---|---|---|---|---|
| 27 | 3 | 22 | 5 | 32 | 2 |
| 28 | 1 | 21 | 3 | 29 | 2 |
| 28 | 2 | 20 | 4 | 31 | 2 |
| 27 | 2 | 27 | 3 | 32 | 3 |
| 26 | 2 | 23 | 3 | 28 | 3 |
| 28 | 2 | 21 | 4 | 30 | 3 |
| 28 | 2 | 19 | 5 | 27 | 2 |
| 27 | 3 | 22 | 3 | 32 | 1 |
| 27 | 2 | 22 | 4 | 30 | 2 |
| 27 | 2 | 20 | 4 | 29 | 1 |

**Table 9.** Python individual runs of the process

| P1 API (s) | P2 API (s) | P3 API (s) |
|---|---|---|
| 2.9 | 4.2 | 1.1 |
| 2.5 | 1.2 | 1 |
| 1.1 | 2.2 | 0.9 |
| 1.3 | 1.2 | 0.9 |

(*continued*)

**Table 9.** (*continued*)

| P1 API (s) | P2 API (s) | P3 API (s) |
| --- | --- | --- |
| 1.3 | 1.5 | 1.7 |
| 1.3 | 1.1 | 0.9 |
| 1.4 | 1.7 | 1 |
| 1.2 | 1.5 | 1 |
| 1.9 | 1.6 | 1.1 |
| 2.1 | 1.6 | 0.9 |

# References

1. van der Aalst, W.M.P., Bichler, M., Heinzl, A.: Robotic process automation. Bus. Inf. Syst. Eng. **60**(4), 269–272 (2018). https://doi.org/10.1007/s12599-018-0542-4
2. Aguirre, S., Rodriguez, A.: Automation of a business process using robotic process automation (RPA): a case study. In: Figueroa-García, J.C., López-Santana, E.R., Villa-Ramírez, J.L., Ferro-Escobar, R. (eds.) WEA 2017. CCIS, vol. 742, pp. 65–71. Springer, Cham (2017). https://doi.org/10.1007/978-3-319-66963-2_7
3. Anagnoste, S.: Robotic automation process - The next major revolution in terms of back office operations improvement. Proc. Int. Conf. Bus. Excell. **11**, 676–686 (2017)
4. Yapa, S.: Reports. In: Customizing Dynamics 365, pp. 205–253. Apress, Berkeley, CA (2019). https://doi.org/10.1007/978-1-4842-4379-4_7
5. Beerbaum, D.: Artificial intelligence ethics taxonomy - robotic process automation (RPA) as business case. SSRN J. (2021)
6. Cewe, C., Koch, D., Mertens, R.: Minimal effort requirements engineering for robotic process automation with test driven development and screen recording. In: Teniente, E., Weidlich, M. (eds.) BPM 2017. LNBIP, vol. 308, pp. 642–648. Springer, Cham (2018). https://doi.org/10.1007/978-3-319-74030-0_51
7. Chakraborti, T., et al.: From robotic process automation to intelligent process automation: – emerging trends –. In: Asatiani, A., et al. (eds.) Business Process Management: Blockchain and Robotic Process Automation Forum, pp. 215–228. Springer International Publishing, Cham (2020)
8. Choudhary, R., Karmel, A.: Robotic process automation. In: Raje, R.R., Hussain, F., Kannan, R.J. (eds.) Artificial Intelligence and Technologies, LNEE, vol. 806, pp. 29–36. Springer Singapore, Singapore (2022). https://doi.org/10.1007/978-981-16-6448-9
9. Czarnecki, C., Fettke, P. (eds.): Robotic Process Automation: Management, Technology, Applications. De Gruyter Oldenbourg, Berlin, Boston (2021)
10. Egger, A., ter Hofstede, A.H.M., Kratsch, W., Leemans, S.J.J., Röglinger, M., Wynn, M.T.: Bot log mining: using logs from robotic process automation for process mining. In: Dobbie, G., Frank, U., Kappel, G., Liddle, S.W., Mayr, H.C. (eds.) ER 2020. LNCS, vol. 12400, pp. 51–61. Springer, Cham (2020). https://doi.org/10.1007/978-3-030-62522-1_4
11. Enriquez, J.G., Jimenez-Ramirez, A., Dominguez-Mayo, F.J., Garcia-Garcia, J.A.: Robotic process automation: a scientific and industrial systematic mapping study. IEEE Access. **8**, 39113–39129 (2020)
12. Eulerich, M., Pawlowski, J., Waddoups, N.J., Wood, D.A.: A framework for using robotic process automation for audit tasks*. Contemp. Act. Res. **39**, 691–720 (2022)

13. P. Fettke, W. Reisig, Systems Mining with Heraklit: The Next Step. ArXiv:2202.01289 [Cs]. (2022)
14. Fielding, R.T.: Architectural Styles and the Design of Network-based Software Architectures. University of California, Irvine (2000)
15. van Hoek, R., Gorm Larsen, J., Lacity, M.: Robotic process automation in Maersk procurement–applicability of action principles and research opportunities. IJPDLM. **52**, 285–298 (2022)
16. G. Inc: Robotic Process Automation (RPA) Software Reviews 2022 | Gartner Peer Insights, Gartner. (n.d.)
17. Kedziora, D., Penttinen, E.: Governance models for robotic process automation: the case of Nordea Bank. J. Inf. Technol. Teach. Cases **11**, 20–29 (2020)
18. Krintz, C., Wolski, R.: Unified API governance in the New API economy. Cut. J. J. Inf. Technol. Manag. **26**, 12–16 (2013)
19. Mohanty, S., Vyas, S.: Intelligent process automation = RPA + AI. In: How to Compete in the Age of Artificial Intelligence, pp. 125–141. Apress, Berkeley, CA (2018). https://doi.org/10.1007/978-1-4842-3808-0_5
20. Mullakara, N., Asokan, A.: Robotic Process Automation Projects: Build real-world RPA solutions using UiPath and Automation Anywhere. Packt Publishing Ltd, Birmingham Mumbai 202AD
21. Naveen Reddy, K.P., Harichandana, U., Alekhya, T., Rajesh, S.M.: A study of robotic process automation among artificial intelligence. IJSRP. **9**, 8651 ((2019))
22. Průcha, P.: Aspect optimization of robotic process automation. In: ICPM 2021 Doctoral Consortium and Demo Track 2021, CEUR Workshop Proceedings. Eindhoven, The Netherlands (2021)
23. Séguin, S., Benkalaï, I.: Robotic process automation (RPA) using an integer linear programming formulation. Cybern. Syst. **51**, 357–369 (2020)
24. Stohr, E.A., Zhao, J.L.: Workflow automation: overview and research issues. Inf. Syst. Front. **3**, 281–296 (2001)
25. Syed, R., et al.: Robotic process automation: contemporary themes and challenges. Comput. Ind. **115**, 103162 (2020)
26. Wewerka, J., Reichert, M.: Towards quantifying the effects of robotic process automation. In: 2020 IEEE 24th International Enterprise Distributed Object Computing Workshop (EDOCW), pp. 11–19. IEEE, Eindhoven, Netherlands (2020)
27. Wewerka, J., Reichert, M.: Robotic process automation - a systematic mapping study and classification framework. Enterp. Inf. Syst. 1–38 (2021)
28. Willcocks, L.P., Lacity, M., Craig, A.: The IT function and robotic process automation. London School of Economics and Political Science, LSE Library (2015)

# Central and Eastern Europe (CEE) Forum

# Preface

## Central and Eastern Europe (CEE) Forum

BPM research in the context of CEE countries opens up endless opportunities for developing possibilities for both the theory and practice of BPM. BPM applications in CEE countries can face numerous challenges, from differences in national culture and business processes resulting from their respective business environment, legal regulation, differences in digital literacy, and more. Organizations from CEE economies sometimes follow practices and models conceived and tested in highly developed countries. Still, they are also obliged to use their own experience and understanding of their local business environment. In this area, research on BPM is still needed to better document, implement, and improve operational business processes in the context of an organization and its environment, its culture and country. A better understanding of BPM in the CEE region would enable practitioners to avoid the issues faced by these countries in the process of BPM adoption and be a valuable contribution to the field of BPM.

We received nine submissions, and the top four highest quality papers were selected for presentation and publication. Each submission was reviewed by at least three Program Committee (PC) members. Papers of Polish authors dominated in this edition.

The literature-based research landscape of BPM in CEE countries was addressed by Renata Gabryelczyk (University of Warsaw, Poland), Edyta Brzychczy, Katarzyna Gdowska, and Krzysztof Kluza (AGH University of Science and Technology, Poland). The authors of this study showed apparent research gaps that should be filled in line with the idea of contextual intelligence to justify the new research questions on BPM that researchers in CEE countries may post.

In another accepted paper, Piotr Sliż (University of Gdansk, Poland) took up the research topic concerning the integration of the concepts and methods of process and project management. The author showed the benefits of this integration and empirically examined the process-project maturity of selected large organizations in Poland.

Another study was conducted by a strong team of co-authors from different universities in Poland: Waldemar Glabiszewski (Nicolaus Copernicus University), Szymon Cyfert (Poznan University of Economics and Business), Roman Batko (AGH University of Science and Technology), Piotr Senkus (independent consultant, formerly UPH), and Aneta Wysokińska-Senkus (War Studies University). The authors proved that the increase in the importance of competencies in knowledge management after the COVID-19 pandemic will significantly change the main components of process orientation in organizations.

In the context of the constantly existing connection of BPM with the planning, design, and implementation of ERP systems, the topic of ERP systems development was considered by the following team of authors: Marek Szelągowski (Polish Academy of Sciences, Poland), Justyna Berniak-Woźny (University of Information Technology and Management, Poland), and, Audrone Lupeikiene (Vilnius University, Lithuania).

Based on a literature review, relevant market reports, and interviews with ERP experts, the authors explored the essence and main approaches to ERP transformation into process-based ERP systems.

We would like to thank the authors and Program Committee members for contributing to the 2nd edition of the Central and Eastern Europe Forum. We are convinced that the CEE Forum will help strengthen the existing body of BPM knowledge in the CEE region. We recommend reading these selected papers to the broader community of BPM researchers, practitioners, and enthusiasts.

September 2022

Vesna Bosilj Vukšić
Renata Gabryelczyk
Mojca Indihar Štemberger
Andrea Kő

# Organization

## Program Chairs

| | |
|---|---|
| Vesna Bosilj Vukšić | University of Zagreb, Croatia |
| Renata Gabryelczyk | University of Warsaw, Poland |
| Mojca Indihar Stemberger | University of Ljubljana, Slovenia |
| Andrea Kő | Corvinus University of Budapest, Hungary |

## Program Committee

| | |
|---|---|
| Agnieszka Bitkowska | Warsaw University of Technology, Poland |
| Edyta Brzychczy | AGH University of Science and Technology, Poland |
| Katarzyna Gdowska | AGH University of Science and Technology, Poland |
| Arkadiusz Jurczuk | Bialystok University of Technology, Poland |
| Marite Kirikova | Riga Technical University, Latvia |
| Krzysztof Kluza | AGH University of Science and Technology, Poland |
| Matija Marić | University of Zagreb, Croatia |
| Ivan Matić | University of Split, Croatia |
| Josip Mesarić | University of Osijek, Croatia |
| Ljubica Milanović-Glavan | University of Zagreb, Croatia |
| Igor Pihir | University of Zagreb, Croatia |
| Natalia Potoczek | Polish Academy of Sciences, Poland |
| Dalia Suša Vugec | University of Zagreb, Croatia |
| Martina Tomičić Furjan | University of Zagreb, Croatia |

# Business Process Management in CEE Countries: A Literature-Based Research Landscape

Renata Gabryelczyk[1]([⊠]) [iD], Edyta Brzychczy[2] [iD], Katarzyna Gdowska[2] [iD], and Krzysztof Kluza[2] [iD]

[1] University of Warsaw, ul. Długa 44/50, 00-241 Warsaw, Poland
r.gabryelczyk@wne.uw.edu.pl
[2] AGH University of Science and Technology, al. Mickiewicza 30, 30-059 Krakow, Poland
{brzych3,kgdowska,kluza}@agh.edu.pl

**Abstract.** This article presents the results of a literature review on Business Process Management in the countries of Central Eastern Europe, authored by researchers affiliated to those countries. In line with the used review protocol, our study comprises 159 journal articles analyzed from a meta-perspective, including 60 empirical articles that underwent content-based analysis. While researching BPM in CEE countries, we diagnosed *Management* as the most-studied phenomenon among the three main areas of BPM alongside *Foundations* and *Engineering*. Among the different characteristics of BPM research diagnosed, we identified, as examples, research developments over time, and, Croatia and Slovenia as the nations most involved in BPM research. In order to identify the research gap regarding the BPM capabilities understudy, and to engage researchers in new trends combining BPM with digitalization, we used the original and updated core elements of the BPM framework. We have shown that *BPM methods* are the most tested, followed by *Governance* and *Strategy Alignment*. As in previous studies, elements of *People* and *Culture* are under researched. The inclusion of enhanced and novel BPM capabilities in light of digitalization is in the very early stages of research in CEE. Our studies show apparent research gaps that should be filled in line with the idea of contextual intelligence, justifying the new questions on BPM that researchers in CEE countries may post. This approach allows the drawing of new conclusions and thus contributes to the development of the BPM body of knowledge.

**Keywords:** Business Process Management · BPM · Central Eastern Europe · CEE · Literature review

## 1 Introduction

The main aim of this paper is the examination of published research on Business Process Management (BPM) by authors affiliated to Central and Eastern European (CEE) countries, and in particular, the exploration of the state of published studies reporting on

A. Marrella et al. (Eds.): BPM 2022, LNBIP 459, pp. 279–294, 2022.
https://doi.org/10.1007/978-3-031-16168-1_18

empirical research results from CEE countries. Our study is therefore aimed at presenting the landscape of BPM research in CEE countries, both filling the existing research gap in the field of BPM research, and fitting into the CEE Forum organized during the international BPM conference. Our motivation to undertake this study was two-fold.

Firstly, with the rapidly growing number of scientific and professional publications, effective knowledge management is crucial to following current trends, participating in scientific discussions in the field, and identifying research gaps [33]. Fully understanding this approach, we decided to apply the literature review method to help scientists from the CEE region diagnose both the well-researched and under-researched BPM aspects.

Secondly, our research challenge fits in with the rhetoric of authors researching various areas of management and management information systems who see many differences between economically and geographically diverse regions and the impact of these differences on the planning, implementation and application of technologies and management concepts [34]. This approach is also in line with the idea of contextual intelligence proposed by Khanna [19], according to whom different research environments require the reconsideration of the validity and generalization of the obtained research results in the various cultural, geographic, social and economic contexts. This is especially true for empirical research. These contextual intelligence imperatives were also demonstrated in earlier, albeit singular, research on BPM in transition economies [10]. Thus, we believe that it is prudent to undertake research on a sample of CEE countries to gain a scientific basis for comparing the obtained results in various contexts of applying, researching and developing the BPM discipline. Another justification for our study is to examine how authors in CEE countries fit into the latest research trends. As the most recent BPM research indicates that "*BPM and digital innovation belong together, like two sides of the same* coin" [21], we decided to check whether authors from CEE countries are meeting this challenge.

We applied the systematic literature review *(SLR)* as the research method best suited to the purpose of this paper. Due to the sheer size of the body of publications, the traditional review approach was superseded by the SLR which enables a more objective-oriented systematic literature review [33, 44]. SLR is a widely used approach for explanatory or critical state-of-the-art analysis of advances in a given field [46], so we therefore found it appropriate for analyzing the BPM research landscape in CEE countries.

The main research question we posted for this study is formulated as follows:

RQ: What is the BPM research landscape in Central and Eastern Europe countries? To find the answer, we structured the following three research questions in more detail based on previous BPM literature reviews conducted without reference to the CEE region [13, 30].

RQ1: What are the main characteristics of publications on BPM in CEE countries from a meta-perspective? RQ2: What are the characteristics of publications on BPM in CEE countries from a content-based perspective? RQ3: What BPM capability areas are explored in empirical research in CEE countries in view of digitalization?

Our article is structured as follows: in the Background section, we present a brief overview of previous research on BPM based on the literature and covering all CEE countries. In the next part, we present the research process that we carried out in accordance with the systematic literature review. We present the results of the obtained research broken down into meta and content-based analyses. The final part covers the conclusion, limitations and plans for the future.

## 2 Background

Over the past three decades, the body of knowledge on Business Process Management includes many papers surveying publications in this field. However, papers authored by researchers affiliated in CEE countries or reporting research conducted on institutions located there are relatively scarce. Therefore, in this section we refer to BPM survey papers that either present referential BPM survey methodology or provide findings on BPM in CEE countries. Houy et al. [13] contributed with a reference framework for a comprehensive analysis of empirical BPM articles where a meta-perspective, content-based, and methodical perspective are used for identifying exciting trends in BPM research. In the mentioned work by Houy et al. [13], the authors affiliated themselves with institutions outside CEE and conducted research on articles indexed in the SCI and Ebsco until 2008. The analysis was not restricted by the authors' countries of affiliation or restricted by the geographical location of their research subject. The CEE is represented by two articles, one each from Slovenia and Estonia.

The aforementioned framework was further developed by Roeser and Kern [30]. It was applied to conduct a systematic literature review of empirical BPM survey papers indexed in selected databases until October 2013. The final selection of papers comprised 51 ranked journal papers in which BPM or BPM-related topics were of primary research interest. The analysis showed that the range and extent of BPM studied differed between regions and countries. For the most part, BPM was studied in Europe (including CEE countries: Croatia and Slovenia), America and Asia. Therefore, further research should examine the relationship between contributions to the BPM body of knowledge and the authors' affiliation and location of the research subject, as this will facilitate examining the effect of the country's specific contextual embeddedness on BPM.

Literature reviews for BPM usually focus on examining the state of the art from the perspective of a selected aspect of BPM, e.g., one of the core elements of BPM such as culture, type of organization under study, BPM software, BPM life cycles or ties of BPM with other concepts or technologies [1, 8, 26, 42, 43, 47]. In this research, geographical criteria related to the authors' affiliation or location of the studied companies are barely used, therefore, it is rare to find analyses dedicated directly to CEE countries. Papers contributing to an improved understanding of BMP in or for CEE are scarce with the majority of them simply covering a selection of CEE countries (e.g., [27, 35]), though some do refer to the entire region (e.g., [9, 10]).

Pilav-Velić & Marjanovic [27] studied key motives for BPM implementation within different business processes and across industry sectors in Bosnia & Herzegovina, while Stojanović et al. [35] conducted a comparative analysis of BPM practice in Slovenia and Serbia. Both formulate recommendations for successful BPM implementation in companies operating in transition economies and emphasize the need for studying BPM in reference to the unique context of transition economies.

For an analysis of the state of BPM research in CEE refer to Gabryelczyk et al. [9] and to Gabryelczyk and Roztocki [10]. Authors in [9] include a literature review of journal publications on BPM by authors affiliated to these countries, published until October 2015, and referring exclusively to research on BPM carried out in the transition countries. The analysis was based on the six core elements of BPM framework. The majority of 29 papers under examination are surveys or case studies focused on *Strategy Alignment*, *Governance*, or *Methods*. Gabryelczyk and Roztocki [10], aiming to develop a BPM framework for transition economies, extended the SRL in November 2016 with previously used criteria. Articles (47) meeting these specific criteria were affiliated in 10 transition countries with the most active Slovenia and Croatia. The majority of the articles were published from 2012 to 2015. The main conclusion from this study was the development of a BPM success framework adapted to the specificity of transition economies.

Thus, the particularly sparse research to date shows both a slow but continuous change in general research culture in CEE countries as well as an increasing interest in BPM at both an academic and applicational level. A significant research gap, therefore, arises here requiring detailed continuous research due to the unique nature of the region.

## 3   Methodology and Research Process

In line with the systematic literature review process by Xiao & Watson [46], we began with the formulation of the research problem and development of a review protocol. This step includes posing research questions, inclusion criteria, search strategies, quality assessment criteria, screening procedures, and strategies for synthesis and reporting.

The research process, including the elements of the review protocol and the outcomes of subsequent steps, is presented in Fig. 1. When formulating the research problem, we focused our research on CEE countries as defined by the OECD comprising Albania, Bulgaria, Croatia, the Czech Republic, Hungary, Poland, Romania, the Slovak Republic, Slovenia, and the three Baltic States of Estonia, Latvia and Lithuania [24], and including some former states of the former socialist countries, i.e., Belarus, Bosnia and Herzegovina, Kosovo, Moldova, Montenegro, Macedonia, Serbia, and Ukraine. Thus, our review included 20 countries.

In the first step of our research process, we used *Scopus* and ISI *Web of Science* databases as meaningful sources of publications. The Scopus database is the largest international abstract and citation database of peer-reviewed literature [48]. To validate the results, we also searched Web of Science, which contains publications that have undergone a review process and provides complete information suitable for analysis [39]. The databases were searched with the phrases *"business process management"* or *"BPM"* in the title, abstract or keywords fields, and a list of the selected countries, with the exclusion of disciplines not related to BPM topics (e.g., Clinical Neurology, Chemistry, Physics). Database searching was carried out in February 2022. As a result, we obtained 426 papers. Since used databases are inclusive, it was necessary to remove duplicate items. As a result, we removed 117 papers in the second step of our process. Following this step, 309 papers remained for inclusion criteria screening.

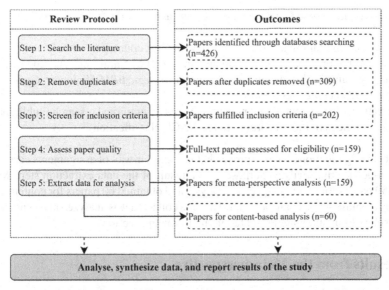

**Fig. 1.** The review process in our study and outcomes of subsequent steps.

The following inclusion criteria was formulated by ourselves: (1) The language of the publication is English, (2) The paper is a journal publication, and (3) The title, abstract, or keywords include *"business process management"* or *"BPM"*. At this step, we divided the whole dataset into four subsets with each subset being checked independently by two researchers in parallel. Inclusion criteria was checked and the title, abstract and keywords carefully read by us to assess if they fit our research parameters. In total, 107 papers failed to meet the inclusion criteria, including 38 papers in which a non-related definition of BPM was used (i.e., beats per minute (heart), balance of payments manual). We obtained 202 relevant papers for the quality assessment step. In the fourth step, we browsed through the full-text articles to assess their quality and eligibility for further analysis. Papers unrelated to the BPM concept were excluded (e.g., BPM appeared only in keywords) and short papers (e.g., research notes). Similarly, as in the previous step, each paper was reviewed by two researchers independently. Disagreements were discussed with the researcher not involved in assessing the disputed paper. After this step, we qualified 159 papers for the data extraction process.

In the data extraction step, we collected data related to meta-perspective analysis and data related to content-based analysis. For the purposes of meta-perspective analysis, we extracted parts of the papers' records (e.g., names of authors, affiliations, publication year, and journal title). We assigned the paper to one of the three main research areas: *computer science, information systems engineering,* and *information system management,* following three main tracks distinguished at the BPM conference: *Foundations, Engineering,* and *Management* (F/E/M). The aforementioned assignments were carried out by two researchers independently. In case of assignment differences, a third researcher was involved in the final decision.

For the content-based analysis purposes, we assumed the following coding:

- presentation of empirical research on BPM in CEE countries,
- relation to the original six core elements of the BPM framework and updated BPM capability framework and their core elements (*Strategic BPM Alignment, Governance, Methods, IT, People and Culture*),
- the relationship to a sector and industry covered by empirical research (public/private and economic activities according to the NACE classification).

The coding mentioned above proceeded in a similar way of assignments carried out for meta-perspective analysis purposes. As a result of the data extraction step, we utilized 159 papers for meta-perspective analysis and obtained 60 papers for content-based analysis (containing 72 contributions). The data collected was analyzed and synthesized. The results of our study are presented in the following sections.

## 4   Results from the Meta-perspective

### 4.1   Review by CEE Countries, Authors and Their Contributions

For the meta-perspective analysis, we considered 159 papers that fulfilled our inclusion criteria. The number of publications is presented by year in Fig. 2.

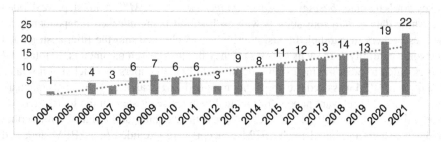

**Fig. 2.** The number of papers by year.

As our query was undertaken in the middle of February, the number of publications (2) found in 2022 is not representative and, therefore, not included in Fig. 2. It is evident that the number of publications has been growing over the years.

Figure 3 Represents the number of papers by CEE countries (the countries without any paper for the meta-perspective analysis were omitted). The papers with authors from various countries were classified as *"Multiple countries"*, so it should be noted that in the case of Slovenia and Croatia, the number of papers including the papers with the cooperation of multiple countries are 32 and 27 respectively. We list those articles in our study as contributions from a given country (including those among all countries whose authors were included in the paper).

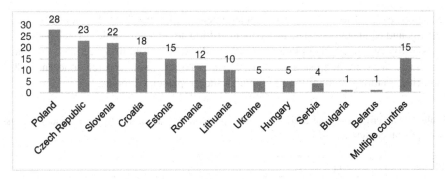

**Fig. 3.** The number of papers by country.

The top five most prominent authors in the surveyed area from the CEE countries (with CEE affiliation) are presented in Table 1. All of these authors are affiliated with institutions in the top five countries based on the number of contributions per country. It can be observed that most authors from our list come from two institutions: the University of Ljubljana (Slovenia) and the University of Zagreb (Croatia). The average number of authors per paper is 3.2, and the median is 3.

**Table 1.** Top authors on BPM in CEE countries.

| # Contributions | Author | Affiliation | Country |
|---|---|---|---|
| 16 | Vesna Bosilj Vukšić | University of Zagreb | Croatia |
| 14 | Mojca Indihar Štemberger | University of Ljubljana | Slovenia |
| 10 | Peter Trkman | University of Ljubljana | Slovenia |
| 9 | Marlon Dumas | University of Tartu | Estonia |
| 9 | Marek Szelągowski | Polish Academy of Sciences | Poland |

The 159 papers are distributed over 106 unique journals, with the majority (80), being solely single paper publications. Sixteen journals published two papers, eighteen published three papers, and the Business Process Management Journal (BPMJ) published 18 papers. It is noted that in comparison with the review from 2016 [9], a significant increase in the case of high-quality journals such as BPMJ (IF 3.464), e.g., [8, 12, 13, 30, 36, 43], Information Systems (IF 2.309), Journal of Competitiveness (IF 4.725) or International Journal of Information Management (IF 14.098) [3, 38]. Thus, the greater number of articles in these journals may indicate an increase in research quality in CEE institutions.

## 4.2 Review Due to Phenomenon: Foundations, Engineering, Management

To find out about CEE contributions to an existing body of knowledge, we assessed each paper before placing them into one of the three categories in terms of the phenomenon according to the Business Process Management conference [45]: Foundations, Engineering, and Management. The assessment criteria are given in Table 2. Each paper was assessed by two researchers independently and we discussed disagreements to obtain consensus. We were interested in identifying areas of the main research concern in CEE countries, enabling recognition of potential underrepresented topics.

**Table 2.** The criteria for the division into the phenomenon according to the Business Process Management conference [45].

| Phenomenon | Criteria |
|---|---|
| Foundations | Papers concerning underlying principles and concepts of BPM systems which introduce novel concepts of BPM systems with the proof-of-concept implementations |
| Engineering | Papers concerning research in engineering aspects of information systems research that are empirically evaluated in a rigorous and preferably reproducible manner |
| Management | Papers concerning BPM concepts and methods that support various issues of business management, such as strategic alignment, governance, methods, information technology, and human aspects, including people and culture, as well as build-on and draw-from real-world organizational endeavors in BPM |

Based on the identification of the category appropriate for each paper, Fig. 4 shows the contributions per country divided into category groups.

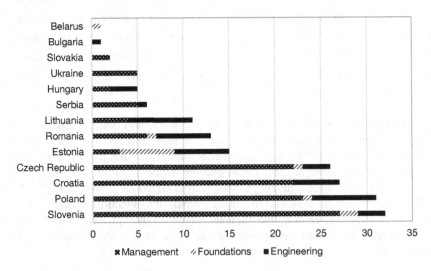

**Fig. 4.** The number of contributions by country and phenomenon.

In most of the presented countries, with the exception of Belarus and Bulgaria, we identified papers with *Management* as the primary phenomenon. There is a clear majority of *Foundations* and *Engineering* contributions found only in Estonia. These contributions are developed mostly in cooperation with researchers from outside the CEE area. Explored topics, especially in the *Management* area, are often the result of an organization's needs in a given CEE country (e.g., ministry [20], healthcare system [28] or retail industry [15]).

## 5 Results from the Content-Based Perspective

### 5.1 Content Related to Empirical Research in CEE Countries

The content-based perspective in our literature review included the analysis of articles covering the results of empirical research conducted in the CEE countries. Empirical research relies on data gathered through evidence and approach, using quantitative and qualitative methods for gathering this evidence [5].

The inclusion criteria for this analysis were met by 60 articles. However, considering multi-author contributions from different countries and/or studies conducted in different CEE countries, the total contributions count is 72. All articles were classified under the meta-perspective in the *"Management"* research area, which results, to an extent, from the definitions adopted for this area from the field of information systems management. All articles have been reviewed in terms of content that allowed us to identify the most involved CEE countries in empirical research and learn about the BPM issues most frequently studied in the specific region.

The countries that provide by far the most numerous empirical research on BPM are Slovenia (20 contributions) and Croatia (18 contributions). BPM researchers from these countries also most often collaborate by providing multi-author articles with affiliations of these two countries, e.g., [3, 12, 16, 36], and research involving organizations from Balkan countries [4, 6, 16].

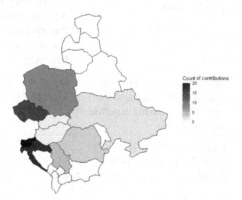

**Fig. 5.** The number of contributions that reports results of empirical research in CEE countries.

Empirical research was also conducted on data obtained in organizations in the Czech Republic (15), Poland (8), Serbia (4), Romania (3), Slovakia (3), Ukraine (2), and Hungary (2). In the CEE countries shown in Fig. 5 in white, we did not diagnose reports from empirical research on BPM.

## 5.2  Content in View of BPM Frameworks

In order to provide an established theoretical basis for examining the contributions of CEE countries in the BPM area, we have adopted the original and updated version of the six core BPM elements framework.

**Table 3.** Definitions and issues related to core elements of BPM in the original and updated BPM framework.

| Core element of BPM | Core elements of BPM framework, definitions by de Bruin and Rosemann 2007 [7] | Enhanced and new BPM capability areas, issues based on Kerpedzhiev et al. 2021 [18] |
| --- | --- | --- |
| Strategic alignment | Continual tight linkage of organisational priorities and enterprise processes, enabling achievement of business goals | Strategic BPM Alignment, Strategic Process Alignment, Process Positioning, Process Portfolio Management |
| Governance | Establishing relevant and transparent accountability and decision-making processes to align rewards and guide actions | Contextual BPM Governance, Contextual Process Governance, Process Architecture Governance, Process Data Governance, Roles and Responsibilities |
| Methods | Approaches and techniques that support and enable consistent process actions and outcomes | Process Content Management, Process Compliance Management, Process Architecture Management, Process Data Analytics, BPM Platform Integration, Multi-purpose Process Design, Advanced Process Automation, Adaptive Process Execution, Agile Process Improvement, Transformational Process Improvement |
| IT | Software, hardware and information management systems that enable and support process activities | |
| People | Individuals and groups who continually enhance and apply their process-related expertise and knowledge | Data Literacy, Innovation Literacy, Customer Literacy, Digital Literacy |
| Culture | Collective values and beliefs that shape process-related attitudes and behaviours | Evidence Centricity, Change Centricity, Customer Centricity, Employee Centricity |

This framework was originally proposed by de Bruin and Rosemann [31] based on BPM critical success factors in order to build the foundations for a BPM maturity model and to identify capability areas [7]. Thereafter, based on the consolidation of literature, triangulation of various research methods and increasing the detail of description, a model was created which distinguished six core elements critical to BPM: strategic alignment, governance, methods, information technology, people, and culture [32]. In this form, the six core BPM elements framework was used in the previous research, e.g., to structure practical BPM case studies from around the world [40, 41] and, in the literature review of BPM research in transition economies [9]. Over time, due to the increasing absorption of new technologies and the rapidly increasing digital transformations, this framework was updated in 2021 in view of new challenges for BPM [18]. Due to the ever-closer connection of methods and IT technologies supporting successful BPM adoption, and, above all, their rapid development, only five main capability areas were distinguished in the new framework, combining the Methods and IT areas (Table 3).

Using the six core elements of the BPM framework allowed us to assess the current state of research in CEE countries and analyze critical research aspects raised by authors in the region. Although many of the articles were of a multi-faceted nature, we identified the dominant core element of BPM in each of them. As a result of the analyses, it transpired that authors researching BPM in CEE countries focus their research interests on *Methods* (27 contributions) defined as tools and techniques that support all activities at different stages of the process life cycle, e.g., for documenting, analyzing, implementing, measuring, and executing a process (Table 4). These are the basic issues for using BPM in an organization, and the high interest in them remains unchanged compared to previous research on BPM elements in transition countries [9, 10]. The results of our research also show a fair number of studies in the areas of *Strategic Alignment* (12 contributions) and *Governance* (14 contributions). Empirical research in CEE often addresses the issues of increasing financial performance through the use of BPM, e.g. [11, 22] and the application of performance criteria within the established responsibility framework [16, 17]. As in previous studies, the areas of *People* and *Culture* remain the least explored.

**Table 4.** Contributions from CEE countries count by six core elements of the BPM framework.

| Country | Strategic alignment | Governance | Methods | IT | People | Culture |
|---|---|---|---|---|---|---|
| Slovenia | 5 | 4 | 7 | 2 | 1 | 1 |
| Croatia | 5 | 5 | 4 | 2 | 0 | 2 |
| Czech Republic | 1 | 4 | 6 | 3 | 1 | 0 |
| Poland | 0 | 0 | 5 | 1 | 1 | 1 |
| Serbia | 1 | 1 | 2 | 0 | 0 | 0 |
| Romania | 0 | 0 | 1 | 1 | 1 | 0 |
| Ukraine | 0 | 0 | 1 | 1 | 0 | 0 |
| Hungary | 0 | 0 | 1 | 0 | 0 | 0 |
| Slovakia | 0 | 0 | 0 | 0 | 1 | 0 |
| Count (72) | 12 | 14 | 27 | 10 | 5 | 4 |

In order to check whether the BPM research in CEE countries takes up the new role of BPM in organizations changing under the influence of digital transformations, we applied the updated BPM capability framework developed by Kerpedzhiev et al. [18]. However, the results of our investigation have shown low research engagement in the enhanced and new BPM capability areas so far. References to novel aspects of BPM in view of digitalization are only acknowledged in 6 articles (8 contributions). Croatia once again leads in two areas: *Methods/IT* and *Governance* (2 contributions). Process Data Governance issues were discussed in two articles jointly with authors from Slovenia [3, 36]. Business Intelligence systems are here considered a novel capability area in the aspect of leveraging process-related data. Croatian author contribution to Methods/IT includes Process Context Selection [29] and Agile Process Improvement [4]. The other two articles were also classified under the *Methods/IT* area. The authors from the Czech Republic took up the topic of Advanced Process Automation [37], while the authors from Romania touched BPM Platform Integration [2].

### 5.3   Content in View of Sector/Industry

The reviews of the articles were also carried out concerning the sector and industry covered by empirical research [14, 30]. The most important characteristics of the obtained sample confirm that 60% (36 papers) of analyzed empirical papers in CEE countries concern research on private sector organizations. In comparison, 18% (11 papers) covered public sector, mainly government agencies. In 10% (6 papers), the survey covered diverse organizations from both sectors. The sector was not specified in 12% of the articles. The results of empirical BPM research in CEE countries confirm earlier findings [13] that the public sector remains much less researched. We extracted data regarding industry according to the Nomenclature of Economic Activities [23]. This analysis shows that 50% (30 papers) contain empirical research conducted on large samples of organizations classified as various economic activities, according to NACE. Among the remaining papers, we can emphasize: O - public administration and defense (10%, 6 papers), C - manufacturing companies (8%, 5 papers).

## 6   Conclusions, Limitations and the Future Research

A systematic literature review performed from a meta-perspective and content-based perspective makes it possible to diagnose research gaps and set directions for further research in the BPM domain [13, 30], while the findings of Pilav-Velić & Marjanovic [27] and Stojanović et al. [35] indicate the need for detailed BPM research in CEE countries due to their unique contextual character of transition economies. The need for monitoring the state of BPM-related research in CEE countries is derived from the hitherto paucity of such research [9, 10] and is in line with the belief that different research environments may result in new research questions leading to significant contributions to the discipline [19]. This was our motivation for surveying the BPM research landscape in CEE countries.

(RQ1) Compared to previous studies [10], we can observe a significant increase in the number of journal papers indexed in *Web of Science* and *Scopus* authored or co-authored by researchers affiliated with CEE institutions. Examining the increase in the number of publications, we assessed the dynamics of growth of such papers over time. We identified the journals with the highest number of articles from the surveyed region, the most active countries, researchers, and research centers. In terms of the number of publications, researchers from Croatia and Slovenia lead the way. Among the works included in the survey, there is noticeable domination of articles classified as *Management*, while *Foundation* and *Engineering* are under researched in the CEE countries apart from Estonia, which stands out in terms of the relative and the absolute number of publications from the Engineering area. (RQ2) As far as the specificity of publications on BPM in CEE countries from the content-based perspective is concerned, it can be observed that two BPM core elements - *People* and *Culture* - are under researched, while the largest number of publications concern *Methods*. The reason for this may be the aforementioned specific character of CEE countries as transition economies, where the organizational culture of enterprises is still focused on achieving a quick performance improvement through the implementation of a new concept, tool or technique rather than investing in the long-term process of introducing organizational changes and training highly qualified personnel. However, the changes in the approach to process management and BPM implementation in organizations are evidenced by many papers referring to *Strategic BPM Alignment* or *Governance*. (RQ3) To investigate which BPM capability areas are explored in empirical research in CEE countries given digitalization, we checked to what extent the examined set of articles can be described by the categories of the updated BPM capability framework proposed by Kerpedzhiev et al. [18]. We discovered that the papers published scarcely fit into the new BPM framework corresponding to the challenges of the digitalization.

In summary, our paper provides an original contribution by identifying under researched and unexplored areas in the field and may suggest what research areas are terra incognita and provide original research topics. Another potential contribution from our study is the assessment in which main tracks distinguished at the BPM conference, i.e., *Foundations, Engineering*, and *Management*, are the authors from CEE countries the most and the least involved in research. Thus, from a theoretical point of view, our work can be considered exploratory research bringing ideas and inspiration for future specific research projects on BPM in CEE countries and potential areas for collaboration with BPM researchers from that region. The obtained results may turn out to be useful when expanding the network of cooperation with scientists from the region or for young BPM scientists when deciding at which institution and under supervision of which mentor to do their PhD studies. In contributing to increasing the body of knowledge on BPM in CEE countries, we also acknowledge the limitations of our study. The main limitation is an unrepresentative set of databases we searched. Our focus was solely on journal papers indexed in the two most vital databases, ignoring other articles, as well as conference proceedings and chapters. Thus, to some extent, our study sample limits the generalization of results. We are justified by selecting databases in which only the best international journals are indexed. Therefore, we can strongly assume that the most substantial BPM publications from CEE countries were published there, especially as the results obtained confirm the previous findings on BPM literature reviews in transition economies (based on multiple databases) [9, 10].

The limitations of the current research give us sound justification to develop our studies in the future. Initially, we plan to expand the literature database and conduct more in-depth studies with a content-based perspective [25], taking into account the findings obtained in the analyzed papers.

# References

1. Anica-Popa, L., Vrincianu, M., Amza, C.: Enhancing business process management by using organisational memory and capitalisation of the cognitive acquis. Transform. Bus. Econ. **9**(1), 472–489 (2010)
2. Barbu, A., Simion, P.C., Popescu, M.A.M., Costea Marcu, I.C., Popescu, M.V.: Exploratory study of the BPM tools used by romanian industrial service companies to increase business performance. TEM J. **9**(2), 546–551 (2020)
3. Bosilj Vukšić, V., Pejić Bach, M., Popovič, A.: Supporting performance management with business process management and business intelligence: a case analysis of integration and orchestration. Int. J. Inf. Manage. **33**(4), 613–619 (2013)
4. Bosilj Vukšić, V., Brkić, L., Tomičić Pupek, K.: Understanding the success factors in adopting business process management software: case studies. Interdiscip. Descr. Complex Syst. **16**(2), 194–215 (2018)
5. Bouchrika, I.: What Is Empirical Research? Definition, Types & Samples. Research.com (2021). https://research.com/research/what-is-empirical-research. Accessed 17 May 2022
6. Brkić, L., Tomičić Pupek, K., Bosilj Vukšić, V.: A framework for BPM software selection in relation to digital transformation drivers. Tehnicki Vjesnik **27**(4), 1108–1114 (2020)
7. de Bruin, T., Rosemann, M.: Using the Delphi technique to identify BPM capability areas. In: ACIS 2007 Proceedings - 18th Australasian Conference on Information Systems, pp. 643–653 (2007)
8. de Morais, R.M., Kazan, S., de Pádua, S.I.D., Costa, A.L.: An analysis of BPM lifecycles: From a literature review to a framework proposal. Bus. Process. Manag. J. **20**(3), 412–432 (2014)
9. Gabryelczyk, R., Jurczuk, A., Roztocki, N.: Business process management in transition economies: current research landscape and future opportunities. In: Proceedings of the 22nd Americas Conference on Information Systems, AMCIS, San Diego (2016)
10. Gabryelczyk, R., Roztocki, N.: Business process management success framework for transition economies. Inf. Syst. Manag. **35**(3), 234–253 (2018)
11. Hernaus, T., Pejić Bach, M., Bosilj Vukšić, V.: Influence of strategic approach to BPM on financial and non-financial performance. Balt. J. Manag. **7**(4), 376–396 (2012)
12. Hernaus, T., Bosilj Vuksic, V., Indihar Štemberger, M.: How to go from strategy to results? Institutionalising BPM governance within organisations. Bus. Process. Manag. J. **22**(1), 173–195 (2016)
13. Houy, C., Fettke, P., Loos, P.: Empirical research in business process management - analysis of an emerging field of research. Bus. Process. Manag. J. **16**(4), 619–661 (2010)
14. Hrabal, M., Tuček, D.: What does it mean to own a process: defining process owner's competencies. FME Transactions **46**(1), 138–150 (2018)
15. Hrosul, V.A., Goloborodko, A., Lehominova, S.V., Kalienik, K.V., Balatska, N.: Modelling balanced criteria system for business business process management. J. Innov. Sustain. RISUS **12**, 139–153 (2021)
16. Indihar Štemberger, M., Bosilj Vukšić, V., Jaklič, J.: Business process management software selection–two case studies. Economic research-Ekonomska istraživanja **22**(4), 84–99 (2009)

17. Jaklič, J., Bosilj Vukšić, V., Mendling, J., Indihar Štemberger, M.: The orchestration of corporate performance management and business process management and its effect on perceived organisational performance. SAGE Open **11**(3), 215824402110401 (2021)
18. Kerpedzhiev, G.D., König, U.M., Röglinger, M., Rosemann, M.: An exploration into future business process management capabilities in view of digitalisation: results from a delphi study. Bus. Inf. Syst. Eng. **63**(2), 83–96 (2021)
19. Khanna, T.: A case for contextual intelligence. Manag. Int. Rev. **55**(2), 181–190 (2015). https://doi.org/10.1007/s11575-015-0241-z
20. Manfreda, A., Buh, B., Indihar Štemberger, M.: Knowledge-intensive process management: a case study from the public sector. Balt. J. Manag. **10**, 456–477 (2015)
21. Mendling, J., Pentland, B.T., Recker, J.: Building a complementary agenda for business process management and digital innovation. Eur. J. Inf. Syst. **29**(3), 208–219 (2020)
22. Milanović Glavan, L., Bosilj Vukšić, V.: Examining the impact of business process orientation on organisational performance: the case of Croatia. Croatian Oper. Res. Rev. **8**(1), 137–165 (2017)
23. NACE: Complete list of all NACE Code (2022). https://nacev2.com/en. Accessed 17 May 2022
24. OECD: Agricultural Policy Monitoring and Evaluation 2020. OECD, Paris (2020)
25. Orlikowski, W.J., Baroudi, J.J.: Studying information technology in organisations: research approaches and assumptions. Inf. Syst. Res. **2**(1), 1–28 (1991)
26. Pejić Bach, M., Bosilj Vukšić, V., Suša Vugec, D., Stjepić, A.M.: BPM and BI in SMEs: the role of BPM/BI alignment in organisational performance. Int. J. Eng. Bus. Manage. **11**, 1847979019874182 (2019)
27. Pilav-Velić, A., Marjanovic, O.: Business process management practices in a small transition economy – current status and research opportunities. In: Proceedings of the 22nd Americas Conference on Information Systems, AMCIS, San Diego (2016)
28. Popesko, B., Tučková, Z.: Utilization of process oriented costing systems in healthcare organizations. Int. J. Math. Model. Methods Appl. Sci. **6**, 200–208 (2012)
29. Rajh, A., Šimundža-Perojević, Z.: Lessons learned from implemented internal and external digitisation processes at the croatian agency for medicinal products and medical devices. Moderna Arhivistika **1**, 197–212 (2018)
30. Roeser, T., Kern, E.M.: Surveys in business process management – a literature review. Bus. Process. Manag. J. **21**(3), 692–718 (2015)
31. Rosemann, M., de Bruin, T.: Towards a business process management maturity model. In: Proceedings of the 13th European Conference on Information Systems, pp. 26–28, ECIS, Regensburg (2005)
32. vom Brocke, J., Rosemann, M. (eds.): Handbook on Business Process Management 1. IHIS, Springer, Heidelberg (2015). https://doi.org/10.1007/978-3-642-45100-3
33. Rowe, F.: What literature review is not: diversity, boundaries and recommendations. Eur. J. Inf. Syst. **23**(3), 241–255 (2014)
34. Roztocki, N., Soja, P., Weistroffer, H.R.: Enterprise systems in transition economies: research landscape and framework for socioeconomic development. Inf. Technol. Dev. **26**(1), 1–37 (2020)
35. Stojanović, D., Tomašević, I., Slović, D., Gošnik, D., Suklan, J., Kavčič, K.: B.P.M. in transition economies: Joint empirical experience of Slovenia and Serbia. Econ. Res. Ekonomska Istraživanja **30**(1), 1237–1256 (2017). https://doi.org/10.1080/1331677X.2017.1355256
36. Suša Vugec, D., Bosilj Vukšić, V., Pejić Bach, M., Jaklič, J., Indihar Štemberger, M.: Business intelligence and organisational performance: the role of alignment with business process management. Bus. Process. Manag. J. **26**(6), 1709–1730 (2020)
37. Šimek, D., Šperka, R.: How robot/human orchestration can help in an HR department: a case study from a pilot implementation. Organizacija **52**(3), 204–217 (2019)

38. Škrinjar, R., Trkman, P.: Increasing process orientation with business process management: critical practices. Int. J. Inf. Manage. **33**(1), 48–60 (2013)
39. Tomaskova, H., Kopecky, M.: Specialisation of business process model and notation applications in medicine—a review. Data **5**(4), 99 (2020)
40. vom Brocke, J., Mendling, J.: Business process management cases–learning from real-world experience. BPTrends Column, Cl. Notes, pp. 1–10 (2016)
41. vom Brocke, J., Jans, M., Mendling, J., Reijers, H.A.: A five-level framework for research on process mining. Bus. Inf. Syst. Eng. **63**(5), 483–490 (2021). https://doi.org/10.1007/s12599-021-00718-8
42. vom Brocke, J., Schmiedel, T.: Towards a conceptualisation of BPM culture: results from a literature review. In: Proceedings of the 15th Pacific Asia Conference on Information Systems: Quality Research in Pacific, PACIS, Brisbane (2011)
43. vom Brocke, J., Sinnl, T.: Culture in business process management: a literature review. Bus. Process. Manag. J. **17**(2), 357–378 (2011)
44. Webster, J., Watson, R.T.: Analysing the past to prepare for the future: writing a literature review. MIS Quar.l **26**(2), xiii–xxiii (2002)
45. Weske, M., Montali, M., Weber, I., Brocke, J.: BPM: Foundations, Engineering, Management. In: Weske, M., Montali, M., Weber, I., vom Brocke, J. (eds.) BPM 2018. LNCS, vol. 11080, pp. 3–11. Springer, Cham (2018). https://doi.org/10.1007/978-3-319-98648-7_1
46. Xiao, Y., Watson, M.: Guidance on conducting a systematic literature review. J. Plan. Educ. Res. **39**(1), 93–112 (2019)
47. Zemguliene, J., Valukonis, M.: Structured literature review on business process performance analysis and evaluation. Entrepren. Sustain. Issues **6**(1), 226–252 (2018)
48. Zerbino, P., Stefanini, A., Aloini, D.: Process science in action: a literature review on process mining in business management. Technol. Forecast. Soc. Chang. **172**, 121021 (2021)

# Process and Project Oriented Organization: The Essence and Maturity Measurement

Piotr Śliż[(✉)] (iD)

University of Gdansk, Sopot, ul. Armii Krajowej 101, 81-824 Sopot, Poland
piotr.sliz@ug.edu.pl

**Abstract.** The main aim of this paper is to present the results of a process-project maturity assessment of large organizations in Poland. The paper consists of two main parts: a theoretical part, which primarily outlines the rationale supporting the prospects and the need for an orientation towards process and project organizations, and an empirical part, presenting an attempt to integrate the MMPM and PMMM maturity models, in order to assess organizational level of process-project maturity. The empirical research carried out on a sample of 90 large organizations shows that vast majority of the organizations surveyed are characterized by low levels of process and project maturity, and 13 of the entities examined can be described, based on the assumptions adopted, as a process-project organization (level 4 of process-project maturity). Further, the research conducted has led to an outline of the factors supporting the recognition of process management as a method fundamental to the designing a process-project organization. Maturity model integration has demonstrated the levels of process and project maturity as well as a statistically positive correlation between the degree of process maturity and project maturity. The original character of this paper primarily concerns the need to fill the literature gap, consisting in the scarcity of publications describing integration of process and project management methods and the deficit of works presenting process-project maturity results.

**Keywords:** Process-project oriented organization · BPM · Process management · Project management · Maturity

## 1 Introduction

The dynamic nature of the socio-economic transformations in the market environment generates a state, in which contemporary organizations focus their activities on attempts to find highly flexible systemic management formulas that enable flexible responses to external (from the environment) and internal (from within the organization) factors. This first of all implies the need to monitor the expectations, needs and satisfaction levels of customers [26], identified as the main accelerators of change in organizations [28]. Moreover, the realities of organizational functioning in the knowledge and information age are shaped by a dynamically expanding set of determinants, which include: the implementation of modern technologies, the expansion of the current set of organization's main resources (human, financial and physical resources) with knowledge

© Springer Nature Switzerland AG 2022
A. Marrella et al. (Eds.): BPM 2022, LNBIP 459, pp. 295–309, 2022.
https://doi.org/10.1007/978-3-031-16168-1_19

[21], the generation of huge volumes of data on the organization's functioning and its environment and feasibility of collecting and exploring thereof [20], the emergence of the opportunities resulting from the process of financial, market, competition, lifestyle, culture, technology, research and knowledge, and legal regulation globalization [30], the ethical significance of business choices [37], as well as the emphasis on the aspect of social responsibility of contemporary organizations and the customers' focus on the ethical dimension of production process implementation [3]. These factors do not constitute a closed catalog but imply a continuous search for and generation of new solutions that eliminate the current state dysfunctions and limits identified so far, simultaneously giving rise to new, hitherto unknown, problems, which creates an impetus for a dynamic search for integrated system formulas within the sphere of organizational management, enabling solving thereof [31]. One solution to the problem, consisting in a search for highly flexible system formulas, entails integration of selected process and project management methods, additionally adopting the assumptions of the ambidexterity concept, identified as the balancing of exploitation and exploration activities [17]. The main operational categories materializing the integration of process and project management on the grounds of ambidexterity entail adoption of such operational categories as: exploitative processes (exploitation) as well as explorative processes and projects (exploration) [4, 31].

The theoretical study revealed a knowledge deficiency primarily associated with the need to integrate business process management (BPM) and project management methods. A cognitive gap, consisting in a deficit of research on simultaneous assessment of the process and project maturity levels in an organization, was also identified. Consequently, the research problem was delineated around the following research question: What process-project organization assumptions should be taken into account in order to assess the level of process-project maturity? Such formulated research problem determined the structure of the main objective. The paper aims to present the essence and assumptions of process-project organization functioning as well as demonstrate a method of process-project maturity measurement. The empirical proceeding to assess the level of process-project maturity was carried out on a non-probabilistically selected sample of 90 large organizations in Poland. To implement the tasks formulated, an opinion poll method was used. Computer-assisted web interviewing (CAWI) research technique was used in the proceedings. The research tool used in the study was survey-type questionnaires.

## 2    Theoretical Background – At the Level of Process and Project Management Permeability

The theoretical research carried out for the purpose of this publication has led an outline of the rationale supporting the possibility and above all the need, from the perspective of business practice representatives, to conceptualize organizations in process-project terms. The following have been considered as such rationales: the fact that business processes and projects occur in every organization [23], the external and internal customer orientation, in both process [11] and project orientation [36] as well as the fact that project management can be viewed as a process [16], the consideration of effective project management as one of the eight success factors of process management implementation in an organization [2, 7], the applicability of common methodologies, techniques and tools used in both discussed concepts [11] and the complementary nature of the process and project approaches [1].

Based on the literature review, barriers, identified as limiting factors (inhibitors) in business process and project management integration, have also been indicated, i.e.: the lack of convergent process and project classification schemes [12], the clear differences between both process and project organizational structures [11], the gaps in the theoretical and practical knowledge on the issues of integrated system formulas [4], the fact that integration of process and project management is achievable at an appropriate level of development dynamics and activity standardization [1], as well as the lack of a concept of process and project maturity assessment models integration; and the deficiencies in the integration of process and project management with such collaborating disciplines as knowledge management, change management or communication management (Cf. [31]).

In the context of the search for an integration plane for the two operational categories described, it is important to highlight the common formal attributes thereof, i.e., purpose, inputs, outputs, allocated resources and customers [24]. Implementation of activities, both in business processes and projects, is oriented on the result generated, desirable from an external and internal customer perspective [5, 24]. Generation of results requires resource use and sharing, with clear emphasis on the knowledge resource, which is used to coordinate the two categories discussed [4]. The clear differences between the operational categories discussed, on the other hand, include the genesis of processes and projects organizations, assuming that processes constitute natural elements of any organization, while projects are brought to life by management, depending on the organization's needs [24]. The essence of the difference between a process and a project entails the nature of the activities performed, which in the case of processes are repetitive, whereas projects involve one-time (unique) activities. This determines the sequence of the activities, which in a business process, as opposed to a project, is continuous in nature. Focusing on the effects generated in both categories, it is worth emphasizing that in processes, they are the same or similar, whereas in projects, they are of diversified character, which results from the nature of the activities performed. Such differences between processes and projects do not prevent a holistic view on the integration of the two categories but delineate the horizon for an individual perspective and a search for management concepts and methods that enable achievement of synergies, from the perspective of the results and the opportunities generated through both.

Turning to the results of the subsequent stage of the theoretical study, the similarities and differences between business process management (BPM) and project management have been outlined. The set of similarities encompasses: the common methodological basis [4, 23] the fact that both process and project management objectives are derived from the organization's strategy [23], the occurrence of processes in project management (executive, auxiliary and managerial) [34], the perception of knowledge as a key resource in an organization [4], the improvement of the processes in projects by application of process management principles and tools [18], and that the fact that projects and processes are defined by managerial standards (ISO 10006, DIN 69 901, BS 6079) [39].

Summing up, the following have been qualified as the key common planes supporting process and project integration, identified as the catalysts enabling process-project organization modeling: the complementary nature of both concepts and management methods [4, 24], the shared customer orientation on the customer, both in external and internal terms [5], the interdisciplinary character of employee teams (process and project teams [5], as well as the focus on high employee empowerment [6], the similarity [31] of the process governance [19] and project governance [1, 35] assumptions; the high utility of the process and project approaches in developing an organization's ability to introduce changes (adaptive and innovative) [25], the strengthening of cross-functional processes [24], the shared methodological layer and the availability of the same techniques and tools in both discussed concepts [11], as well as the fact that integration of the process and project management concepts can serve as a starting point for development of much more elaborate conceptions that are based on e.g., process-project-product [27] or process-project-knowledge triads [4]. Considering organizations in process-project terms, potential threats, identified as inhibitors of process and project management integration, should be indicated, which include: the diverse objectives of business processes and projects, the diverse methodologies for recording the course of business process and project activities, the occurrence of intra-group conflicts in both process and project teams, but also intergroup conflicts between process and project teams [31]. The literature addresses and discusses such issues as: the increase in the costs associated with organization functioning under process and project management concept integration and coordination; the disruptions in the diffusion of knowledge between process and project teams, resulting from a functional organization division [24], i.e., functioning within the sphere of processes, projects and functions [23], the need to distribute the resources between the areas of process and project activity; the difficulty of designing a coherent system consolidating the behavior of the employees working within the process and project sphere; the intra-organizational problems at the interface of business processes and projects and the lack of studies on process-project organization management [24], the lack of unified methodology for organization maturity assessment from the perspective of integrated process-project orientation implementation assessment; as well as the cost of and the time needed for information system and communication tool unification in an organization [31].

Based on the subject literature, an assumption was adopted, to be empirically verified, supporting the recognition of business process management as a basis for organization modeling within a process-project dimension. The following premises were considered: an enterprise entails a set of processes [5] occurring in any organization, which, as objects of an organization's structure [22], can coexist with such categories as projects (development of a common ground for project management) [23], procedures (within the dimension of process stiffening factor reduction or increase) and functions (coexistence of functional and process management), processes are at the center of today's and tomorrow's competition [38], processes are identified as critical organizational resources [29], which are mainly meant to generate added value for the customers [13], different criteria can be used for process typology (e.g., auxiliary and supporting processes) [33], processes, just like process organizations, can be evaluated from the perspective of various measures, i.e., maturity [10] or efficiency [8, 9].

An organization, in process-project view, entails such a state of the system, in which the benefits resulting from process and project solution implementation, as well as from the integration, assuming ambidexterity, of the concepts and methods of such category management, are discounted consciously. It is thus a complex system that is based on a business process and project symbiosis in exploitation and exploration activities, to achieve a synergy effect (Cf. [31]). A catalog of factors was adopted as essential constituents shaping process-project organizations. Operation based on the ambidexterity concept assumptions and extension of the process typology to include exploitative and explorative processes are the first two of those determinants. Focus on designing interdisciplinary process and project teams is another. The set also includes provision of an environment conductive to, as per contextual ambidexterity, the employees' execution of both exploitation and exploration activities. This, however, primarily requires implementation of a competence market within the organization, to acquire information on the competences needed for process and project execution already at the stage of process design or project planning. Furthermore, it is essential for the management to create an environment enabling emergence of employee initiatives within the organization. Such initiatives improve the processes involving reconfiguration of the project management methodology towards authorial solutions adapted to the specificity of a given organization. It is also imperative that the management implements a process-project organizational structure developed on the assumptions of a matrix, in which the auxiliary processes are in the vertical layer, while the basic and project processes are in the horizontal layer. The essence of this assumption is based on a mechanism in which auxiliary process managers are expected to provide, on market terms, the intangible and material resources necessary for implementation of the effects generated in the main processes and projects. The process owners and project managers, in the horizontal layer, are in turn responsible for generating effects consistent with the customer needs and expectations [31]. A conceptual diagram of a company organizational structure from a process-project perspective is shown in Fig. 1.

The basis for a structure outlined as such entails implementation of market principles into the organization, which constitutes an important determinant of a process-project organization [31].

## 3 Methods and Models

### 3.1 Study Scope and Characteristics of the Organizations Included in the Empirical Investigation

The scope of the empirical investigation conducted should be allocated within the field of three approaches: object-oriented, subject-oriented, spatial and temporal. Within the object scope of the study, four problem groups intersect, i.e., identification of the levels of process and project maturity, using the MMPM and PMMM models, and the resultant identification of the level of process-project maturity. Due to the limitations of this work, the characteristics of the two models have been presented in detail in the publications: [14, 15, 31, 32].

The object scope of the empirical study entails public and private sector large organizations (more than 250 employees). The entities surveyed were selected according to

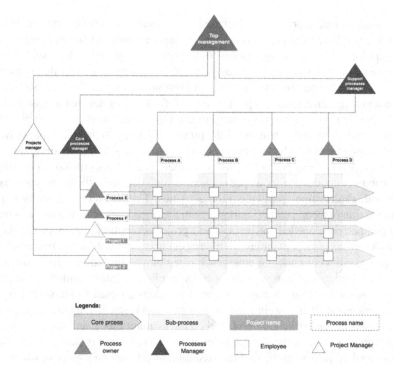

**Fig. 1.** Conceptual diagram of organizational structure in the company examined. Source: Own elaboration based on the assumptions presented in: [31].

a territorial criterion, the sector, the dominant type of activity (production, trade or services) and the range of operation (local, national or international). The temporal scope allowed inclusion of organizations, regardless of the period of their functioning on the market.

The empirical investigation was carried out using non-probabilistic (non-random) sampling with purposive selection. The selection criteria involved operating on the territory of Poland, organization size (large organizations), activity range (local, national and international), the dominant type of activity, according to the Polish Classification of Activities. Organizations operating in all sectors of the economy were included in the survey. As a result, 90 organizations were surveyed using an opinion poll method incorporating the CAWI technique. The response rate was 29.20%.

## 3.2 Research Proceedings

The research was divided into five stages, which are characterized in Table 1.

The survey was conducted in Poland, in 2021–2022, using the CAWI technique. The survey questionnaire was delivered to the respondents via Google Forms.

**Table 1.** Characteristics of research proceedings

| Stage | Description |
|---|---|
| Stage 1 | Review of the subject literature - assessment of the state of knowledge on process and project management concept and method integration as well as process-project maturity |
| Stage 2 | Definition of the criteria and selection of the process and project maturity models. Adjustment of the measurement scales in both maturity models and digitization of the research tool (questionnaire), enabling implementation of the survey, using the CAWI technique |
| Stage 3 | Selection of the research method and the sampling technique. Preparation of a registry of organization and implementation of a pilot study to assess the understanding of the questions and answers contained in the survey questionnaire |
| Stage 4 | Implementation of the survey, using a designed, objectified process and project maturity assessment |
| Stage 5 | Data collection. Selection and quality assessment of the data obtained. Questionnaire selection. Data analysis. Hypothesis verification. Formulation of the conclusions, limitations and directions for further research |

Source: own elaboration

### 3.3 Assumptions of the Integrated Process-Project Maturity Assessment Model

Efforts to increase the level of process-project maturity should be focused on integrating the process and project architecture in response to the turbulent nature of the business environment. Dedicated or integrated (process and project) maturity models are used to determine the level of process-project maturity.

Process-project maturity occurs in an organization when the management consciously discounts the benefits resulting from the implementation and integration of process and project solutions at levels of the functioning system and the organizational structure. Maturity model integration, in turn, is defined as consolidation (integration) of the maturity assessment areas (shared areas), in order to obtain a complex diagnosis of the phenomenon analyzed. In the subject area discussed, maturity model integration is understood as planned activities involving consolidation of two or more maturity models (process and project) into a whole, allowing assessment of the degree of process management and project management element implementation. Such consolidation, in this sense, also involves elimination of repetitive (duplicated) assessment areas (defined as shared areas). Organizational structures, maturity models, and excellence models may constitute the subject of consolidation [31].

Process-project maturity is gradual in nature and can be described by five maturity levels (Table 2).

It should be emphasized here that an organization can qualify for a higher level of process-project maturity, if it meets the lower-level evaluation criteria (e.g., for an organization to qualify as level three, it must achieve a minimum of level two process and project maturity). Assessment of the level of process-project maturity is based on the results of sub-assessments from the MMPM and PMMM models.

**Table 2.** Characteristics of process-project maturity levels

| Level | Level characteristic |
|---|---|
| Level 1. Organization based on a functional management formula | Organizational structure in functional (silo) in nature. A classic (silo) system formula of organization management dominates. Project activities are implemented on an ad-hoc basis, e.g., to obtain funding from external sources. Implementation of both processes and projects is chaotic (ad hoc implementation). This is caused a lack or low level of process and project formalization (lack of process and project documentation) and a lack of the organizational roles characteristic for process and project management |
| Level 2. Organization with ad hoc application of process and project elements ($MMPM = 2$, $PMMM = 2$) | Within the sphere of structural solutions, the organization is based on a functional structure with identified processes and teams established to implement projects. The exploitation and exploration layers are not specified. 'Shared language' is used to identify the sets of activities within the process and project categories. This means that processes and projects are defined correctly. The main processes are identified and formalized as descriptive or graphic process documentation. Within the sphere of the functioning system, projects are established, in addition to processes |
| Level 3. Process and project oriented organization ($MMPM = 3$, $PMMM = 3$) | The organization seeks opportunities to discount the dynamism of processes, while changes occur in the structural sphere, initiating implementation of a matrix structure, with project organization by separate, authorized units. The organization employs a measurement system for the processes identified and formalized. Such activities enable decision making based on the measurement results obtained. The exploitation layer is delineated clearly, while the management activities are primarily aimed at increasing the efficiency and quality of the process effects (results) generated. Parallelly, projects are initiated within the process architecture sphere, which are implemented on the basis of formalized process documentation. To execute projects, project manager roles have been established in the organization |

*(continued)*

**Table 2.** (*continued*)

| Level | Level characteristic |
|---|---|
| Level 4. Process organization with a high level of project management maturity (MMPM = 4, PMMM = 4) | The organizational structure, depending on the size and type of the genotype activity, is characterized by a matrix arrangement. Efforts are made to implement a process structure with project organization or a process-project structure. The organization is characterized by a high (minimum fourth) level of process maturity. Achievement of this level indicates that the organization has reached a system state, in which the management consciously discounts the benefits of process and project solution implementing and recognizes the need for the grouping and integration thereof within the exploitation and exploration layers |
| Level 5. Process and project organization (MMPM = 5, PMMM = 5) | A process and project structure has been implemented in the organization. Each project is implemented in a process convention. The organization is characterized by the highest level of process and project maturity. The organization's level of process and process maturity and its level of project management maturity are at level five. The structure of the organization is built based on a matrix structure or a process or project structure |

Source: own elaboration, based on: [31]

Table 3, in turn, presents the extremes of a process and project organization, defining the characteristics indicating its immaturity and maturity.

Summing up, the path to process-project maturity, especially for organizations managed according to a functional system formula, should be defined as transformation towards simultaneous implementation of solutions that increase the level of process and project maturity in the organization, in combination with deployment of operational categories within the exploitation and exploration layers and implementation of activities that are aimed at integration thereof within a single organization. Such activities should be carried out in an evolutionary manner, choosing a strategy that enables planning and gradual implementation of various process and project components.

Table 4 outlines the nine steps in the design of the integrated process-project maturity assessment model used in the implementation of the empirical investigation presented in this publication.

**Table 3.** Immaturity and maturity characteristics of a process-project organization

| Features of process-project immaturity (levels 1–3) | Features of process-project maturity (levels 4–5) |
|---|---|
| The dominant systemic management formula in the organization is the functional approach (no symptoms indicating either process solution implementation or an orientation towards process transformation) | Processes and projects are the key object of a flexible organizational structure (process structure with project organization or process-project structure) |
| Functional, strongly petrified and highly centralized organizational structure | The exploitation layer is characterized by a high level of process maturity in the organization (minimum level four, according to the MMPM2 model), while the exploration layer is characterized by a high level of project maturity |
| Employee improvisation as regards business process and project implementation | |
| Executive improvisation as regards process and project management integration | |
| The ability to simultaneously execute the organization's processes and projects results from individual employee initiatives, rather than from the operating system of the entire organization | Process and project implementation is carried out via specially appointed roles functioning within the sphere of a process-project structure |
| | The organization employs a system of process and project effect (result) measurement |
| Projects are implemented in an ad hoc manner and project management methodologies are not used | The organization implements a market relations mechanism, at the level of both operational and exploratory layers, as well as customer-supplier relations mechanism, at the level of processes and projects |
| No system of process and project evaluation metrics | |
| Lack of market relations mechanism implementation within the organization (in particular with regard to the relations at the level of the exploitation and exploration layers and at the level of operational categories, e.g., main processes - auxiliary processes, internal projects - main processes) | When assessing the level of professional project management, both project maturity and excellence models are used |
| The organization's ad hoc improvements and optimization activities are oriented at the exploitation layer only | |

Source: [31]

Contemporary organizations, in order to achieve supremacy on the market, should focus on both business processes and projects. As a result, efforts should be made in the sphere of organizational management, to integrate business process and project management methods. In order to assess the degree to which the assumptions of both methods are implemented, an integrated model is needed. A proposal of such a model is presented in this paper, based on two empirically verified maturity models MPMM [32] and PMMM [14, 15].

**Table 4.** Design stages of an integrated process-project maturity assessment model

| Step | Scope | Activities |
|------|-------|-----------|
| Step 1 | Definition of the integration assumptions | Function: descriptive<br>Assessment scope: organization maturity level<br>Type of assessment: assessment based on the symptoms of process capability and project solution implementation declaration<br>Research method: CAWI opinion poll<br>Assessment: objectivized<br>Assessment specificity: universal model |
| Step 2 | Overview of the maturity models available in the process and project management literature | Identification, review and selection of the models fulfilling the assumptions defined in stage 1 |
| Step 3 | Maturity model selection | Based on the literature review and secondary research, considering, inter alia, the level of model operationalization in research, the MMPM and PMMM models were selected |
| Step 4 | Adjustment of the model to the specificity of the process-project maturity issue under examination | Measurement scale change from a 7-point to a 5-point scale |
| Step 5 | Definition of the type of integration | Integration, at the level of organization maturity, using a 5-point scale (−2 to 2 range). Results integration to assess process and project maturity |
| Step 6 | Model testing | Testing of integrated models on a sample of 10 organizations. Verification of the understanding of survey questions and answers |
| Step 7 | Model modification | Correction of the questions and answers marked as incomprehensible by the respondents |
| Step 8 | Definition of the target group, the measurement method and technique | Target group: large organizations. The questionnaire was addressed to management representatives, process owners and project managers<br>Sampling technique: non-sampling<br>Method: opinion poll |
| Step 9 | Model implementation | Implementation of the survey using an integrated model |

Source: own elaboration, based on: [31]

# 4 Results

## 4.1 Process-Project Maturity Level Assessment in the Surveyed Group of Large Organizations Operating in Poland

Initially, an attempt was made to assess the level of process-project maturity in the surveyed entities operating in Poland. The primary research objective was to identify the degree of process and project management element implementation, using an integrated model of process-project maturity assessment consisting of two components: the multicriteria model of process maturity (MMPM) model [32] and the project management maturity model (PMMM) [14, 15]. Detailed model assumptions have been presented in the publications characterizing the two concepts [15, 32] Aggregate results of the process and project maturity assessment are shown in Table 5.

**Table 5.** Process and project maturity assessments (N = 90)

| Maturity level | Level 1. (MMPM) | Level 2. (MMPM) | Level 3. (MMPM) | Level 4. (MMPM) | Level 5. (MMPM) | Total |
|---|---|---|---|---|---|---|
| Level 1. (PMMM) | 21 | 9 | 3 | 1 | 1 | 35 |
| Level 2. (PMMM) | 5 | 11 | – | 1 | – | 17 |
| Level 3. (PMMM)1 | 1 | 4 | 15 | – | – | 20 |
| Level 4. (PMMM) | 1 | – | c | 7 | – | 8 |
| Level 5. (PMMM) | 2 | 2 | – | 2 | 4 | 10 |
| Total | 30 | 26 | 18 | 11 | 5 | **90** |

Source: own elaboration, based on the research carried out

As the data in Table 5 shows, the vast majority of organizations (21), examined on the basis of the assumptions of the MMPM and PPMM maturity models, using the available research questionnaires and the assumptions presented in Tables 2 and 3 of the process-project maturity model, were classified as level 1 organizations. It is worth noting that only 13 organizations in the entire sample were identified as a process-project organization.

The results show that despite the optimistic opinions regarding the rise of interest in process and project management, voiced by researchers, the practical dimension of management concept and method implementation is at a relatively low level of maturity in the research streams described. It is worth emphasizing here that the results presented in Table 5 are in line with the conclusions arising from other empirical investigations concerning the low level of process and project maturity in Polish organizations. Undoubtedly, the research results presented broaden this research field and set directions for further studies on identifying the factors supporting and limiting implementation of process and project orientation.

### 4.2 Research Hypothesis Verification

Further in the study, an attempt was made to verify the research hypothesis formulated. The aim of the statistical analysis was to verify the relationship between the levels of process and project maturity. The hypothesis was verified using the Spearman's rank correlation coefficient, for which a null hypothesis of no correlation between the variables under examination was formulated. The p-value was 0.003, which allows the null hypothesis rejection in favor of an alternative hypothesis about the variables' correlation. Based on the results of the statistical analysis, a conclusion was made that the hypothesis is true – a statistically significant positive, but low, Spearman's rank correlation $= 0.41$ ($p < 0.05$) occurs, indicating a clear positive relationship between the variables of process maturity level and project maturity level. This indicates that, in the analyzed group of organizations, an increase in the level of process maturity positively affects an increase in the level of project maturity, which is consistent with the formulated presumption that the basis for a process-project organization entails implementation of BPM assumptions.

## 5   Conclusion

The main axis of this paper fits the need to fill the cognitive gap associated with the small number of publications addressing, on the theoretical and empirical grounds, the issue of process and project management method integration and organizational process-project maturity level assessment. The study conducted fills this gap, even if partially, by propounding an integration of two models of process (MMPM) and project (PMMM) maturity, as a result of which the category of process-project maturity has been presented and defined.

As with any research of this type, this study, too, is burdened with limitations, consisting of non-probabilistic sampling technique, thus limited the conclusions to the group of organizations under examination only. Moreover, it should be emphasized that the assumptions of a process and project organization do not constitute a closed catalog but form a basis for a broader discussion on the integration of selected process and project management concepts and methods, with particular emphasis on agile methodologies. This provides a basis for further research and qualitative proceedings. As a result of the survey, an additional gap was identified, consisting in the small number of publications addressing the category of exploration processes and the concept of explorative BPM. The direction of further research has been determined. Another goal set by the Author is to assess the relationship between the levels of process and project maturity assessment and the level of selected ICT technology implementation in an organization. This direction, in exploratory layer context, seems cognitively interesting, as it will identify which ICT technologies (e.g., artificial intelligence, Internet of Things or cloud computing) can be of potential support in the achievement of higher levels of maturity. The author also aims to seek solutions to exemplify process and project oriented organizations (organizations at high levels of process and project maturity).

# References

1. Ahrens, V.: Complementarity of project and process management. Arbeitspapiere der Nordakademie (2018)
2. Bai, C., Sarkis, J.: A grey-based DEMATEL model for evaluating business process management critical success factors. Int. J. Prod. Econ. **146**, 281–292 (2013). https://doi.org/10.1016/j.ijpe.2013.07.011
3. Bezençon, V., Blili, S.: Ethical products and consumer involvement: what's new? Eur. J. Mark. **44**, 1305–1321 (2010). https://doi.org/10.1108/03090561011062853
4. Bitkowska, A.A., Wydawnictwo C.H.: Beck: Od klasycznego do zintegrowanego zarzadzania procesowego w organizacjach. Wydawnictwo C. H. Beck, Warszawa (2019)
5. Brilman, J.: Nowoczesne koncepcje i metody zarzadzania. Polskie Wydaw. Ekonomiczne, Warszawa (2002)
6. Czubasiewicz, H., Grajewski, P.: Koncepcja empowermentu w zarządzaniu organizacjami. SiP, pp. 153–173 (2019). https://doi.org/10.33119/SIP.2018.162.11
7. Gabryelczyk, R.: Exploring BPM adoption factors: insights into literature and experts knowledge. In: Ziemba, E. (ed.) AITM/ISM -2018. LNBIP, vol. 346, pp. 155–175. Springer, Cham (2019). https://doi.org/10.1007/978-3-030-15154-6_9
8. Grajewski, P.: Polskie Wydawnictwo Ekonomiczne: Organizacja procesowa. Polskie Wydawnictwo Ekonomiczne, Warszawa (2016)
9. Grajewski, P., Sliż, P., Trenbrink, C.: Assessing efficiency and effectiveness in the automotive after-sales processes. Int. J. Serv. Oper. Manage. (online first) (2021)
10. Harter, D.E., Krishnan, M.S., Slaughter, S.A.: Effects of process maturity on quality, cycle time, and effort in software product development. Manage. Sci. **46**, 451–466 (2000). https://doi.org/10.1287/mnsc.46.4.451.12056
11. Harvey, J., Aubry, M.: Project and processes: a convenient but simplistic dichotomy. IJOPM. **38**, 1289–1311 (2018). https://doi.org/10.1108/IJOPM-01-2017-0010
12. Hobbs, B., Aubry, M., Thuillier, D.: The project management office as an organisational innovation. Int. J. Project Manage. **26**, 547–555 (2008). https://doi.org/10.1016/j.ijproman.2008.05.008
13. Ittner, C.D., Larcker, D.F.: The performance effects of process management techniques. Manage. Sci. **43**, 522–534 (1997). https://doi.org/10.1287/mnsc.43.4.522
14. Kerzner, H., Kerzner, H.: Advanced Project Management: Best Practices on Implementation. Wiley, Hoboken, N.J (2004)
15. Kerzner, H.: Strategic Planning for Project Management Using a Project Management Maturity Model. Wiley, New York (2001)
16. Koskinen, K.U.: Organizational learning in project-based companies: a process thinking approach. Proj. Manag. J. **43**, 40–49 (2012). https://doi.org/10.1002/pmj.21266
17. March, J.G.: Exploration and exploitation in organizational learning. Organ. Sci. **2**, 71–87 (1991). https://doi.org/10.1287/orsc.2.1.71
18. Marciszewska, A., Nowosielski, S.: Podejście procesowe w usprawnianiu zarządzania projektami. Prace Naukowe Uniwersytetu Ekonomicznego we Wrocławiu. **169**, 73–83 (2011)
19. Markus, M.L., Jacobson, D.D.: Business process governance. In: vom Brocke, J., Rosemann, M. (eds.) Handbook on Business Process Management, vol. 2, pp. 201–222. Springer, Heidelberg (2010). https://doi.org/10.1007/978-3-64201982-1_10
20. McAfee, A., Brynjolfsson, R.: Big data: the management revolution. Harv. Bus. Rev. **90**, 60–68 (2012)
21. Nag, R., Gioia, D.A.: From common to uncommon knowledge: foundations of firmspecific use of knowledge as a resource. AMJ. **55**, 421–457 (2012). https://doi.org/10.5465/amj.2008.0352

22. Nanz, G.: zfo-Toolkit-Prozess-Alignment-Von der Unternehmensstrategie zum Geschäftsprozess. eitschrift Fuhrung und Organisation, vol. 81, (2012)
23. Nowosielski, S.: Procesy a projekty w organizacji. Ekonomika i Organizacja Przedsiębiorstwa, pp. 140–150 (2017)
24. Nowosielski, S.: Procesy i projekty w organizacji. O potrzebie i sposobach współdziałania. SiP, pp. 109–129 (2019). https://doi.org/10.33119/SIP.2018.169.8
25. Osbert-Pociecha, G.: Zdolność organizacji do zmian–dlaczego zmiany wymagają podejścia procesowego i/lub projektowego. Prace Naukowe Uniwersytetu Ekonomicznego we Wrocławiu. 10, 87–96 (2017)
26. Prahalad, C.K., Ramaswamy, V.: The Future of Competition: Co-Creating Unique Value with Customers. Harvard Business Press, Boston, Massachusetts (2004)
27. Reiß, M.: Integriertes Projekt-, Produkt- und Prozeßmanagement (1992). https://doi.org/10.18419/OPUS-5580
28. Saarijärvi, H., Kannan, P.K., Kuusela, H.: Value co-creation: theoretical approaches and practical implications. Eur. Bus. Rev. 25, 6–19 (2013). https://doi.org/10.1108/095553413112 87718
29. Seethamraju, R.: Business process management: a missing link in business education. Bus. Process. Manag. J. 18, 532–547 (2012). https://doi.org/10.1108/14637151211232696
30. Sengupta, S., Mohr, J., Slater, S.: Strategic opportunities at the intersection of globalization, technology and lifestyles. Handbook Bus. Strategy 7, 43–50 (2006). https://doi.org/10.1108/10775730610618602
31. Sliz, P.: Organizacja procesowo-projektowa: istota, modelowanie, pomiar dojrzalosci. Difin, Warszawa (2021)
32. Śliż, P.: Concept of the organization process maturity assessment. J. Econ. Manage. 33, 80–95 (2016)
33. Sułkowski, Ł.: Status poznawczy zarządzania procesowego. In: Podejście procesowe w zarządzaniu, pp. 45–49. SGH (2004)
34. Trocki, M., Grucza, B., Ogonek, K.: PWE: Zarządzanie projektami. Warszawa (2003)
35. Trocki, M.: Project Governance–kształtowanie ładu projektowego organizacji. Studia i Prace Kolegium Zarządzania i Finansów 159, 9–23 (2018)
36. Turner, J.R., Keegan, A.: The management of operations in the project-based organisation. J. Chang. Manag. 1, 131–148 (2000). https://doi.org/10.1080/714042464
37. Vranceanu, R.: The ethical dimension of economic choices. Bus. Ethics 14, 94–107 (2005). https://doi.org/10.1111/j.1467-8608.2005.00394.x
38. Willaert, P., Van den Bergh, J., Willems, J., Deschoolmeester, D.: The process-oriented organisation: a holistic view developing a framework for business process orientation maturity. In: Alonso, G., Dadam, P., Rosemann, M. (eds.) BPM 2007. LNCS, vol. 4714, pp. 1–15. Springer, Heidelberg (2007). https://doi.org/10.1007/978-3-540-75183-0_1
39. Wyrozębski, P.: Bariery w wykorzystaniu wiedzy projektowej w organizacjach. In: Wkład nauk ekonomicznych w rozwój gospodarki opartej na wiedzy, pp. 289–302. Warszawa (2014)

# The Competencies for Knowledge Management and the Process Orientation in the Post-Covid Economy. Generation Z Students Perspective

Waldemar Glabiszewski[1]([✉]) [iD], Szymon Cyfert[2] [iD], Roman Batko[3] [iD], Piotr Senkus[1,2,3,4] [iD], and Aneta Wysokińska-Senkus[4] [iD]

[1] Nicolaus Copernicus University, 87-100 Torun, Poland
`waldemar.glabiszewski@umk.pl`
[2] Poznan University of Economics and Business, 61-875 Poznan, Poland
`s.cyfert@ue.poznan.pl`
[3] AGH University of Science and Technology, 30-059 Krakow, Poland
[4] War Studies University Warsaw, 00-910 Warsaw, Poland
`a.wysokinska-senkus@akademia.mil.pl`

**Abstract.** Taking into account the changes occurring in the aftermath of the Covid-19 crisis and taking the perspective of students from generation Z currently entering the labour market, the aim of this article is to assess the increase in the importance of competencies for knowledge management on the impact of process orientation. The empirical research was conducted using an online survey method at two leading universities in Poland in economic education, and complete data were collected from 711 generation Z students. They were analysed using the structural equation modelling method. The results show that the increase in the importance of competencies for knowledge management following the Covid-19 crisis will affect the change of the five of six process orientation components, i.e. defining business processes more clearly, allocating resources to a greater extent based on business processes, broader setting of specific performance targets for different business processes, measuring the outcome of different business processes more closely and more unambiguous designation of process owners. Only in the case of rewarding employees based on the implementation of business processes in which they are involved, students from generation Z do not see the impact of the increasing importance of competencies for knowledge management.

**Keywords:** Process orientation · Knowledge management · Competencies · Covid-19 · Generation Z

## 1 Introduction

Changes resulting from the Covid-19 crisis have a huge impact on the functioning of enterprises, their formation, management and the results they achieve [1, 16, 31]. The crisis caused by the Covid-19 pandemic is forcing companies to rebuild their business

A. Marrella et al. (Eds.): BPM 2022, LNBIP 459, pp. 310–325, 2022.
https://doi.org/10.1007/978-3-031-16168-1_20

models and reformulate their place and role in society [6]. Ritter and Pedersen [26], among the six business models they identified, point to the existence of an antifragile business model, in which the process management concept is used. Research has shown that the process approach in times of crisis provides a higher degree of flexibility because of their ability to adapt to a rapidly changing environment and function better in a new environment. Ritter and Pedersen [26] assume that a process approach to its management increases organisation's ability to cope with a crisis. Furthermore, mistakes made during previous crises should become the subject of an organisational learning process in order to prevent similar mistakes in subsequent crises. In this context, the crucial importance of the knowledge management concept should be emphasised. According to Schiuma et al. [28], the knowledge management process should be considered essential to guide an organisation through a crisis. Hence, it seems that in crisis conditions it becomes desirable to develop competencies for knowledge management, assuming that these will increase the tendency to process orientation, which is desirable in these conditions.

The above considerations provide a basis for analysis competencies for knowledge management and process orientation, and furthermore to link them together in the context of changes occurring in the aftermath of the Covid-19 crisis. Taking into account the changes caused by the coronacrisis from the perspective of Generation Z students, two important questions arise (1) will competencies for knowledge management grow in importance? (2) and will increase in the importance of competencies for knowledge management affect the propensity to adopt a process orientation? The available research results indicate that generation Z has some important characteristics differentiating it from earlier generations, including a high level of entrepreneurship (17). Therefore, it should be assumed that the representatives of Generation Z are distinguished by greater propensity to start their own business. Hence, we treated the intention to establish our own company in the near future as a moderating variable in our research.

Thus the aim of the article is to assess the increase in the importance of competencies for knowledge management on the change of process orientation. The main hypothesis was formulated in the research, according to which the increase in the importance of competencies for knowledge management changing under the influence of coronacrisis will influence the change of individual components of process orientation. As process orientation was defined in the paper by six variables, six specific hypotheses were verified, stating the influence of competencies for knowledge management on each of its components separately i.e.: (H1) the need to define business processes more clearly so that most employees have a clear understanding of these processes, (H2) the need to allocate resources to a greater extent based on business processes, (H3) the need to set specif-ic performance targets for different business processes, (H4) the need to measure the outcome of different business processes more closely, (H5) the need to clearly designate process owners to take responsibility, (H6) the need to reward employees more based on the performance of the business processes in which they are involved.

Both the competencies for knowledge management and process orientation are complex constructs and require unambiguous definition. Therefore, for the purposes of the study, we adopted the measurement scales from previously conducted studies [8, 10].

Our article is organised as follows. First, by discussing the nature and scope of competencies for knowledge management and process orientation, we identify and define the dimensions of the relationships under study that we are interested in. Second, we situate our substantive problem in the conditions of the post-covid economy. Third, we identify it from the perspective of Generation Z students just entering the labour market. Fourth, we present our research method and results. Fifth, based on the empirical findings and the results of a critical review of the literature, we conduct a discussion that leads to a key conclusion that the changing competencies for knowledge management in the aftermath of the Covid-19 crisis have an impact on the importance of the individual components of the orientation process (five out of six).

## 2   Theoretical Background

The literature on the importance of competencies in the field of knowledge management and process orientation streams into three broad categories: works that perceive knowledge management processes as necessary to deal with a crisis; works that focus on the skills and abilities that allow organizations to function efficiently in times of uncertainty caused by a crisis; works that emphasise importance of competences in the field of knowledge management facilitating business process management (BPM).

Ritter and Pedersen who wrote about the importance of management processes in coping with a crisis claim that "argue that a process approach based on the idea that exposure to a crisis seems to trigger the need to be prepared for similar events makes the organisation achieve the ability to cope with the crisis. In particular, mistakes made during previous crises are important for learning and preventing similar mistakes in the face of the current crisis" [26, p. 219].

The authors also emphasize the importance of anticipating how a crisis develops and understanding how decisions made during a crisis affect outcomes beyond its end.

A study by Clauss et al. [9] has shown that focusing on temporary business model innovation can help organizations gradually change existing business models quickly by considering the limited resources at their disposal.

Schiuma et al. [28] argues that the knowledge management can be considered essential to guide organisations through a crisis. It can help implement change management, optimise operations and, most importantly, support organisational learning mechanisms that can turn into innovations that strengthen organisational systems that respond to a complex socio-economic landscape. Knowledge management processes can drive workplace protection, ensuring employees are secure and engaged in finding new solutions. Furthermore, knowledge management processes play a key role in developing supply chains' resilience to crises and in reorganising operations. As such, knowledge management processes can provide valuable insights to help leaders and managers equip the organisation with capability to respond effectively to changes in a coronacrisis situation. Moreover, a knowledge management process perspective can be helpful in moving beyond the current crisis and in designing resilient organisations prepared for future crises. One of the key management implications of the pandemic is that organisations have learned that conventional strategic thinking may not be helpful in planning for the future. Indeed, it is more effective to combine traditional business strategies with new thinking. For this reason, leaders need new skills and capabilities to successfully lead in the post-pandemic era.

The skills and abilities that allow for efficient functioning in times of uncertainty caused by a crisis are highlighted by Leinwand, Mani and Sheppard [18]. Researchers after conducting a study among companies such as Microsoft, Cleveland Clinic and Philips, defined the skills and abilities that are useful for success under uncertainty. The findings show that leaders in these companies sought to be proficient in a wide range of traits and did not rely solely on their strengths. They learned to collaborate with others of different backgrounds and mindsets, and they emphasised working together to lead the company despite all differences. Leaders operating in a complex and transforming business environment need to be good strategists, with a clear vision of where their company will be when the crisis is over. Leaders must also have the ability to execute strategy, be accountable for the company's transformation, translate strategy into concrete execution steps and see them through to completion, and make quick operational decisions. The digital age requires leaders to not only make bold decisions in times of uncertainty, but also to try to admit what they do not know and hire people with potentially very different skills, experience, and capabilities. Leaders need to be open to new technologies and new ways of doing things that differ from existing methods.

Another group of competencies in the field of knowledge management in the era of the coronavirus crisis was presented by Ahmad and van Looy [1], who wrote about the challenges of business process management (BPM) in the context of new technologies that change the way people work in organisations. Although digital innovations can increase process efficiency and firm productivity, employees do not always accept the associated changes in work. The authors therefore examined the increased digitisation efforts during the Covid-19 pandemic, during which employees were forced to radically rethink their previous way of working, relying heavily on technology for communication and other business tasks. The subject of this article was the changes in ways of working that were caused by the pandemic and how employees coped with them, as well as the exploratory skill sets needed to adapt to the changes. For the purpose of the study, the authors identified four groups of business process management (BPM) skills from the literature, such as sense of creativity, sense of opportunism, sense of flexibility and sense of adaptation to change.

Researchers pay attention not only to the positive implications of knowledge management. The risk associated with knowledge management is also becoming an important issue in the face of the coronavirus. Durst and Zięba [14] defined "knowledge risk" as a measure of the probability and severity of the adverse effects of any activities involving knowledge or related in some way with knowledge, which may affect the functioning of the organization at any level. Organizations are exposed to the risk of knowledge and its possible consequences under conditions of an external and dynamic crisis such as the COVID-19 crisis. The rise in home workers also increases the risk of cybercrime due to increased use of technology. The risk is related, for example, to the employees' lack of sufficient knowledge about cybersecurity and the use of devices that are not properly secured. The current COVID-19 situation has increased some of these threats in companies.

The research undertaken in the above-mentioned publications allow to conclude that the competences in the field of knowledge management will gain importance and will influence the tendency to adopt process orientation.

Shujahat et al. when defining knowledge management as a discipline and function in which knowledge is created, acquired, shared, coded and used through an enabling environment to increase innovation and organizational performance, highlight two main components of knowledge management: the knowledge management environment and the knowledge management processes. They emphasize that the most important component of knowledge management is the knowledge management process, which creates knowledge and can exist independently of the formal organizational support for knowledge management [30]. Oliva and Kotabe note that good knowledge management in organizations depends on good practices that go through basic processes [24]. Archer-Brown and Kietzmann also show the process nature of activities related to knowledge management, arguing that knowledge should be perceived both as 'stock' and 'flow', because of its dynamic nature and how it is generated, transferred and improved [2].

Discussing the benefits of using the knowledge management process, Shahzad et al. suggest that knowledge management enables organizations to respond to changes, improves the sustainability of operations and a competitive advantage, which provides credibility for shareholders and influences customer confidence [28]. Ode and Ayavoo assume that the effective application of knowledge reduces costs and increases the efficiency of the organization [23], while Shujahat et al. show the relationship between the knowledge management process and the innovation and productivity of employees [30].

Sarka et al. show the need to study knowledge management in different needs and expectations of the generational cohorts [27]. In particular, important seem to be the not fully recognized needs and expectations of the Z generation, currently entering the labour market, who were born and brought up in specific conditions. Findings Barhate and Dirani suggests that easy access to technology enables Gen Z to learn [3]. On the other hand, Dolot, pointing to the importance of feedback on the results of work, emphasizes the importance of mobility and the opening of generation Z to new technologies, which are its natural environment [13]. Similarly, Dogra and Kaushal note that the latest phase of dynamic technology influenced Generation Z, which acquired new skills by showing an interest in gaining knowledge due to social and technological changes [12]. According to Jayathilake et al. Z-generation workers and their way of thinking and working are significantly different from previous generations. They are characterized by a much lower level of loyalty to employers, although their knowledge and skills contribute to the achievement of organizational goals. In this context, Jayathilake et al. draw attention to the benefits of using knowledge about digital technology and its distribution through reverse mentoring, influencing the improvement of social relations in the enterprise [16]. Bencsik et al. suggest that generation Z has features important from the perspective of knowledge management that distinguish this generation from the previous ones [4].

Qandah et al. suggest that gaining a competitive advantage requires organizations to identify what kind of knowledge is needed, how to acquire it and apply it effectively and efficiently, which influences the importance of knowledge management ability [25]. Similarly, Desouza and Awazu point out that effective knowledge management in a competitive business environment requires an organization to have the skills to create, transmit, store, retrieve, apply, segment and destroy knowledge [11]. Pointing to the process-based nature of knowledge management, Mota Veiga et al. emphasize the importance of skills related to the creation, transfer, integration and application of knowledge

[20]. In attempting to understand dynamic capabilities through knowledge management, Nielsen defines three capabilities critical to knowledge management: knowledge development, knowledge (re)combination, and knowledge use [21]. Conchado et al. indicate generic competencies for knowledge management acquired through higher education: analytical thinking, ability to rapidly acquire new knowledge, mastery of own field or discipline, knowledge of other fields or disciplines [10].

## 3  Empirical Analysis

### 3.1  Data Collection, Variables and Method

The data presented in this paper were got during the second stage of an empirical study carried out as part of a research project entitled 'Competencies for process improvement with the use of ICT tools', funded by NCBiR (the National Centre for Research and Development in Poland). The main goal of the research proceedings was to assess the influence of competencies for knowledge management on the management processes in the post-covid economy, from the perspective of students representing Generation Z and entering the labour market. In particular, the study attempts to answer the question of whether the increase in the importance of competencies for knowledge management predicted in the aftermath of the Covid-19 crisis will impact the change of individual components of process orientation.

In the study we adopted the measurement scales from previous studies for all the constructs of our model. Each part of the questionnaire is described below.

Competencies for knowledge management were assessed by participants using a 4-item scale developed by Conchado et al. [10]. Sample items are 'As a result of the Covid-19 crisis, the importance of ability to rapidly acquire new knowledge will increase', 'As a result of the Covid-19 crisis, the importance of analytical thinking will increase'. The scale ranges from $1 = $ Very Strongly Disagree to $7 = $ Very Strongly Agree. All the measurement items in the construct proposed by Conchado et al. are listed in Table 1.

**Table 1.** Components of competencies for knowledge management.

| Symbol | Item name |
| --- | --- |
| $x_1$ | Ability to rapidly acquire new knowledge |
| $x_2$ | Analytical thinking |
| $x_3$ | Knowledge of other fields or disciplines |
| $x_4$ | Mastery of your own fields or disciplines |

Source: (Conchado et al., 2015)

Process orientation was assessed by participants using a 6-item scale developed by Chen et al. [8]. Sample items are 'After the Covid-19 crisis is over, companies should define their business processes more clearly so that most employees have a clear understanding of these processes', 'After the Covid-19 crisis is over, companies should allocate resources to a greater extent based on business processes'. The scale ranges from 1 = Very Strongly Disagree to 7 = Very Strongly Agree. All the measurement items in the construct proposed by Chen et al. are listed in Table 2.

**Table 2.** Components of process orientation.

| Symbol | Item name |
| --- | --- |
| $y_1$ | In our firm, business processes are sufficiently defined so that most employees have a clear understanding of these processes |
| $y_2$ | Our firm allocates resources based on business processes |
| $y_3$ | Our firm sets specific performance goals for different business processes |
| $y_4$ | Our firm measures the outcome of different business processes |
| $y_5$ | Our firm clearly designates process owners to assume responsibilities |
| $y_6$ | Employees are rewarded based on the performance of business processes in which they are involved |

Source: (Chen et al. 2009)

In order to obtain data, we used the method of an internet survey. The original survey questionnaire was addressed to students of the Nicolaus Copernicus University in Toruń which has the status of a research university and the Poznań University of Economics. Both universities in the field of education at the faculties of economics are among the best universities in Poland. In order to make sure that the respondents are representatives of the Z generation, the questionnaire was asked about their age, assuming the year 1997 as a border for generations Y and Z. If the respondent indicated that he was born before 1997, he was not subjected to further research. The respondents were bachelor's students in the 3rd year and master's studies in the 1st and 2nd years and in their curriculum, they already had lectures on knowledge management and process management, which allowed them to properly assess studied phenomena. Knowing the opinion of the Z generation, which is just entering the labor market, is important because although this generation is not the dominant cohort yet, their beliefs and views will soon change the labor market. What's more, their current theoretical knowledge gained during their studies, in many cases supported by pre-predictive experience, allows them to formulate accurate opinions that are worth listening to.

The study was conducted between January and March 2021. The online questionnaire was correctly and fully completed in terms of the diagnosed items by 711 students. Their characteristics are presented in Table 3.

Table 4 presents the description and scale of all analysed variables.

## 4   Results

In order to test the hypothesis that the increase in the importance of competencies for knowledge management (x), which changed in the aftermath of the Covid-19 crisis, influence the change of the individual process orientation components ($y1,\ldots, y6$), structural

**Table 3.** Structure of the sample.

| Characteristics | % in sample |
|---|---|
| *Form of studies* | |
| Stationary | 77.8 |
| Extramural | 22.2 |
| *Programme of studies* | |
| Finance and Accounting | 35.6 |
| Management | 39.2 |
| Economics | 20.0 |
| Social Policy | 5.2 |
| *Gender* | |
| Female | 57.23 |
| Male | 42.77 |
| *Intention to set up own business in the near future* | |
| Students planning to start their own business | 53.0 |
| Students not planning to start their own business | 47.0 |

equation models were estimated. The aim of the research procedure was to identify the influence of competencies for knowledge management on each analysed component of process orientation separately. Therefore, six SEM models were developed. As an explanation, each model takes into account the factor X, which is a value formed by the four variables presented in Tables 1 and 4.

The factor X was the latent variable in all models, therefore its reliability was tested. The Cronbach's alpha statistic for this factor was 0.705 [7], which means it is acceptable, and for all measures x1–x4 describes the same construct. Furthermore, whether the respondent is planning or is not planning to start his own company in the near future was considered as a moderator variable in each model. The estimated models, which are the basis for the verification of hypotheses, are presented in Scheme 1.

The models were estimated based on 711 observations in SPSS AMOS v.16 using a maximum likelihood method. The significant level was set at 0.05. In the first step the models for all dataset were estimated. In the second step the models for two groups distinguished based on whether students plan to start their own business in the near future or not were calculated and compared. The results obtained for base models are presented in Table 5. Table 6 also includes the results of the external model (confirmatory factor analysist), which is a common part for all six models.

All estimated models have a very good values for fit measures statistics – IFI > 0.9 and RMSEA < 0.8 [5]. The influence of the factor X for the variables Y1, Y2, Y3, Y4

**Table 4.** Description of variables.

| Description | Label | Type |
|---|---|---|
| *Explanatory variables* | | |
| Knowledge management | $x$ $(x_1,..., x_4)$ | |
| -As a result of the Covid-19 crisis, the importance of ability to rapidly acquire new knowledge will increase | $x_1$ | Ordinal (1–7) |
| -As a result of the Covid-19 crisis, the importance of analytical thinking will increase | $x_2$ | Ordinal (1–7) |
| -As a result of the Covid-19 crisis, the importance of knowledge of other fields or disciplines will increase | $x_3$ | Ordinal (1–7) |
| -As a result of the Covid-19 crisis, the importance of mastery of your own fields or disciplines will increase | $x_4$ | Ordinal (1–7) |
| *Explained variables* | | |
| After the Covid-19 crisis is over, companies should define their business processes more clearly so that most employees have a clear understanding of these processes | $y_1$ | Ordinal (1–7) |
| After the Covid-19 crisis is over, companies should allocate resources to a greater extent based on business processes | $y_2$ | Ordinal (1–7) |
| After the Covid-19 crisis is over, companies should be more likely to set specific performance targets for different business processes | $y_3$ | Ordinal (1–7) |
| Following the end of the Covid-19 crisis, companies should measure the outcome of different business processes more closely | $y_4$ | Ordinal (1–7) |
| After the Covid-19 crisis is over, companies should clearly designate process owners to take responsibility | $y_5$ | Ordinal (1–7) |
| After the Covid-19 crisis is over, employees should be rewarded more based on the performance of the business processes in which they are involved | $y_6$ | Ordinal (1–7) |
| *Moderator variable* | | |
| Intention to set up own business in the near future | X95 | Dichotomous |

and Y5 are statistical significant. It means that the competencies for knowledge management (being a component of the ability to rapidly acquire new knowledge, analytical thinking, knowledge of other fields or disciplines and mastery of your own fields or disciplines) after the Covid-19 crisis are over will be affect on greater tendency to use process orientation, i.e., Y1 – defining their business processes more clearly so that most employees have a clear understanding of these processes, Y2 – allocating resources to a greater extent based on business processes, Y3 – broader setting of specific performance targets for different business processes, Y4 – measuring the outcome of different business processes more closely, Y5 – more unambiguous designation of process owners.

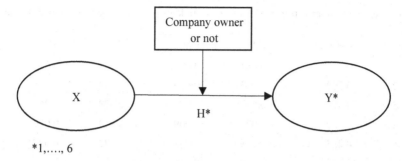

**Scheme 1.** Estimated SEM models.

**Table 5.** Results obtained from the base SEM models.

| Model | Parameter | C.R | P Value | IFI | RMSEA |
|-------|-----------|------|---------|-----|-------|
| X->Y1 | 0.329 | 0.056 | 0.000 | 0.961 | 0.060 |
| X->Y2 | 0.235 | 0.053 | 0.000 | 0.984 | 0.037 |
| X->Y3 | 0.463 | 0.173 | 0.000 | 0.982 | 0.044 |
| X->Y4 | 0.200 | 0.054 | 0.000 | 0.912 | 0.079 |
| X->Y5 | 0.174 | 0.058 | 0.000 | 0.931 | 0.077 |
| X->Y6 | 0.085 | 0.052 | 0.078 | 0.975 | 0.045 |

**Table 6.** External SEM model (Factor analysis) – common for every six model.

| Relation | Parametr | S.R | P value |
|----------|----------|------|---------|
| X1->X | 0.485 | 0.051 | |
| X2->X | 0.585 | 0.054 | 0.000 |
| X3->X | 0.580 | 0.053 | 0.000 |
| X4->X | 0.469 | 0.051 | 0.000 |

Only the impact of competencies for knowledge management (X) on the need to reward employees based on the results of the business processes in which they are involved (Y6) was insignificant. All data loadings in confirmatory factor analysis were also statistical significant (Table 6).

Table 7 contains the results obtained in the group of students planning to start their own business. Table 8, in turn, shows the group of respondents who are not planning to start their own business.

All estimated models have very good values for fit measures statistics – IFI > 0.9 and RMSEA < 0.8. It can be noted that in the second group of respondents (Table 8), the impact of factor X on Y5 is also statistically insignificant. Furthermore, in the group of students planning to start their own business in the near future, the impact of the variable

**Table 7.** Results obtained from SEM models – students planning to start their own business.

| Model | Parameter | C.R | P Value | IFI | RMSEA |
|-------|-----------|-----|---------|-----|-------|
| X->Y1 | 0.283 | 0.081 | 0.000 | 0.937 | 0.076 |
| X->Y2 | **0.298** | 0.065 | 0.000 | 0.969 | 0.053 |
| X->Y3 | **0.527** | 0.062 | 0.000 | 0.978 | 0,050 |
| X->Y4 | 0.189 | 0.073 | 0.009 | 0.918 | 0.079 |
| X->Y5 | **0.223** | 0.079 | 0.002 | 0.932 | 0.077 |
| X->Y6 | 0.067 | 0.068 | 0.308 | 0.966 | 0.052 |

**Table 8.** Results obtained from SEM models – students who are not planning to start their own business.

| Model | Parameter | C.R | P Value | IFI | RMSEA |
|-------|-----------|-----|---------|-----|-------|
| X->Y1 | **0.373** | 0.081 | 0.000 | 0.995 | 0.023 |
| X->Y2 | 0.164 | 0.079 | 0.021 | 0.997 | 0.022 |
| X->Y3 | 0.378 | 0.079 | 0.000 | 0.992 | 0.028 |
| X->Y4 | **0.224** | 0.079 | 0.002 | 0.924 | 0.076 |
| X->Y5 | 0.116 | 0.076 | 0.097 | 0.940 | 0.073 |
| X->Y6 | 0.107 | 0.074 | 0.125 | 0.997 | 0.022 |

X seems to be stronger on the need to allocate resources based on business processes (Y2) and the tendency to set specific performance goals for various business processes (Y3) than in the second group of the respondents. On the other hand, however, the influence of the change in competencies for knowledge management anticipated as a result of the Covid-19 crisis on defining your processes more clearly (Y1) and on measuring results of different business processes more accurately (Y4) is more meaningful in the group of respondents who are not planning to start their own business.

## 5 Discussion

The article poses a general hypothesis, according to which the increase in the importance of competencies for knowledge management changing under the influence of coronacrisis will influence the change of individual components of process orientation. As process orientation was defined by six variables, six specific hypotheses were verified, stating the influence of competencies for knowledge management on each of its components separately i.e.:

H1: The increase in the importance of competencies for knowledge management changing under the influence of the Covid-19 crisis will trigger the need to define business processes more clearly so that most employees have a clear understanding of these processes.

H2: The increase in the importance of competencies for knowledge management changing under the influence of the Covid-19 crisis will trigger the need to allocate resources to a greater extent based on business processes.

H3: The increase in the importance of competencies for knowledge management changing under the influence of the Covid-19 crisis will trigger the need to set specific performance targets for different business processes.

H4: The increase in the importance of competencies for knowledge management changing under the influence of the Covid-19 crisis will trigger the need to measure the outcome of different business processes more closely.

H5: The increase in the importance of competencies for knowledge management changing under the influence of the Covid-19 crisis will trigger the need to clearly designate process owners to take responsibility.

H6: The increase in the importance of competencies for knowledge management changing under the influence of the Covid-19 crisis will trigger the need to reward employees more based on the performance of the business processes in which they are involved.

Diagnosis in this respect was made from the perspective of Generation Z students entering into the labour market.

The results show the increase in the importance of competencies for knowledge management following the Covid-19 crisis will affect the change of the five of six process orientation components. Therefore, companies should clearly strengthen their process orientation under the new post-crisis operating conditions and in the face of supplying the labour market with representatives of generation Z. Only in the case of rewarding employees based on the implementation of business processes in which they are involved, students from Generation Z do not see the impact of the increasing importance of competences for knowledge management. The lack of this relation intrigues and induces to deepen the research in order to justify such a state of affairs, since it influences the effective motivation of employees representing this generation. Perhaps young people expect additional remuneration for the very development of their competences, and not only for the effects of their application.

We observed the strongest statistically dependencies for the relationship, an increase in the importance of competencies in knowledge management on the change in setting specific performance goals for different business processes and on the change sufficiently defining business processes, so that most employees have a clear understanding of these processes.

Our findings also suggest differences in the impact of an increase in the importance of competencies for knowledge management on individual the process orientation components in the case of students who plan to set up their own business in the near future and students who do not plan to do so. Students planning to run their own business see stronger dependencies than students not planning to, in terms of the impact of competencies for knowledge management on allocating resources based on business processes (Y2), setting specific performance goals for different business processes (Y3) and clearly designating process owners to assume responsibilities (Y5), students not planning to run their own business see stronger dependencies than students planning to run their own business in relation to the impact of competencies for knowledge on sufficiently defining business processes, so that most employees have a clear understanding of these processes (Y1), measuring the outcome of different business processes (Y4). The differences between the two student groups identified in the study outline a clearly emerging approach to the impact of competencies for knowledge management on individual areas of process orientation. Future companies owners, most likely because of their business approach, recognize the importance of competencies for knowledge management for 'soft' elements of the process approach (allocation of resources, goals performance setting, process owners designating), which can be considered more advanced activities in construction a process-oriented organization. On the other hand, students not planning to run their own business see the benefits associated with the impact of competencies for knowledge management on 'hard' elements (processes defining, business processes outcome measuring), which can be considered a starting point in building a process-oriented organization. What from a slightly different perspective is also visible in the studies cited by Clauss et al. [9], who noted that rapid change in existing business models should be accompanied by consideration of the limited resources available to entrepreneurs.

At this point, it is worth referring to research results Leinwand, Mani and Sheppard [18] arguing that the competencies they define as important for success in times of uncertainty caused by the Covid-19 pandemic, i.e., broad-based proficiency (i.e., knowledge of other fields and disciplines in addition to proficiency in one's own fields and disciplines), the ability to function in a rapidly changing business environment (ability to acquire knowledge quickly), including new technologies, as well as awareness of the preferences of individual customers, local communities and ecosystems (ability to think analytically) will impact intensifying the process orientation of companies.

Findings van Looy [31], suggest, that under the influence of the pandemic, organisations have adopted alternative ways of operating, and their processes have become more digitised, simplified and more efficient, which means that defining them well makes them more understandable for employees, and companies are more willing to set performance targets for various business processes. Moreover, maintaining a balance between exploitation (continuous improvement of business processes) and exploration (which can be understood as acquiring knowledge from other domains, adapting to changes, new business processes), leads to increased business efficiency.

# 6 Conclusion

In this article, we have assessed the influence of competencies for knowledge management importance on the change of process orientation in the post-covid economy, from the perspective of Generation Z students entering the labour market. Based on a discussion in knowledge management and process orientation, we proposed six hypotheses that the increase of importance of competencies for knowledge management, because of the Covid-19 crisis, will have affect the change six extracted of the process orientation components. Using Structural Equation Modelling, we identify the relationships between competencies for knowledge management and process orientation from the perspective of Generation Z students entering the labour market.

The results support the argument for the relationship between competencies for knowledge management and process orientation, and suggest that from the perspective of Generation Z students the increasing importance of competencies for knowledge management will not only translate into one area of the process orientation (rewarding employees based on the performance of the business processes in which they are involved). The above observation can be considered as a premise for strengthening in the curricula both issues related to process orientation and the development of competencies for knowledge management, which, as research has shown, impact process orientation. It is also worth conducting in the future research to answer the question why Generation Z students do not see seemingly obvious relationships between competencies for knowledge management and rewarding employees based on the performance of the business processes in which they are involved.

The results of our research have theoretical implications. This study enriches process management literature by demonstrating that competencies for knowledge management are important variables influencing process orientation. The study also helps to better understand the attitudes of Generation Z, in this case pointing to their views on changes in the process approach in knowledge management competences.

Also, the research results have some practical implications, which are significant from the perspective of the development of the process orientation. Our findings show that from the perspective of Generation Z, which is just entering the labour market, competencies for knowledge management have no effect on rewarding employees based on the performance of the business processes in which they are involved. The above observation should induce managers to revise their views on the applied motivational stimuli, prompting employees from Generation Z to engage in activities aimed at building a process oriented organization.

This study has limitations which may give rise to possible future research. First, the study was conducted on a large group of students studying economics, and therefore with a satisfactory level of knowledge of both the process approach and knowledge management, which allowed them to complete the questionnaire correctly. However, it would be worth examining whether their views on influence of competencies for knowledge management on the management process in the post-covid economy differ from students of other faculties, as well as from other Generations (in particular, Generation Y). Second, in the survey we used existing measurement scales, however we are aware that the list of measurements developed for process orientation and competencies for knowledge management may not be exhaustive or complete from the perspective Z Generation. Therefore, in future research, it would be worth expanding the variables studied, referring them to the specificity of Generation Z.

# References

1. Ahmad, T., Van Looy, A.: Development and testing of an explorative BPM acceptance model: insights from the COVID-19 pandemic. PLoS ONE **16**(11) (2021). https://doi.org/10.1371/journal.pone.0259226
2. Archer-Brown, C., Kietzmann, J.: Strategic knowledge management and enterprise social media. J. Knowl. Manag. **22**(6), 1288–1309 (2018)
3. https://doi.org/10.1108/JKM-08-2017-0359/FULL/PDF

4. Barhate, B., Dirani, K.M.: Career aspirations of generation Z: a systematic literature review. Eur. J. Train. Develop. **46**(1–2), 139–157 (2022). https://doi.org/10.1108/EJTD-07-2020-0124/FULL/PDF

5. Bencsik, A., Molnar, P., Juhasz, T., Machova, R.: Relationship between knowledge sharing willingness and life goals of generation Z. In: Proceedings of the European Conference on Knowledge Management, vol. 1, pp. 84–94. ECKM, Padua (2018)

6. Bollen, K.A.: Structural Equations with Latent Variables. Wiley, New York (1989). https://doi.org/10.1002/9781118619179

7. Carroll, A.B.: The pyramid of corporate social responsibility: toward the moral management of organizational stakeholders. Bus. Horiz **34**, 39–48 (1991). https://doi.org/10.1016/0007-6813(91)90005-G

8. Cortina, J.M.: What is coefficient alpha? An examination of theory and applications. J. Appl. Psychol. **78**(1), 98–104 (1993). https://doi.org/10.1037/0021-9010.78.1.98

9. Chen, H., Tian, Y., Daugherty, P.J.: Measuring process orientation. Int. J. Logist. Manage. **20**(2), 213–227 (2009). https://doi.org/10.1108/09574090910981305

10. Clauss, T., Breier, M., Kraus, S., Durst, S., Mahto, R.V.: Temporary business model innovation – SMEs' innovation response to the covid-19 crisis. R&D Manage. (2021). https://doi.org/10.1111/radm.12498

11. Conchado, A., Carot, J.M., Bas, M.C.: Competencies for knowledge management: development and validation of a scale. J. Knowl. Manag. **19**(4), 836–855 (2015). https://doi.org/10.1108/JKM-10-2014-0447/FULL/PDF

12. Desouza, K.C., Awazu, Y.: Segment and destroy: the missing capabilities of knowledge management. J. Bus. Strateg. **26**(4), 46–52 (2005). https://doi.org/10.1108/02756660510608558/FULL/PDF

13. Dogra, P., Kaushal, A.: An investigation of Indian generation Z Adoption of the voice-based assistants. VBA **27**(5), 673–696 (2021). https://doi.org/10.1080/10496491.2021.1880519

14. Dolot, A.: The characteristics of Generation Z. E-Mentor **74**, 44–50 (2018). https://doi.org/10.15219/EM74.1351

15. Durst, S., Zieba, M., Gonsiorowska, M.: Knowledge Risks in the COVID-19 Pandemic. IFKAD 2021: Managing Knowledge in Uncertain Times, Rome, Italy 1–3 Sept 2021. https://www.researchgate.net/publication/354695051_Knowledge_Risks_in_the_COVID-19_Pandemic

16. Gaviria-Marin, M., Merigó, J.M., Baier-Fuentes, H.: Knowledge management: a global examination based on bibliometric analysis. Technol. Forecast. Soc. Chang. **140**, 194–220 (2019). https://doi.org/10.1016/J.TECHFORE.2018.07.006

17. Jayathilake, H.D., Daud, D., Eaw, H.C., Annuar, N.: Employee development and retention of Generation-Z employees in the post-COVID-19 workplace: a conceptual framework. Benchmarking **28**(7), 2343–2364 (2021). https://doi.org/10.1108/BIJ-06-2020-0311/FULL/PDF

18. Lanier, K.: 5 things HR professionals need to know about Generation Z. Strateg. HR Rev. **16**(6), 288–290 (2017). https://doi.org/10.1108/shr-08-2017-0051

19. Leinwand, P., Mani, M.M., Sheppard, B.: 6 Leadership Paradoxes for the Post-Pandemic Era. Harvard Business Review (2021). https://hbr.org/2021/04/6-leadership-paradoxes-for-the-post-pandemic-era

20. Liu, C.J., Wang, W.: Knowledge creation in pandemic time: challenges and opportunities. Issues Inform. Syst. **22**(1), 306–319 (2021). https://doi.org/10.48009/1_iis_2021_306-319

21. Mota Veiga, P., Fernandes, C., Ambrósio, F.: Knowledge spillover, knowledge management and innovation of the Portuguese hotel industry in times of crisis. Journal of Hospitality and Tourism Insights. (ahead-of-print) (2022). https://doi.org/10.1108/JHTI-08-2021-0222/FULL/PDF

22. Nielsen, A.P.: Understanding dynamic capabilities through knowledge management. J. Knowl. Manag. **10**(4), 59–71 (2006). https://doi.org/10.1108/13673270610679363/FULL/PDF

23. Nonaka, I., Takeuchi, H.: The Knowledge-creating Company: How Japanese Companies Create the Dynamics of Innovation. Oxford University Press, New York (1995)

24. Ode, E., Ayavoo, R.: The mediating role of knowledge application in the relationship between knowledge management practices and firm innovation. J. Innov. Knowl. **5**(3), 210–218 (2020). https://doi.org/10.1016/J.JIK.2019.08.002

25. Oliva, F.L., Kotabe, M.: Barriers, practices, methods and knowledge management tools in startups. J. Knowl. Manag. **23**(9), 1838–1856 (2019). https://doi.org/10.1108/JKM-06-2018-0361/FULL/PDF

26. Qandah, R., Suifan, T.S., Masa'deh, R., Obeidat, B.Y.: The impact of knowledge management capabilities on innovation in entrepreneurial companies in Jordan. Int. J. Organ. Anal. **29**(4), 989–1014 (2020). https://doi.org/10.1108/IJOA-06-2020-2246/FULL/PDF

27. Ritter, T., Pedersen, C.L.: Analyzing the impact of the coronavirus crisis on business models. Ind. Mark. Manage. **88**, 214–224 (2020). https://doi.org/10.1016/j.indmarman.2020.05.014

28. Sarka, P., Heisig, P., Caldwell, N.H.M., Maier, A.M., Ipsen, C.: Future research on information technology in knowledge management. Knowl. Process. Manag. **26**(3), 277–296 (2019). https://doi.org/10.1002/KPM.1601

29. Schiuma, G., Jackson, T., Lönnqvist, A.: Managing knowledge to navigate the coronavirus crisis. Knowl. Manag. Res. Pract. **19**(4), 409–414 (2021). https://doi.org/10.1080/14778238.2021.1992711

30. Shahzad, M., Qu, Y., Zafar, A.U., Rehman, S.U., Islam, T.: Exploring the influence of knowledge management process on corporate sustainable performance through green innovation. J. Knowl. Manag. **24**(9), 2079–2106 (2020). https://doi.org/10.1108/JKM-11-2019-0624/FULL/PDF

31. Shujahat, M., Sousa, M.J., Hussain, S., Nawaz, F., Wang, M., Umer, M.: Translating the impact of knowledge management processes into knowledge-based innovation: the neglected and mediating role of knowledge-worker productivity. J. Bus. Res. **94**, 442–450 (2019). https://doi.org/10.1016/J.JBUSRES.2017.11.001

32. Van Looy, A.: How the COVID-19 pandemic can stimulate more radical business process improvements: using the metaphor of a tree. Knowl. Process Manage. **28**, 107–116 (2021). https://doi.org/10.1002/kpm.1659

# The Future Development of ERP: Towards Process ERP Systems?

Marek Szelągowski[1] ⓘ, Justyna Berniak-Woźny[2](✉) ⓘ, and Audrone Lupeikiene[3] ⓘ

[1] Systems Research Institute, Polish Academy of Sciences, Newelska 6, 01-447 Warsaw, Poland
`marek.szelagowski@dbpm.pl`
[2] University of Information Technology and Management, Sucharskiego 2, 35-225 Rzeszów, Poland
`jberniak@wsiz.edu.pl`
[3] Institute of Data Science and Digital Technologies, Vilnius University, Akademijos 4, 08412 Vilnius, Lithuania
`audrone.lupeikiene@mif.vu.lt`

**Abstract.** Organizations operating in Industry 4.0 and 5.0 use both ERP and BPMS systems. As recently as 10–15 years ago, the reasons behind using these two classes of systems were different. ERPs were used to manage the organization's resources, and BPMS – to support the implementation of business processes, often understood as work or document flows. However, as a result of digital transformation, both business needs as well as ERP and BPMS vendors responding thereto made these two classes of systems overlap to an increasing degree. Thus, the aim of this article is to answer the question: Are we heading towards process-based ERP systems or is the future in the flexible, open integration of postmodernERP and iBPMS? The authors conducted a narrative literature review and content analysis of 88 ERP systems offered on the Polish market. As a result, 11 ERP systems containing functionalities specific to BPMS were identified. Further, to define the essence of the transformation of ERP into process-based ERP systems, 5 expert interviews were conducted, which allowed for the formulation of two approaches to this transformation: the integration of ERP systems with iBPMS as an external subsystem taking over the implementation of selected business processes based on metadata and data of the ERP system; or process management within the ERP system by enabling the configuration of selected processes in ERP subsystems or modules based on a repository of process models, e.g. in BPMN.

**Keywords:** Enterprise Resource Planning (ERP) · postmodernERP · Business Process Management System (BPMS) · intelligent BPMS (iBPMS)

## 1 Introduction

For almost 30 years, Enterprise Resource Planning (ERP) systems were considered to be the main systems supporting management in organizations [1]. However, the increasingly broad use of process-based methodologies and hyperautomation techniques in management forces organizations to also use Business Process Management Systems

© Springer Nature Switzerland AG 2022
A. Marrella et al. (Eds.): BPM 2022, LNBIP 459, pp. 326–341, 2022.
https://doi.org/10.1007/978-3-031-16168-1_21

(BPMS). The vast majority of organizations that already use ERP systems have to decide whether or not and to what extent BPMS should be implemented or whether the ERP system should be changed to a process-based one. ERP vendors are faced with even more significant decisions. Should "process" functionalities be built into an existing ERP system with a view to preparing integration mechanisms enabling the on-demand addition of BPMS elements, including selected hyperautomation techniques such as process mining, robotic process automation (RPA), or artificial intelligence (AI)? Both for systems vendors and the users themselves, these are strategic decisions that are difficult to make, essential from the perspective of the competitive ability of the organization, and involve long-term significant human resources. Thus, the aim of this paper is to answer the research question: "Are we heading toward process-based ERP systems or is the future in the flexible, open integration of postmodernERP and iBPMS?".

The paper begins with the outline of the methodology. Parts 3 and 4 present the results of the literature review relating to the current status and development trends of ERP and BPMS. Part 5 compares the requirements, development drivers, and architectures of both system classes. Then, the results of the ERP systems analysis supplemented with the experts' interviews are presented and discussed. The last part presents the conclusions of the research.

## 2 Methodology

Studies on the research topic have been performed in three stages. The first step consisted of a narrative literature review held on the basis of the resources available in scientific databases, such as the repositories of SpringerLink, Emerald, ScienceDirect, Proquest, and Google Scholar. The main topics of interest were critical success factors, drivers of the evolution and architectures of ERP and BPMS. The summation of the literature research formed the basis of the next stage of the study. In the second stage, the authors analyzed online resources pertaining to the ERP systems available on the Polish market, with a focus on the possibilities of their use with a view to supporting business process management. The authors have based this stage on Qualitative Content Analysis – a research methodology of systematic analysis and interpretation of contents of texts – in this case, the ERP systems offered on the Polish market [2]. In the last stage of the study, the authors used partly structured expert questionnaires with representatives of 5 selected ERP system vendors. For each of the questionnaires, the same scenario was used, which allowed the authors to easily compare the results [3].

## 3 Enterprise Resources Planning Systems

Since the mid-1990s, ERP systems, which integrate support for various different areas of operation and business processes [1, 4], have become the standard regardless of the industry. The coherent combination of managing sales, production, human resources, and finances allowed for more efficient planning and monitoring of ongoing operations. However, since the mid-2000s, it has started to become increasingly clear that the monolithic architecture of ERP systems has certain limitations: it is unable to tailor the system's operations to the business processes of the organization, lacks the flexibility of process performance, lacks standard integration mechanisms with external systems and databases, and suffers from vendor lock-in, which often results in dismissing the needs of the users and high systems maintenance costs. Both the pressures of business and growing technological possibilities from the late 2000s onward have led to significant changes in the architecture of ERP systems and the emergence within the ERP system architecture of a module responsible for the integration of its various other modules, but also enabled the efficient integration of the system with external software and data sources [5]. In acknowledgment of the revolutionary nature of the introduced changes, in 2014 Gartner proposed the creation of a new class of "postmodernERP" systems, characterized by a modular internal architecture and the readiness to be integrated with external functionalities and modules [6]. The resulting composite IT architecture enables the users to quickly adapt or expand in accordance with the changing needs of business without being limited to the offer of a single vendor, a single software standard, or a single group of business processes.

The evolution of ERP systems began with inventory databases, which were later enriched with the planning and registry of operations (transactions) with the management of increasingly complex business processes [7]. In effect, even postmodernERPs remain transactional systems, that is, systems intended to register and monitor transactions instead of designing and executing end-to-end business processes. The support of business processes requires their strict integration with BPMS or the inclusion of process management tools within the architecture of the ERP system itself. In both cases, from the perspective of the user, this requires in the minimum the capability to design end-to-end business processes and to hold transactions configured in specific modules of the postmodernERP system from the level of the executed business processes. Vendors of ERP systems undertake to develop their offer in terms of embedding process management or enabling the strict integration of ERP with selected BPMS or Business Process Analysis (BPA) systems.

## 4  Business Process Management Systems

Business processes can be considered the arteries of modern organizations [8], as they represent the specific way in which organizational work is structured and executed, with a view to creating value and supporting business strategy implementation. BPMS, which combined information technology and knowledge in the field of management sciences and were applicable to operational business processes [9], support holistic management and increase the flexibility of implemented processes. They are defined as an application infrastructure supporting BPM projects and programs that support the entire process life cycle, from identification, through modeling, design, implementation and analysis, to continuous improvement [10]. BPMS allow organizations to increase the flexibility of business processes in a diverse application landscape [11]. However, due to the growing volume of data and the increasingly complex decision-making process resulting from the growing dynamics of the business ecosystem, BPMS have reached their limits. With the advent of Industry 4.0, traditional structured business processes have been largely replaced by dynamic ones: either partially structured or unstructured [12]. According to Olding and Rozwell [13], traditional, structured BPs encompass only about 30% of processes in organizations operating in Industry 4.0.

The answer to the changes taking place were intelligent BPMS (iBPMS) a type of high-performance (low-code/no-code) application development platforms that enable dynamic changes to operating models and procedures, documented as models, directly driving the execution of business operations [14]. Such platforms serve as a single tool allowing for the easy leverage of the analytics and intelligence of BPM through the use of the cloud, Internet of Things (IoT) integration, message-oriented middleware, business activity monitoring, the use of artificial intelligence (AI), and much more. In turn, business users make frequent (or ad hoc) process changes in their operations, regardless of technical resources managed by IT, such as integration with external systems and security administration. iBPMS also enable "citizen developers" and professional developers to collaborate to improve and transform business processes. They allow new, emergent practices to quickly scale across a function or enterprise. Although they take into account aspects of business transformation and digitization, changes in the requirements related to Industry 5.0 and the need for seamless collaboration between people and machines are driving the further evolution of BPM software [15]. The purpose of the changes is to provide a tool building a sustainable competitive position on the market.

## 5 Postmodern ERP and iBPMS – Differences and Similarities

### 5.1 Goals and CSFs for the Implementation of ERP and BPMS

In Industry 4.0 and 5.0, the measure of success of an organization is its ongoing efficiency and the development potential of its products and services, as well as the capability to use and develop its own intellectual capital [16].

In the literature from the last 10 years, there exist multiple publications on the requirements of Critical Success Factors (CSFs) for ERP systems and BPMS. All point to the fact that the success of implementing ERP or BPMS is dependent not on a single factor, but on the synergy of several or even several dozen CSFs [17–19]. They also lead to the observation that the goals and CSFs of the implementation of ERP and BPMS are if not identical, then at least largely overlapping and complementary. In Industry 4.0, it is expected that the implementation of ERP or BPMS will result in an increase in efficiency and effectiveness, as well as the flexibility of business processes [20]. However, from the point of view of Industry 5.0, in order to unleash the potential of both classes of systems and unlock the innovation of employees, the need to change the work culture and the empowerment of employees should also be taken into account.

### 5.2 Trends in the Development of Postmodern ERP and iBPM Systems

The results of the literature review highlight several concerns that determine the driving forces behind the development of ERP and iBPM systems. The most important among them are the needs of organisations operating in global, ever-changing business ecosystem of Industry 4.0/5.0, as well as technological possibilities available to system vendors. The system users' requirements resulting from nature of their work and social culture cannot be ignored either. The drivers pursued lead to appropriate trends in the development of ERP and iBPMS, which are characterised in more detail in terms of foreseen requirements having a key impact on the further development of systems [21, 22].

**(D1) The Constant Efforts of Enterprises to Improve Productivity and Efficiency**
The main driver of the practical use of both ERP and BPM systems is the pursuit of reducing costs and increasing the efficiency/productivity of the business [23]. On the one hand, 72% of organisations indicate that cost reduction is their goal of implementing and using a BPMS [24]. On the other hand, ERP systems vendors declare that organsations use their systems to integrate the management of business processes [25]. Beyond the 2000s, emphasis for both class of systems has shifted from supporting internal management to leveraging value in real time [1, 26]. Nowadays, production, provision of services, and decision-making are federated within and between different organisations and divisions. According to Bailey et al. [27], by 2026, more than 50% of large organisations will compete as collaborative digital ecosystems rather than discrete firms. Therefore, improving productivity and efficiency should be analysed not only in the local context, but also in the context of the global business ecosystem. This means that in order for ERP and BPMS to be useful for cross-functional integration and value creation, they must be implemented in a technological ecosystem that covers and integrates the entire business ecosystem.

The foreseen requirements having a key impact on the further development of systems are as follows: (1) The need to support a business in such a way that it could systematically explore new opportunities, adapt, and fundamentally transform itself; (2) The need to support processes of highest maturity levels and of different natures; (3) The need to enable the management of end-to-end processes covering networks of different types of organisational units; (4) The need to align business processes with a strategic level; (5) The need to ensure systems quality characteristics, such as interoperability, performance, and scalability; (6) The need to create preconditions for cooperation with other systems types to fully automate end-to-end processes.

**(D2) Abrupt Changes in Work and Social Culture**
The real enterprise environment is highly dynamic and deals with a large number of various exceptions. The COVID-19 pandemic has demonstrated the reality of unforeseen disruption. According to Chong et al. [28], organisations that are able to adapt to such challenges are resilient, and characteristics of resilience include the development of local networks of teams and business units. This driver clearly indicates the importance of tools for the real-time management and improvement of business processes. Such a situation significantly accelerated changes in the work culture and made it possible to implement new business models based on digitisation [29]. This in turn resulted in the necessity to maintain a permanently higher rate of adoption of remote work and digital touchpoints [30]. By necessity, in many organisations technology has become the key to every interaction [28]: The foreseen requirements having a key impact on the further development of systems are as follows: (1) The need to support a digital-first, remote-first business model; (2) The need to support decisions on business innovations, including new business models and agility; (3) The need to support different types of process variability, run-time process variability, and their management in real time;

(4) The need to enable holding business activities anywhere, exploiting the potential of mobile technologies.

### (D3) Technological Changes

ERP and BPMS vendors today have at their disposal opportunities provided by rapidly evolving and emerging new information technologies. These technologies originate from different fields, including cyber-physical systems, Internet of Things (IoT), cloud computing, hyperautomation, service-oriented paradigm, industrial information integration [31], to list just some of the more important ones. They are the major force behind a technological shift toward supporting new models of business.

The foreseen requirements having a key impact on the further development of systems are as follows: (1) The need for the orchestrated use of multiple technologies; (2) The need to extend the variety of supported technologies and simplify them to expand the scope of business automation; (3) The opportunity for on-demand access to required services (i.e., required infrastructure, platforms, software) and for building resilient, flexible, and agile application architectures thanks to the availability of cloud technology [32]; (4) The need to develop customer-facing systems by blurring business and technological aspects: (5) The opportunity to create flexible, adjustable, composable systems even faster thanks to the use of principles of service-oriented architecture; (6) The creation of preconditions to extend the digital workforce with smart things and cyber-physical systems.

### (D4) Rapidly Growing Data Streams and Data Heterogeneity

Some departments or even entire organisations (for example, insurance companies) have always been data driven. Nowadays, businesses make extensive use of data because of the potential they provide. This requires ensuring that large amounts of structured, semi structured, and unstructured data can be stored and processed, including in their native form. In the context of Industry 4.0/5.0, business data that flow through business processes and are exchanged among the different types of actors are highly heterogeneous. The steps of business processes are carried out not only by traditional workers, but by various internet-connected devices as well. In addition, data should be available as soon as they are created and acquired. According to Guay [33], without appropriate data management, the expected business value of postmodern ERP systems will not materialize. The same can be said for iBPMS. The foreseen requirements having a key impact on the further development of systems are as follows: (1) The need to enable collaboration between machinery and people in running the business activities by enabling data exchange; (2) The need to extend the system's data infrastructure to cover not only traditional data bases and warehouses, but also data lakes, repositories, mobile data bases, etc.; (3) The need to ensure data quality, integrity, and security; (4) The creation of preconditions for the real-time and embedded analytics; (5) The creation of preconditions for end-to-end processes and for overall business visibility; (6) The creation of preconditions for processes mining and optimization.

### (D5) Prerequisites for Increasing the Intelligence of ERP and iBPM Systems

AI-enabled solutions are implemented in different fields, changing the work of entire organisations and their employees. The research indicates that the development of AI has made it possible to automate complex business process that until then could be executed only by humans. Advancements in machine learning, robotics, knowledge representation, automated reasoning and data analysis, planning and scheduling, computer vision, and natural language processing make the prerequisites for extensive hyperautomation, which is among the most important strategic technology trends [30, 32]. Thus, ERP and iBPM systems will be increasingly extensively rely on AI-based solutions combined with the digital workforce to improve business efficiency and workflow. The foreseen requirements having a key impact on the further development of systems are as follows: (1) The need to automate an increasing number of processes and remove the need for human intervention; (2) The need to shift ERP and iBPMS workplaces to a heterogeneous workforce, where people, as well as robots and intelligent things interact with the system; (3) The opportunity to develop an AI-driven user experience providing the users with more useful content; (4) The opportunity to use automated reasoning and inferred data to interpret documents written in a natural language when replacing people in the performance of tasks; (5) The opportunity for advanced and extensive business analytics and for its automatisation.

### 5.3  Architectures of Postmodern ERP and iBPM Systems

The term ERP is a generalized and an abstract term, because the products of specific providers can differ in many particular aspects. Some ERPs support only some operational and financial processes. They vary in functionality, data representation schemes, operation modes, and in many other details. An iBPMS is a solution for management of structured and unstructured business processes. To highlight their distinctive features, it can be said that BPM systems help enterprises optimize, implement, and automate flows of business activity to achieve business goals. iBPMS go one step further, i.e. "i" refers to the intelligence and advanced capabilities of these systems. The common denominator of both class of systems is not limited to principal architectural solutions. In practice, iBPMS increasingly supplement or even overlap with the typical functionalities of postmodern ERP systems. Typically, postmodern systems are characterized as federated and loosely coupled when the functionality is sourced as cloud services or via business process outsourcers [1, 33]. All these features are quality characteristics (i.e. adaptability, scalability, integration feasibility to name just a few) and can be implemented via design approaches, system IS architecture styles, and design patterns.

### (A1) Functional Architecture of PostmodernERP Systems

PostmodernERP has taken shape through several stages of development. The system consists of many functional subsystems or modules that share a database. As a rule, every functional subsystem/module focuses on one business area, such as human resources, sales and distribution, procurement, asset management, manufacturing, finances, and planning. As the system evolved, additional capabilities were integrated. The extension of the system can be considered from two dimensions: (1) horizontal – where functionality is extended by adding domain-specific constituents, i.e. by integration with subsystems or modules of the same category; (2) vertical – where ERP evolves thanks to new technological capabilities, i.e. by adding functionality to enable advanced capabilities such as intelligent automation, advanced analytics, and real time activities.

The horizontal dimension can be adapted to the needs of the business through special-ized subsystems, such as supply chain management (SCM), supplier relationship man-agement (SRM), product lifecycle management (PLM), or business warehouse (BW), have been created to expand some ERP functions or to implement new functionality. As a result, the boundaries of ERP were rethought in two ways: (1) these subsystems, namely warehouse management (WM), SRM, and CRM, were in fact included in core ERP [34, 35], (CRM and SRM are the examples in Fig. 1); or (2) they were developed as independent subsystems or modules and could be integrated among themselves and with the core ERP system (PLM, BW and SCM are the examples in Fig. 1).

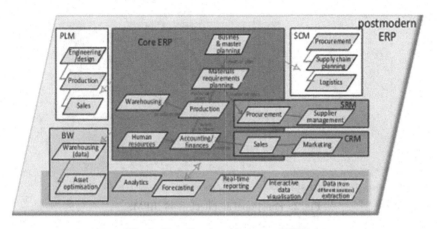

**Fig. 1.** Functional architecture of ERP.

This second option allows users to purchase and configure systems from modules that meet their needs. However, additional integrations increase the complexity of the system as a whole. As a rule, the core ERP serves as the central point of the integrated constituents. Considering the aspect of vertical extension, functional modules cover and extend the activities traditionally performed by people. Analytics comprises predictive, embedded, and real-time analytics in addition to classic data warehouse-based analysis. Some modules can be named as software agents, which perform tasks ranging from routine repetitive tasks to complex solutions.

The vertical dimensions of the expansion of ERP systems offers an increasing num-ber of new possibilities thanks to the fact that constantly emerging and improved new technologies allow for the automation of an increasing number of activities that were previously performed only by people. To ensure a truly live business, some functional modules, such as planning, procurement, or manufacturing [34, 36] must have their real-time execution counterparts, or, generally speaking, they should integrate the digital world. In other words, a module to process different types of data (e.g. unstructured, binary) from sensors, social networks, the IoT should appear on a vertical scale.

## (A2) Functional Architecture of Postmodern iBPM Systems

The different types of architectures for BPMS have received quite a lot of attention from, among others, Arsanjani et al. [37] and Pourmirza et al. [38]. The functional architecture of iBPMS, analogously to postmodernERP architecture, can be considered as an extension of its predecessor from horizontal and vertical points of view (Fig. 2).

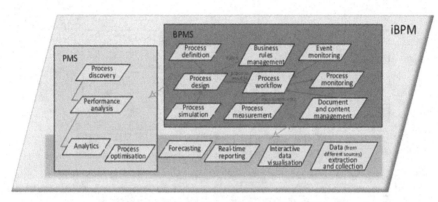

**Fig. 2.** Functional architecture of iBPMS.

The functional modules of an iBPMS support business process identification, engineering, execution, monitoring, and measurement. Process engineering includes process model development, optimizing, evaluating, and quality assurance. Multiple alternatives should be generated, studied, and analyzed in simulation and replaying on historical data studies in order to engineer the best possible business processes. In general, iBPMS takes manual processes and transforms them into digital processes that operate intra- and inter-enterprise systems. The business rules management module focuses on defining and storing rules which control business processes, while the content management module – on storing and securing documents, images, and other types of information entities. iBPMS extends the functionality of its predecessors by highly complex event monitoring and processing, increasing the ability of a business to identify opportunities or adapt to unexpected situations. In the context of integration with specialized systems, process mining systems (PMS) are worth mentioning. In addition to typical types of process mining [39], a PMS can be used to detect routine work in processes that can be automated [9].

In the vertical dimension, iBPMS implement end-to-end process automation via hyperautomation, including mimicking the behavior of workers. The process analytics functional module adds advanced predictive and real-time analytics, in which big data are used as well. Analytics also includes customer records on social networks, which enable both the definition and execution of more dynamic process discovery [40, 41]. Nowadays, iBPMS link workers, machines, and the IoT to ensure support for intra- and inter-processes [42]. One consequence of this is the creation of a functional module to process large amounts of different types of data in real time. In addition, the functional architecture of BPMS is extended by a real-time decision making module.

## 6 ERP Systems Evolution – Vendors Perspective

In order to confront the results of literature research with business practice, the authors have analyzed ERP systems offered by vendors in Poland. In total, 88 such systems were identified. Following the analysis of the content of the offers from the perspective of using the solutions to support business processes, 10 vendors were identified for a total of 11 ERP systems, which are undoubtedly already designed in accordance with the principles of composite architecture and which enable the use of business process models. These are: Infor (Infor LN), Sygnity Business Solutions (Quatra MAX), Oracle (Oracle e-Business Suite and Oracle ERP Cloud (Fusion)), SAP (SAP S/4HANA) Comarch (Comarch ERP Egeria), Soneta (enova365 platinum version), IFS (IFS CLOUD), BPSC (Impuls EVO), SIMPLE (SIMPLE ERP), and Gardens (GardensERP).

In the last stage of the study, the authors applied 2 step expert interviews. First, based on the literature review and content analysis results, the authors developed and administered partly structured expert questionnaires to representatives of 5 ERP systems vendors, who accepted invitations to participate in the study. The results of the questionnaires were presented in Table 1.

**Table 1.** Process-based functionalities of selected ERP systems.

| | System name | Infor LN | Oracle ERP Cloud (Fusion) | SAP S/4HANA | Comarch ERP Egeria | enova365 |
|---|---|---|---|---|---|---|
| 1 | Process modeling | Yes, own DEM notation | Yes, with Oracle BPM Cloud | Yes, with SAP Signavio | Yes, with Camunda | Yes, own workflow description notation |
| 2 | Importing process from Business Process Analysis (BPA) | Yes | Yes, with Oracle BPM Cloud | Yes with SAP Signavio | Yes, with Camunda | No |
| 3 | Process execution in accordance with predefined models (changes to the model change the means of execution) | Yes | Yes | Yes | Yes | Yes |
| 4 | Adding or ommiting tesks or subprocesses in the course of execution | Yes, modifications, versions | - | Yes | Yes | NO |
| 5 | Launching tasks in other systems in the process view | Yes | Yes | Yes | Yes | NO |
| 6 | Controlling of executed processes | Yes | Yes, but transactional not process-based | Yes | Yes | Yes |

The studies show that typically process-based functionalities are already present in postmodernERP systems. These systems allow for the modeling of business processes (e.g. INFOR or enova) or are strictly integrated with iBPMS applications (e.g. Comarch, Oracle, or SAP). All vendors who participated in the study offer the possibility of executing processes in accordance with predefined models (changes to the model lead to changes in execution) and control over ongoing and finished processes. It should be noted that almost all systems allow for the execution of processes not just in a way which is fully compliant with the predefined sequence of actions, but which also allows for the possibility to adapt the process the needs of the specific execution context. This is a key feature which enables the execution within these systems of fundamental processes, which are decisive with regard to the results and the competitive position of the organisation and the vast majority of which require, in Industry 4.0/5.0, the dynamic adaptation of the process to the needs of the clients or the broader business environment [13, 14]. In most of the analysed cases, there is also the possibility to launch tasks in other systems in the course of process execution. Both these features considerably raise the flexibility and possibilities of the integration of ERP systems.

In the second step, the authors conducted in-depth interviews with experts participating in the study. The interviews were aimed at understanding the essence of the applied ERP system development approaches towards business process management. In the course of expert interviews, the respondents have provided a broader description of the offered ERP systems from the perspective of their present possibilities in the scope of process management.

The INFOR LN 10.7 system from Infor has a composite architecture, which enables the modeling of processes of any nature, as well as data flow, including the integration with external software. It has its own notation, which is similar to BPMN. In the course of work, the processes available to users have the form of active diagrams, which offer the possibility of maintaining the system and executing processes in accordance with a predefined sequence of tasks and decisions or through the direct selection of actions from the process diagram level. The system enables the users to launch tasks in other applications in the course of process execution. Data on the ongoing and finished processes may be presented in the form of diagrams containing the full information on the process executors, the state of completion, the time of completion, and the data processed. The system includes built-in tools from the areas of RPA, process mining, and ML/AI, but also allows for integration with external tools.

Oracle Fusion applications are implemented through Oracle Business Process Management and depending on the executed process may be modified in accordance with client requirements. New business process models may be designed and implemented with the help of the Oracle Process Cloud Service, which also provides the choice of the method of contact with process and task executors. In the course of work within the system, all actions within processes are logged and controlled, which facilitates undertaking actions and reporting problems or identifying delays, but also allows for the analysis of the executed tasks and processes.

The architecture of the Comarch ERP Egeria 8 system was based on microservices. The system allows for the execution of business processes in accordance with patterns implemented therein and updated by the developer on an ongoing basis in response to legal changes. At the same time, the system allows the users to configure their own unique processes and implement them in iBPMS Camunda, strictly integrated with Egeria. Another possibility is the integration with external document management and workflow class software with the use of the functionalities of both systems.

SAP S/4HANA from SAP – S/4HANA consists of domain-specific application written in ABAP code and an additional layer of the SAP Fiori application, which service predefined business roles. SAP offers pre-prepared business process patterns modeled in BPMN along with instructions for configuring the correct parameters in the SAP S/4HANA system and the SAP Signavio subsystem, which enable work with processes throughout their entire lifecycle – from design and modeling, through management and ongoing execution, up to evaluating their business efficiency.

## 7  Discussion

Industry 4.0 is characterized by the convergence of technologies that improve the efficiency and effectiveness of business processes [21]. ERP systems enable the integration of business processes and ongoing access to integrated data throughout the enterprise [7]. The implementation of the postmodernERP system provides organizations with benefits as a catalyst for business innovation, a platform for business process efficiency, a tool for standardizing processes, and by saving IT costs. One of the most important decision groups in ERP implementation are decisions regarding the configuration of the organization's business processes [43], i.e. decisions directly linking ERP systems with the functional scope of iBPMS. From this perspective, it is not surprising that CSFs and drivers for the development of postmodernERP and iBPMS systems are almost totally overlapping. In Industry 4.0 and the emerging Industry 5.0, both classes of systems require: (1) support in achieving current results, incl. Through the effectiveness of business processes, a system of continuous monitoring and improvement, effective management of organizational change, including the implementation of business process improvements; (2) development support based on employee involvement and participation, organizational culture, awareness, and understanding of process management.

For both classes of systems, compliance with the above CSFs requires: (1) ensuring system-level feasibility, best described by the CSF "System Architecture for Flexibility and Integration to Generally Accepted Standards"; (20 ensuring the actual implementation of BPM at the organizational level, including changes in the organizational culture, best described by the CSFs "Appropriate Implementation Strategy" and"Organizational Culture."

This is clearly indicated not only by the D1 driver "Continuous efforts of enterprises to improve their productivity and efficiency," but by the analysis of all other drivers presented in Sect. 5.2. Only a combination in the development of both classes of systems of "technological" (Industry 4.0) and "cultural" (Industry 5.0) views can ensure a balanced and sustainable competitive position of organizations using these systems.

As shown in the paper, the architectural requirements for both classes of systems are essentially the same. They can be summarized in two main points: (1) composite architecture enabling the integration of modules and even external subsystems and their data, in accordance with the requirements of planning, implementation and analysis of business processes; (2) flexibility to adapt to the organization's business processes, regardless of their nature.

Theoretically, these requirements can be met in three ways: (1) integration of the ERP system with iBPMS, as an external subsystem taking over the implementation of selected business processes based on the metadata and data of the ERP system; (2) process operation of the ERP system, by enabling the configuration of selected processes in selected modules based on a repository of process models, e.g., in BPMN; (3) building the full functionality of the ERP system using iBPMS.

The authors reject the third option of preparing an application as impractical. A system built in such a way would require the preparation of a database layer and a presentation layer, analogous to ERP systems. In addition, a significant part of the processes supported by ERP systems is static, often defined by law, and it is much more effective to "program" them in the application. In practice, as the analysis of architectural requirements and possibilities has shown, there are only two ways leading to the same goal, which is the process operation of the ERP system.

# 8   Conclusions

The aim of the article was to answer the research question: "Are we heading toward process-based ERP system or is the future in the flexible, open integration of postmodernERP and iBPMS?". The complementary and overlapping functionalities of postmodernERP and iBPMS mean that both systems are at present dedicated to the same group of users. This fact, along with the similarity of the CSFs and drivers of development of both classes of systems, as well as identical architectures and the use of the same ICT solutions, de facto determines the strict integration of both classes, and, in the future – their combination into a single class of systems. They are merely two points of departure, from which further development leads to the same end point, namely the process-based operations of an ERP system or, broadly looking, an enterprise information system (EIS). To answer the posed research question, we are undoubtedly going in the direction of process-based ERP systems. However, this "process-based" nature can be achieved by the two paths presented in the article.

This new direction of the development of postmodernERP will undoubtedly become a crucial topic of further research on the development of systems, encompassing e.g. tracking the directions of the development of iBPMS and postmodernERP, the identification of the limitations of thereof, as well as the combination of both classes of systems into a single class, not to mention tracking the proliferation and the effects of using techniques from the area of hyperautomation and the analysis of changes to implementation methodologies.

The limitation of this research is its focus on the systems offered on the Polish market and on the vendor perspective. In the course of further work, the authors intend to extend their research to all European Union countries and for research from the perspective of companies using both systems. This will enable them to formulate a final answer to the question about the future of postmodernERP and iBPMS.

# References

1. Katuu, S.: Enterprise resource planning: past, present, and future. New Rev. Inf. Network. **25**(1), 37–46 (2020). https://doi.org/10.1080/13614576.2020.1742770
2. Mayring, P.: Qualitative content analysis. Forum: Qual. Soc. Res. **1**(2) (2000). http://www.utsc.utoronto.ca/~kmacd/IDSC10/Readings/text%20analysis/CA.pdf. Accessed 21 May 2022
3. Bogner, A., Littig, B., Menz, W. (eds.): Interviewing Experts. Palgrave Macmillan London (2009). https://doi.org/10.1057/9780230244276
4. Lee, A., Chen, S.-C., Kang, H.-Y.: A decision-making framework for evaluating enterprise resource planning systems in a high-tech industry. Qual. Technol. Quant. Manage. **17**(3), 319–336 (2020). https://doi.org/10.1080/16843703.2019.1626073
5. Lupeikiene, A., Dzemyda, G., Kiss, F., Caplinskas, A.: Advanced planning and scheduling systems: modeling and implementation challenges. Informatica **25**(4), 581–616 (2014). https://doi.org/10.15388/Informatica.2014.31
6. Hardcastle, C.: Postmodern ERP Is Fundamentally Different From a Best-of-Breed Approach. Gartner Research, ID: G00264620 (2014)
7. Nazemi, E., Tarokh, M., Djavanshir, G.: ERP: a literature survey. Int. J. Adv. Manuf. Technol. **61**(9–12), 999–1018 (2012). https://doi.org/10.1007/s00170-011-3756-x
8. Ilahi, L., Ghannouchi, S.A., Martinho, R.: BPFlexTemplate: A Business Process template generation tool based on similarity and flexibility. Int. J. Inf. Syst. Project Manage. **5**(3), 67–89 (2017). https://doi.org/10.12821/ijispm050304
9. van der Aalst, W.M.P.: Process mining and RPA: how to pick your automation battles? In: Czarnecki, C., Fettke, P. (eds.) Robotic Process Automation: Management, Technology, Applications. De Gruyter STEM, pp. 223–240 (2021). https://doi.org/10.1515/9783110676693-012
10. Fanning, K., Centers, D.: Intelligent business process management: hype or reality? J. Corpor. Account. Finan. **24**(5), 9–14 (2013). https://doi.org/10.1002/jcaf.21870
11. Koopman, A., Seymour, L.F.: Factors impacting successful bpms adoption and use: a south african financial services case study. In: Nurcan, S., Reinhartz-Berger, I., Soffer, P., Zdravkovic, J. (eds.) BPMDS/EMMSAD -2020. LNBIP, vol. 387, pp. 55–69. Springer, Cham (2020). https://doi.org/10.1007/978-3-030-49418-6_4
12. Kemsley, S.: The changing nature of work: from structured to unstructured, from controlled to social. In: Rinderle-Ma, S., Toumani, F., Wolf, K. (eds.) BPM 2011. LNCS, vol. 6896, pp. 2–2. Springer, Heidelberg (2011). https://doi.org/10.1007/978-3-642-23059-2_2
13. Olding, E., Rozwell, C.: Expand Your BPM Horizons by Exploring Unstructured Processes. Gartner Technical Report G00172387 (2015)
14. Szelągowski, M.: Dynamic BPM in the Knowledge Economy: Creating Value from Intellectual Capital. LNNS, vol. 71 (2019). https://doi.org/10.1007/978-3-030-17141-4
15. Szelągowski, M., Lupeikiene, A.: Business process management systems: evolution and development trends. Informatica **31**(3), 579–595 (2020). https://doi.org/10.15388/20-INFOR429
16. Ozdemir, V., Hekim, N.: Birth of industry 5.0: making sense of big data with artificial intelligence, "The Internet of Things" and next-generation technology policy. OMICS A J. Integrat. Biol. **22**(1), 65–76 (2018). https://doi.org/10.1089/omi.2017.0194
17. Gabryelczyk, R.: Is BPM truly a critical success factor for ERP adoption? An examination within the public sector. Procedia Comput. Sci. **176**, 3389–3398 (2020). https://doi.org/10.1016/j.procs.2020.09.058
18. Rosemann, M., vom Brocke, J.: The six core elements of business process management. In: vom Brocke, J., Rosemann, M. (eds.) Handbook on Business Process Management 1. IHIS, pp. 105–122. Springer, Heidelberg (2015). https://doi.org/10.1007/978-3-642-45100-3_5

19. Ganesh, K., Mohapatra, S., Anbuudayasankar, S., Sivakumar, P.: Enterprise Resource Planning. Fundamentals of Design and Implementation. Springer Cham (2014). https://doi.org/10.1007/978-3-319-05927-3

20. Karim, J., Somers, T., Bhattacherjee, A.: The impact of ERP implementation on business process outcomes: a factor-based study. J. Manag. Inf. Syst. **24**(1), 101–134 (2007). https://doi.org/10.2753/MIS0742-1222240103

21. Katuu, S.: Trends in the enterprise resource planning market landscape. J. Inf. Organ. Sci. **45**(1), 55–75 (2021)

22. Gartner.    https://www.gartner.com/en/information-technology/insights/top-technology-trends. Accessed 1 June 2022

23. Fiodorov, I., Sotnikov, A., Telnov, Y., Ochara, N.M.: Improving Business Processes Efficiency and Quality by Using BPMS (2021). https://doi.org/10.51218/1613-0073-2830-327-336

24. Procesowcy: Dojrzałość procesowa polskich organizacji podsumowanie IV edycji badania dojrzałości procesowej organizacji funkcjonujących w Polsce (2020). https://procesowcy.pl/portfolio-items/dojrzalosc-procesowa-polskich-organizacji-2020/?portfolioCats=19

25. Oracle: https://www.oracle.com/erp/what-is-erp/#link1. Accessed 10 May 2022

26. Gartner: Strategic Roadmap for Postmodern ERP. ID G00384628 (2019)

27. Bailey, S., et al.: Predicts 2022: Supply Chain Strategy. Reprint, Gartner (2021)

28. Chong, E., Handscomb, C., Williams, O., Hall, R., Rooney, M.: Agile resilience in the UK: Lessons from COVID-19 for the "next normal." McKinsey & Company (2020). https://www.mckinsey.com/business-functions/people-and-organizational-performance/our-insights/agile-resilience-in-the-uk-lessons-from-covid-19-for-the-next-normal. Accessed 21 April 2022

29. Rachinger, M., Rauter, R., Müller, C., Vorraber, W., Schirgi, E.: Digitalization and its influence on business model innovation. J. Manuf. Technol. Manag. **30**(8), 1143–1160 (2018)

30. Gartner Top strategic technology trends for 2021 (2021). https://www.gartner.com/smarterwithgartner/gartner-top-strategic-technology-trends-for-2021. Accessed 10 May 2022

31. Xu, L., Xu, E., Li, L.: Industry 4.0: state of the art and future trends. Int. J. Prod. Res. **56**(8), 2941–2962 (2018)

32. Gartner: Top strategic technology trends for 2022. https://www.gartner.com/en/information-technology/insights/top-technology-trends. Accessed 21 April 2022

33. Guay, M.: Postmodern ERP Strategies and Considerations for Midmarket IT Leaders (2016). http://proyectos.andi.com.co/camarabpo/Webinar%202016/Postmodern%20ERP%20strategies%20and%20considerations%20for%20midmarket%20IT%20leaders-%20Gartner.pdf. Accessed 21 April 2022

34. SAP: https://www.sap.com/products/enterprise-management-erp.html. Accessed 10 May 2022

35. INFOR: Enterprise Resource Planning (ERP). Software for improving and managing business operations. https://www.infor.com/nordics/solutions/erp. Accessed 21 April 2022

36. SAP/IBM: The Next-Generation Supply Chain. Digital Transformation Made Possible by IBM and SAP. https://www.ibm.com/downloads/cas/WGZLQOX3. Accessed 21 April 2022

37. Arsanjani, A., Bharade, N., Borgenstrand, M., Schume, P., Wood, J.K., Zheltonogov V.: Business process management design guide: using IBM business process manager. IBM **272** (2015)

38. Pourmirza, S., Peters, S., Dijkman, R., Grefen, P.: A systematic literature review on the architecture of business process management systems. Inf. Syst. **66**, 43–58 (2017)

39. van der Aalst, W.M.P.: Process Mining: Data Science in Action. Springer, Berlin (2016)

40. TIBCO: https://www.tibco.com/reference-center/what-is-ibpms. Accessed 21 April 2022

41. Brambilla, M., Fraternali, P., Ruiz, C.K.V.: Combining social web and BPM for improving enterprise performances. In: Proceedings of the 21st International Conference Companion on World Wide Web - WWW 2012, Companion, pp. 223–226 (2012). https://doi.org/10.1145/2187980.2188014
42. TALLYFY: The Role of Business Process Management (BPM) within the Internet of Things (IoT) (2021). https://tallyfy.com/bpm-internet-of-things-iot/. Accessed 21 Apr 2022
43. Mahendrawathi, E., Zayin, S., Pamungkas, F.: ERP post implementation review with process mining: a case of procurement process. In: Procedia Computer Science 4th Information Systems International Conference 2017, vol. 124, pp. 216–223. ISICO 2017. Bali, Indonesia (2018). https://doi.org/10.1016/j.procs.2017.12.149

# Author Index

Acitelli, Giacomo   200
Agostinelli, Simone   200
Aksar, Burak   123
Axmann, Bernhard   185

Batko, Roman   310
Berniak-Woźny, Justyna   326
Borghoff, Vincent   170
Brzychczy, Edyta   279

Capece, Michela   200
Casanova-Marqués, Raúl   100
Chakraborti, Tathagata   123
Cika, Petr   100
Cyfert, Szymon   310

del-Río-Ortega, Adela   154
Di Ciccio, Claudio   51
Dzurenda, Petr   100

Egger, Christian   138
Enríquez, José González   185

Fdhila, Walid   68
Flechsig, Christian   138
Fuggitti, Francesco   123

Gabryelczyk, Renata   279
Gdowska, Katarzyna   279
Glabiszewski, Waldemar   310

Harmoko, Harmoko   185

Isahagian, Vatche   123, 246

Jiménez Ramírez, Andrés   154
Judmayer, Aljosha   68

Kedziora, Damian   154
Kluza, Krzysztof   279
Köpke, Julius   84

Lupeikiene, Audrone   326

Marangone, Edoardo   51
Mashkif, Nir   215
Mecella, Massimo   200
Modliński, Artur   154
Muthusamy, Vinod   246

Nečemer, Michael   84

Oved, Alon   215

Plattfaut, Ralf   170
Pogiatzis, Antreas   36
Průcha, Petr   260

Ramírez, Andrés Jiménez   185
Ricci, Sara   100
Rizk, Yara   123, 246
Rybinski, Fabian   231

Samakovitis, Georgios   36
Schüler, Selina   231
Senkus, Piotr   310
Shlomov, Segev   215
Skrbek, Jan   260
Sliż, Piotr   295
Stiehle, Fabian   5
Stifter, Nicholas   68
Szelągowski, Marek   326

Talamadupula, Kartik   246
Tonga Naha, Rodrigue   21

Venkateswaran, Praveen   246
Völker, Maximilian   138

Weber, Ingo   5, 51
Weske, Mathias   138
Wysokińska-Senkus, Aneta   310

Yaeli, Avi   215

Zeltyn, Sergey   215
Zhang, Kaiwen   21

Author Index

Printed in the United States
by Baker & Taylor Publisher Services